# Advances in Intelligent Systems and Computing

Volume 1140

AF844814

The series "Advances in Intelligent Systems and Computing" contains publications on theory, applications, and design methods of Intelligent Systems and Intelligent Computing. Virtually all disciplines such as engineering, natural sciences, computer and information science, ICT, economics, business, e-commerce, environment, healthcare, life science are covered. The list of topics spans all the areas of modern intelligent systems and computing such as: computational intelligence, soft computing including neural networks, fuzzy systems, evolutionary computing and the fusion of these paradigms, social intelligence, ambient intelligence, computational neuroscience, artificial life, virtual worlds and society, cognitive science and systems, Perception and Vision, DNA and immune based systems, self-organizing and adaptive systems, e-Learning and teaching, human-centered and human-centric computing, recommender systems, intelligent control, robotics and mechatronics including human-machine teaming, knowledge-based paradigms, learning paradigms, machine ethics, intelligent data analysis, knowledge management, intelligent agents, intelligent decision making and support, intelligent network security, trust management, interactive entertainment, Web intelligence and multimedia.

The publications within "Advances in Intelligent Systems and Computing" are primarily proceedings of important conferences, symposia and congresses. They cover significant recent developments in the field, both of a foundational and applicable character. An important characteristic feature of the series is the short publication time and world-wide distribution. This permits a rapid and broad dissemination of research results.

** **Indexing: The books of this series are submitted to ISI Proceedings, EI-Compendex, DBLP, SCOPUS, Google Scholar and Springerlink** **

More information about this series at http://www.springer.com/series/11156

Roman Szewczyk · Cezary Zieliński ·
Małgorzata Kaliczyńska
Editors

# Automation 2020: Towards Industry of the Future

## Proceedings of Automation 2020, March 18–20, 2020, Warsaw, Poland

Springer

*Editors*
Roman Szewczyk
ŁUKASIEWICZ Research Network
Industrial Research Institute
for Automation
and Measurements PIAP
Warsaw, Poland

Cezary Zieliński
ŁUKASIEWICZ Research Network
Industrial Research Institute
for Automation
and Measurements PIAP
Warsaw, Poland

Małgorzata Kaliczyńska
ŁUKASIEWICZ Research Network
Industrial Research Institute
for Automation
and Measurements PIAP
Warsaw, Poland

ISSN 2194-5357 ISSN 2194-5365 (electronic)
Advances in Intelligent Systems and Computing
ISBN 978-3-030-40970-8 ISBN 978-3-030-40971-5 (eBook)
https://doi.org/10.1007/978-3-030-40971-5



This Springer imprint is published by the registered company Springer Nature Switzerland AG
The registered company address is: Gewerbestrasse 11, 6330 Cham, Switzerland

# Foreword

Global industry, and especially the European one, is currently facing the most significant transition since the Industrial Revolution at the beginning of the nineteenth century. In the face of already observed climate change, the pledge of conversion to zero-emission industry before 2050 is most welcome; however, attaining this target will be the most challenging and expensive transformation undertaken by our technology-oriented civilization within this century. Realization of this venture requires progress in many fields, but especially in industrial automation, robotics and measurement techniques, in association with cyber-physical systems employing artificial intelligence. Those will be the key elements reducing costs and enabling European society to keep its quality of life. Both information technology and physical devices will create synergic systems enabling the increase of industrial efficiency and reduction of resource utilization. The goal is to at least sustain, or even better, to increase the quality of our lives in the new reality of zero-emission industry.

As a result, it is expected that the next 30 years will be the period, when we will face radical innovations in production processes, transportation management and social life. To manage challenges associated with the transformation to zero-emission industry, the joint efforts of both scientists and engineers are necessary. For this reason, scientific community has to accept the challenge for further developing the above-mentioned fields of study and engineering. As this process is currently ongoing, the new results of research in this area are reported in this volume.

We strongly believe that the solutions and guidelines presented in this volume will be useful for both researchers of technical sciences and engineers solving problems in the world of ongoing radical changes.

December 2019

Roman Szewczyk
Cezary Zieliński
Małgorzata Kaliczyńska

# Contents

**Robotics**

# About the Editors

**Prof. Roman Szewczyk** received both his Ph.D. and D.Sc. in the field of mechatronics. He specializes in modelling of properties of magnetic materials as well as in sensors and sensor interfacing, in particular magnetic sensors for security applications. He leads the establishment of the sensing unit for the mobile robot developed for the Polish Police Central Forensic Laboratory as well as the unit for methods of non-destructive testing based on the magnetoelastic effect. He has been involved in over ten European Union-funded research projects within the FP6 and FP7 as well as projects financed by the European Defence Organization. Moreover, he leads two regional- and national-scale technological foresight projects and was active in the organization and implementation of technological transfer between companies and research institutes. He is Secretary for Scientific Affairs in the ŁUKASIEWICZ Research Network – Industrial Research Institute for Automation and Measurements PIAP. He is also Associate Professor at the Faculty of Mechatronics, Warsaw University of Technology, and Former Vice-Chairman of the Academy of Young Researchers of the Polish Academy of Sciences.

**Prof. Cezary Zieliński** received his M.Sc. Eng. degree in control in 1982, Ph.D. degree in control and robotics in 1988, the D.Sc. (habilitation) degree in control and robotics in 1996, all from the Faculty of Electronics and Information Technology, Warsaw University of Technology, Warsaw, Poland, and Full Professorship in 2012. Currently, he is Full Professor in both the ŁUKASIEWICZ Research Network – Industrial Research Institute for Automation and Measurements PIAP and Warsaw University of Technology. Since 2007, he has been Member of the Committee for Automatic Control and Robotics, the Polish Academy of Sciences. He is Head of the Robotics Group in the Institute of Control and Computation Engineering working on robot control and programming methods. His research interests focus on robotics in general and in particular include: robot programming methods, formal approach to the specification of architectures of multi-effector and multi-receptor systems, robot kinematics, robot position–force control, visual servo control and design of digital circuits. He is Author/Co-author of over 200 reviewed conference and journal papers as well as books concerned with the above-mentioned research subjects.

**Dr. Małgorzata Kaliczyńska** received her M.Sc. Eng. degree in cybernetics from the Faculty of Electronics, Wrocław University of Technology, and her Ph.D. degree in the field of fluid mechanics from the Faculty of Mechanical and Power Engineering in this same university. Now, she is Assistant Professor in the ŁUKASIEWICZ Research Network – Industrial Research Institute for Automation and Measurements PIAP and Editor of the scientific and technological magazine "Measurements, Automation, Robotics". Her areas of research interest include distributed control systems, Internet of things, Industry 4.0, information retrieval and webometrics.

# Control and Automation

# Model Reference QSM Control Strategy for Higher Relative Degree Outputs

Katarzyna Adamiak(✉)

Institute of Automatic Control, Łódź University of Technology,
18/22 Stefanowskiego Street, 90-924 Łódź, Poland
katarzyna.adamiak@edu.p.lodz.pl

**Abstract.** This paper deals with the problem of sliding mode control for discrete time, linear, time-invariant disturbed systems. Our aim is to improve the robustness of the system by minimization of the disturbance impact. We adopt the switching type definition of the quasi-sliding mode and propose a new, model reference based control strategy with higher relative degree sliding variables. Our idea is to first generate the desired evolution of the switching variable with the application of the conventional switching type reaching law developed by Gao *et al.* Next, we propose new model following, higher relative degree reaching laws for real disturbed systems. We consider sliding variable of relative degree one and two. We prove that the combination of the model reference strategy with higher relative degree output concept allows to ensure all the properties of the quasi-sliding mode as defined in the seminal paper of Gao *et al.* and reduces the width of the quasi-sliding mode band. Therefore, our strategy ensures the improvement of the system robustness. Finally, we conclude our paper with a simulation example.

**Keywords:** Discrete time systems · Higher relative degree · Model reference control · Reaching law · Sliding mode control

## 1 Introduction

Sliding mode control is nowadays one of the most promising design methodologies in the control theory. Its concept originates from the theory of variable structure systems. Sliding mode was first studied by Emelyanov [1] and later developed by Utkin [2, 3] and Itkis [4]. The main idea of the sliding mode control is to force the representative point of the system to approach and remain on a predesigned sliding surface in the state space. Therefore, the dimensions of the dynamical problem are reduced and the variable structure systems become insensitive to disturbances and model uncertainties [5].

However, high frequency switching of the control signal, required for the ideal sliding mode usually causes implementation difficulties [6, 7]. This observation led to the introduction of discrete time sliding modes. The first definition was brought by Milosavljević [8], who determined the necessary condition for the existence of the quasi-sliding mode. Later his idea was further developed by many researchers [9–12]. In this work we will refer to the seminal definition brought in 1995 by the work of Gao, Wang and Homaifa [13]. Pursuant to their study the quasi-sliding mode exists when the


R. Szewczyk et al. (Eds.): AUTOMATION 2020, AISC 1140, pp. 3–13, 2020.
https://doi.org/10.1007/978-3-030-40971-5_1

representative point of the system moves monotonically from any initial position to the sliding hyperplane and crosses it in finite time. Once the representative point of the system has crossed the sliding plane for the first time it must cross it again in each consecutive step and the absolute value of the sliding variable must not exceed an *a priori* known constant. Therefore, in discrete time case the representative point of the system does not remain on the sliding surface, but only within a band around the switching plane. These three postulates constitute the fundamental definition of switching type quasi-sliding mode, which will be adopted further in this paper.

The work of Gao *et al.* also proposed a reaching law approach to the design of quasi-sliding mode controller, which will be used in this study. The reaching law approach assumes defining an external function that will describe the trajectory of the sliding variable in the entire course of motion. The first reaching function consisted of a proportional term, determining the pace of the convergence to the sliding plane and a switching element, directly responsible for the width of the quasi-sliding mode band [14]. Afterwards, multiple researchers have worked on methods of narrowing this band and therefore improving the robustness of the systems. Many new switching type [15–17] and nonswitching type [18–20] reaching laws have been introduced ever since.

Until these days, the control community focuses on different aspects of the quasi-sliding mode. One of the main areas of contemporary research is the employment of higher relative degree sliding variables [21, 22]. The relative degree describes the time delay between the application of the control signal and its effect in system's output. The relative degree 1 sliding variable is used when the control signal is designed to directly influence the value of the output in the next step. The idea of higher relative degree ($r$) applies to a situation when the control signal in the discrete time instant $k$ does not influence the sliding variable in the next instants $k + 1, k + 2, \ldots, k + r - 1$, but in the instant $k + r$. It has been shown that with a specific choice of the sliding surface, such an approach results in a reduction of the quasi-sliding mode band width in comparison to the conventional SMC methods.

Based on the previous studies in this field, this paper proposes a new model following higher relative degree reaching law for discrete time disturbed systems. In order to reduce the influence of external disturbances on the system, we combine the model reference quasi-sliding mode control with higher relative degree output approach. We prove that our control strategy guarantees the properties of the quasi-sliding mode as defined by Gao *et al.* and offers a significant reduction of the width of the ultimate band.

## 2 Problem Statement

In this paper, we consider a discrete time, linear, time-invariant system:

$$\boldsymbol{x}(k+1) = \boldsymbol{A}\boldsymbol{x}(k) + \boldsymbol{b}u(k) + \boldsymbol{b}f(k), \tag{1}$$

where $\boldsymbol{x}$ is the $n \times 1$ state vector, $u(k)$ is the control signal, $\boldsymbol{A}$ and $\boldsymbol{b}$ are the state matrix and the input distribution vector respectively and $f(k)$ represents a scalar disturbance.

The disturbance is limited as follows: $|f(k)| \leq f_{\max}$. We assume that the system is described in the so called Frobenius form, so:

$$\boldsymbol{A} = \begin{bmatrix} 0 & 1 & 0 & \cdots & 0 \\ 0 & 0 & 1 & \cdots & 0 \\ \vdots & \vdots & \vdots & \ddots & \vdots \\ 0 & 0 & 0 & \cdots & 1 \\ -a_0 & -a_1 & -a_2 & \cdots & -a_{n-1} \end{bmatrix} \quad \text{and} \quad \boldsymbol{b} = \begin{bmatrix} 0 \\ 0 \\ \vdots \\ 0 \\ 1 \end{bmatrix}. \tag{2}$$

Any controllable system:

$$\boldsymbol{\eta}(k+1) = \boldsymbol{\Phi}\boldsymbol{\eta}(k) + \boldsymbol{\Gamma}u(k) + \boldsymbol{\Gamma}f(k) \tag{3}$$

may be transformed to the Frobenius form with the transformation: $\boldsymbol{A} = \boldsymbol{T}\boldsymbol{\Phi}\boldsymbol{T}^{-1}$ and $\boldsymbol{b} = \boldsymbol{T}\boldsymbol{\Gamma}$, where $\boldsymbol{T} = \boldsymbol{S_\Phi}\boldsymbol{S}_A^{-1}$ and $\boldsymbol{S_\Phi}$ and $\boldsymbol{S}_A$ are the appropriate controllability matrices: $\boldsymbol{S_\Phi} = [\boldsymbol{\Gamma}\ \boldsymbol{\Phi\Gamma}\ \ldots\ \boldsymbol{\Phi}^{n-1}\boldsymbol{\Gamma}]$ and $\boldsymbol{S}_A = [\boldsymbol{b}\ \boldsymbol{Ab}\ \ldots\ \boldsymbol{A}^{n-1}\boldsymbol{b}]$.

The aim of our control scheme is to drive the system from any initial position $\boldsymbol{x}(0) = \boldsymbol{x}_0$ to the demand state $\boldsymbol{x}_d$. For that purpose, in the next chapter we propose a model reference-based quasi-sliding mode control strategy with relative degree 1 and 2 sliding variables.

# 3 Model Reference QSM Control Strategy

## 3.1 Reference Model

In this study we introduce a trajectory following quasi-sliding mode control strategy for disturbed systems with higher relative degree outputs. First, we generate the desired evolution of the sliding variable. We use the plant's mathematical model, described as:

$$\boldsymbol{x}_m(k+1) = \boldsymbol{A}\boldsymbol{x}_m(k) + \boldsymbol{b}u_m(k), \tag{4}$$

where $\boldsymbol{x}_m$ is the $n \times 1$ model's state vector, $u_m(k)$ is the model's control signal. The initial conditions of the model are the same as the initial conditions of the system, $\boldsymbol{x}_m(0) = \boldsymbol{x}_0$. The model is independent of any disturbance. As we want to specifically generate the evolution of the demand sliding variable in each step, we apply the relative degree 1 output. Therefore, we define the desired sliding variable as:

$$s_d(k) = \boldsymbol{c}_1[\boldsymbol{x}_d - \boldsymbol{x}_m(k)], \tag{5}$$

where $\boldsymbol{c}_1$ is an $1 \times n$ vector chosen so that $\boldsymbol{c}_1\boldsymbol{b} \neq 0$. We select the sliding surface as $s_d(k) = 0$. We choose to control the model according to the conventional reaching law based sliding mode control strategy introduced by Gao *et al.* [13]. We use the reaching function:

$$s_d(k+1) = (1-q)s_d(k) - \varepsilon \operatorname{sgn}[s_d(k)], \tag{6}$$

where $q$ and $\varepsilon$ are real positive constants and $q < 1$. The signum function equals 1 for variables greater or equal to 0 and −1 for variables smaller than 0. Considering (5) and (6) the control signal for the model becomes:

$$u_m(k) = (\boldsymbol{c}_1\boldsymbol{b})^{-1}\{\boldsymbol{c}_1\mathbf{x_d} - \boldsymbol{c}_1\boldsymbol{A}\boldsymbol{x}_\mathbf{m}(k) - (1-q)s_d(k) + \varepsilon \operatorname{sgn}[s_d(k)]\}. \tag{7}$$

Relation (6) implies the following two properties:

(1) if $s_d(k) \geq 0$, then $s_d(k+1) \leq s_d(k) - \varepsilon$,
(2) if $s_d(k) < 0$, then $s_d(k+1) \geq s_d(k) + \varepsilon$.

Therefore, it can be concluded that not later than in step $k_{0d} + 1 = \text{ceil}[|s_d(0)|/\varepsilon + 1]$ the sign of $s_d(k_{0d} + 1)$ becomes opposite to the sign of $s_d(0)$. Furthermore, considering (6), we conclude that for any $k \geq k_{0d}$,

$$\operatorname{sgn}[s_d(k)] = -\operatorname{sgn}[s_d(k+1)] \tag{8}$$

and notice that for any $k \geq k_{0d} + 1$ the demand sliding variable satisfies:

$$s_d(k) \leq \varepsilon. \tag{9}$$

## 3.2 Reaching Law for RD1 System

Having generated the desired evolution of the sliding variable, in this section we propose model following reaching law for the system (1) with the relative degree 1 sliding variable. Therefore, we define the sliding variable as:

$$s_1(k) = \boldsymbol{c}_1[\boldsymbol{x}_d - \boldsymbol{x}(k)] \tag{10}$$

and choose the sliding surface $s_1(k) = 0$. Moreover, we define the impact of disturbance on the system as $d_1(k) = \boldsymbol{c}_1\boldsymbol{b}f(k)$, which for any $k \geq 0$ satisfies:

$$|d_1(k)| \leq d_{\max} = \boldsymbol{c}_1\boldsymbol{b}f_{\max}. \tag{11}$$

Conventionally, such a system may be controlled with the reaching law of Gao [13]:

$$s_1(k+1) = (1-q)s_1(k) - \varepsilon \operatorname{sgn}[s_1(k)] + d_1(k), \tag{12}$$

where $0 < q < 1$ and $\varepsilon > 0$. It has been shown in [14] that, if the control parameters satisfy $\varepsilon > (2-q)d_{\max}/q$, then the reaching law ensures finite time convergence to the sliding plane. In the sliding phase the sliding variable satisfies: $|s_1(k)| \leq \varepsilon + d_{\max} = \delta$. In order to reduce the band's width, we propose a new model following reaching law:

$$s_1(k+1) = s_d(k+1) + d_1(k). \quad (13)$$

According to (13), the evolution of the sliding variable $s_1$ is required to reach the desired value $s_d$ with accuracy to the disturbance. Traditional reaching laws calculate $s_1(k+1)$ taking into account $s_1(k)$. Consequently, the disturbance effect is accumulated in the sliding variable and transferred to the value of this variable in the next step. Therefore, the sliding variable bears the impact of all disturbance values from $d_1(0)$ till $d_1(k)$. On the contrary, the model following method uses the reference value to obtain $s_1(k+1)$. As the reference profile does not hold any disturbance effect, the impact of the disturbance on the sliding variable is limited to one control step.

Considering (1), (10) and (13), we get the control signal:

$$u(k) = (\boldsymbol{c}_1\boldsymbol{b})^{-1}[\boldsymbol{c}_1\boldsymbol{x}_\mathbf{d} - \boldsymbol{c}_1\boldsymbol{A}\boldsymbol{x}(k) - s_d(k+1)]. \quad (14)$$

Now, we prove that with appropriate choice of the control parameters the reaching law (13) ensures all the properties of the quasi-sliding mode in the understanding of Gao.

**Theorem 1.** If $\varepsilon > d_{\max}/q$ then $\mathrm{sgn}[s_1(k)] = -\mathrm{sgn}[s_1(k+1)]$ for any $k \geq k_0 = k_{0d} + 2$, and for any $k \geq k_{0d} + 1$ the sliding variable belongs to the interval:

$$|s_1(k)| \leq \varepsilon + d_{\max} = \delta_1. \quad (15)$$

**Proof.** We may rewrite the reaching law (13) as:

$$s_1(k+2) = s_d(k+2) + d_1(k+1). \quad (16)$$

Considering (6) we get:

$$s_1(k+2) = (1-q)\{(1-q)s_d(k) - \varepsilon\mathrm{sgn}[s_d(k)]\} - \varepsilon\mathrm{sgn}[s_d(k+1)] + d_1(k+1). \quad (17)$$

Taking into account (8), for any $k \geq k_{0d}$ we obtain

$$\mathrm{sgn}[s_1(k+2)]|s_1(k+2)| = \mathrm{sgn}[s_d(k)]\left[(1-q)^2|s_d(k)| + q\varepsilon\right] + d_1(k+1). \quad (18)$$

We notice that the term in the brackets in (18) is always positive. As, $d_1(k+1)$ satisfies (11) we get:

$$\mathrm{sgn}[s_1(k+2)]|s_1(k+2)| \leq \mathrm{sgn}[s_d(k)]\left[(1-q)^2|s_d(k)| + q\varepsilon\right] + d_{\max}. \quad (19)$$

Therefore, if $\varepsilon > d_{\max}/q$, then:

$$(1-q)^2|s_d(k)| + q\varepsilon > d_{\max}. \quad (20)$$

Considering (8) and (20), for any $k \geq k_{0d} + 2$, we obtain:

$$\operatorname{sgn}[s_1(k)] = \operatorname{sgn}[s_d(k-2)] = -\operatorname{sgn}[s_d(k-1)]. \tag{21}$$

Taking into account that, starting from $k_{0d}$, the sign of the reference sliding variable changes in all the following instants, we conclude that the change of the sign of $s_1(k)$ occurs in all the following instants, starting from $k_0 = k_{0d} + 2$. Moreover, from (9), (11) and (13), for any $k \geq k_0 + 1$ we get (15), which ends the proof. ∎

We have shown that the reaching law (13) satisfies all the properties of the quasi-sliding mode as defined in [13] and determined the quasi-sliding mode band width. This band depends on the values of $d_{max}$ and $\varepsilon$. As $(2 - q) \in (1, 2)$, we conclude that:

$$\varepsilon > \frac{(2-q)d_{\max}}{q} > \frac{d_{\max}}{q}, \tag{22}$$

so the value of $\varepsilon$ in the original control strategy of Gao must be greater than in the model reference control method. Consequently,

$$\delta_1 < \delta \tag{23}$$

and the sliding variable in the Gao's strategy belongs to a greater interval $[-\delta, \delta]$ than in the model following strategy $[-\delta_1, \delta_1]$. Therefore, the application of the model following reaching law improves the robustness of the system.

### 3.3 Reaching Law for RD2 System

The width of the ultimate band may be further reduced by the application of relative degree 2 sliding variable. Therefore, we select the second order sliding variable as:

$$s_2(k) = \mathbf{c}_2[\boldsymbol{x}_d - \boldsymbol{x}(k)], \tag{24}$$

so that $\boldsymbol{c}_2\boldsymbol{b} = 0$ and $\boldsymbol{c}_2\boldsymbol{A}\boldsymbol{b} \neq 0$. Consequently, the control signal in instant $k$ does not influence $s_2(k + 1)$, but $s_2(k + 2)$. Calculating $s_2(k + 1)$ from (1), (24) and considering $\boldsymbol{c}_2\boldsymbol{b} = 0$ we get:

$$s_2(k+1) = \boldsymbol{c}_2\boldsymbol{x}_\mathbf{d} - \boldsymbol{c}_2\boldsymbol{A}\boldsymbol{x}(k). \tag{25}$$

Therefore, the control signal does not influence $s_2(k + 1)$. Now, let us calculate $s_2(k + 2)$. According to (1) and (24):

$$s_2(k+2) = \mathbf{c}_2[\boldsymbol{x}_d - \boldsymbol{x}(k+2)] = \boldsymbol{c}_2\boldsymbol{x}_d - \boldsymbol{c}_2\boldsymbol{A}^2\boldsymbol{x}(k) - \boldsymbol{c}_2\boldsymbol{A}\boldsymbol{b}u(k) - \boldsymbol{c}_2\boldsymbol{A}\boldsymbol{b}f(k). \tag{26}$$

From (26) we conclude that $u(k)$ directly influences $s_2(k + 2)$. Next, we denote the disturbance impact on the RD2 system as $d_2(k) = \boldsymbol{c}_2\boldsymbol{A}\boldsymbol{b}f(k)$, which for any $k \geq 0$ satisfies:

$$|d_2(k)| \leq d_{2\max} = \mathbf{c}_2 \boldsymbol{A} \boldsymbol{b} f_{\max}. \tag{27}$$

We choose the sliding surface $s_2(k) = 0$ and propose a new reaching law for system (1)

$$s_2(k+2) = s_d(k+2) + d_2(k). \tag{28}$$

From (1), (24) and (28) we obtain the control signal:

$$u(k) = (\boldsymbol{c}_2 \boldsymbol{A} \boldsymbol{b})^{-1} \left[ \boldsymbol{c}_2 \boldsymbol{x}_{\mathbf{d}} - \boldsymbol{c}_2 \boldsymbol{A}^2 \boldsymbol{x}(k) - s_d(k+2) \right]. \tag{29}$$

**Theorem 2.** If $\varepsilon > d_{2\max} / q$, then $\mathrm{sgn}[s_2(k)] = -\mathrm{sgn}[s_2(k+1)]$ for any $k \geq k_0 = k_{0d} + 2$, and for any $k \geq k_{0d} + 1$ the sliding variable belongs to the interval:

$$|s_2(k)| \leq \varepsilon + d_{2\max} = \delta_2. \tag{30}$$

**Proof.** The proof is analogical to the proof of Theorem 1. The change of sign of the reference sliding variable $s_d$ is guaranteed for any $k \geq k_{0d}$. Therefore, if in the consecutive steps $\mathrm{sgn}[s_2(k+2)] = \mathrm{sgn}[s_d(k)]$, then the sign of $s_2(k)$ will change as well. Let us express $s_2(k+2)$ using (6) and (28):

$$s_2(k+2) = (1-q)\{(1-q)s_d(k) - \varepsilon\ \mathrm{sgn}[s_d(k)]\} - \varepsilon\ \mathrm{sgn}[s_d(k+1)] + d_2(k). \tag{31}$$

As for any $k \geq k_{0d}$ (8) holds, we get:

$$s_2(k+2) = \mathrm{sgn}[s_d(k)] \left[ (1-q)^2 |s_d(k)| + q\varepsilon \right] + d_2(k). \tag{32}$$

We notice that the term in the brackets on the right-hand side of (32) is always greater than or equal to $q\varepsilon$. From (27) and (32) we conclude that if $\varepsilon > d_{2\max} / q$, then $\mathrm{sgn}[s_2(k+2)] = \mathrm{sgn}[s_d(k)]$ and, taking into account (8), for any $k \geq k_0 = k_{0d} + 2$

$$\mathrm{sgn}[s_2(k)] = -\mathrm{sgn}[s_2(k+1)]. \tag{33}$$

Considering (9), (27) and (28) we get $|s_2(k)| \leq \varepsilon + d_{2\max}$, which ends the proof. ■

**Theorem 3.** If $|\boldsymbol{c}_2 \boldsymbol{A} \boldsymbol{b}| < |\boldsymbol{c}_1 \boldsymbol{b}|$, then

$$d_{2\max} < d_{\max} \tag{34}$$

and

$$\delta_2 < \delta_1 < \delta. \tag{35}$$

**Proof.** Equation (34) is directly derived from (11) and (27). Since $\delta_1 = \varepsilon + d_{\max}$ and $\delta_2 = \varepsilon + d_{2\max}$, taking into account (23) and (34) we obtain (35). ■

We conclude that the model following RD2 control strategy ensures all properties of the quasi-sliding mode and reduces width of the quasi-sliding mode band compared to RD1 model reference strategy and to the conventional control strategy of Gao.

## 4 Simulation Example

In this chapter we verify our model reference sliding mode control strategy with a simulation example. We take into account a third order plant, which represents a model of a DC motor. The system is described by a continuous time transfer function $G(s) = \frac{24}{s^3 - 13.06s^2 - 3.66s}$. The system is controlled through sample and hold devices and therefore requires a discrete time representation. We discretize the model with the sampling time 1 and transform it into the Frobenius form. Following the transformation method described in chapter 2 we obtain a state space model:

$$\boldsymbol{x}(k+1) = \begin{bmatrix} 0 & 1 & 0 \\ 0 & 0 & 1 \\ 0 & -0.751 & 1.751 \end{bmatrix} \boldsymbol{x}(k) + \begin{bmatrix} 0 \\ 0 \\ 1 \end{bmatrix} [u(k) + d(k)]. \tag{36}$$

The initial conditions of the system are $\boldsymbol{x}_0 = [10\ 10\ 30]^{\mathrm{T}}$ and we aim to drive the system to the desired state $\boldsymbol{x}_d = [0\ 0\ 0]^{\mathrm{T}}$. The disturbance $d(k)$ changes between its extreme values $\pm 1$. In order to present the properties of the new model following strategy we compare the fundamental control strategy of Gao *et al.* [13] with the model following strategy with RD1 and RD2 sliding variables. For the original strategy we choose the sliding vector $\boldsymbol{c}_1 = [0\ 0\ 1]$, which ensures stability of the closed-loop system. As $d_{\max} = 1$, for the Gao's strategy we set $q = 0.6$ and $\varepsilon = 2.34$ to satisfy the switching condition and minimize the quasi-sliding band width. On the other hand, for the model following strategy we consider two cases. We apply the relative degree 1 sliding variable with $\boldsymbol{c}_1 = [0\ 0\ 1]$ in the model. Next we choose the sliding variable of RD1 with $\boldsymbol{c}_1$ and of RD2 with $\boldsymbol{c}_2 = [0\ 0.1\ 0]$ for the real disturbed system. Therefore, $d_{1\max} = 1$ and $d_{2\max} = 0.1$. For a fair comparison we set $q = 0.6$. For RD1 system we choose $\varepsilon = 1.67$, according to Theorem 1 and for RD2 system $\varepsilon = 0.167$, as defined by Theorem 2. The results of our simulations are presented in Figs. 1, 2 and 3. Plots obtained with the Gao's strategy have been shown in blue, while the results of our control method with RD1 sliding variable were plotted in green and with RD2 sliding variable are presented with red.

Figure 1 shows the evolution of the sliding variables. It is clearly visible that in the RD1 systems the sliding variables change from the beginning of the control process, while in the RD2 case the sliding variable begins to change after 2 control steps. The original control strategy resulted in the quasi-sliding mode band width of 3.34, while in our method in RD1 system this width has been reduced to 2.67 and in RD2 system to 0.267. The price to pay for this improvement is a slight prolongation of the reaching phase. In the model reference RD2 strategy the sliding plane is crossed for the first time two time instants later than in the Gao's case.

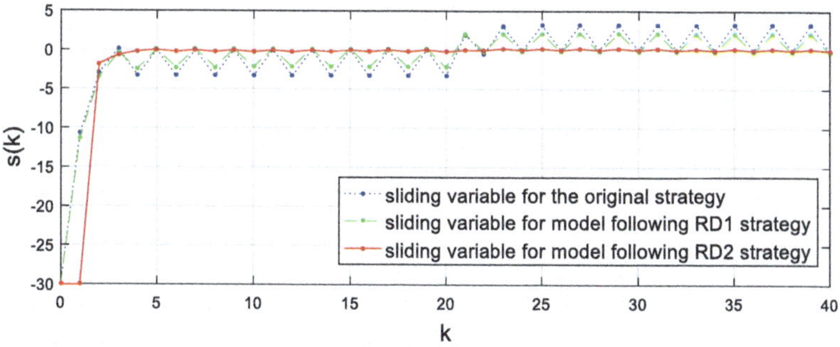


**Fig. 1.** Evolution of the sliding variable.

In Fig. 2 we may see the evolution of the control input. The model reference strategy allowed to reduce the control values in the sliding phase. Finally, Fig. 3 presents the evolution of the first state variable. It is clear that with the new control strategy the error has been limited and therefore the robustness of the system has improved. The integral absolute error (IAE) for the first state variable for the Gao's control method is 122.92, while for the model following strategy it was reduced to 108 for RD1 case and 115.88 in RD2 case.

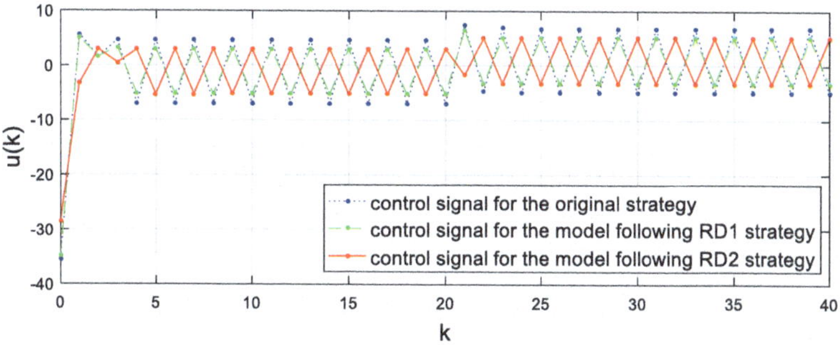


**Fig. 2.** Evolution of the control signal.

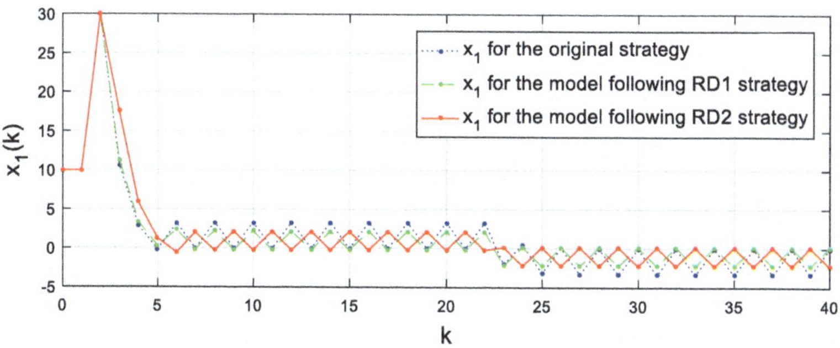


**Fig. 3.** Evolution of the first state variable $x_1$.

## 5 Conclusions

The paper presents a new quasi-sliding mode control strategy for discrete time, disturbed systems. We aimed to improve the system's robustness by application of a new model following control strategy with the sliding variables of relative degree 1 and 2. First, we generated the desired sliding variable based on the model of the system with the conventional reaching law of Gao *et al.* Next, we proposed a new trajectory following reaching law for the real disturbed system. We showed that both for RD1 and RD2 system, with an appropriate choice of the control parameters, the model following strategy guarantees all the properties of the quasi-sliding mode as defined by Gao. As the model is not influenced by any disturbances, the strategy results in a reduction of the width of the quasi-sliding mode band in comparison to the conventional control method. Lastly, we verified our results with a simulation example.

## References

1. Emelyanov, S.V.: Variable-Structure Control Systems. Nauka, Moscow (1967). (in Russian)
2. Utkin, V.I.: Variable structure systems with sliding modes. IEEE Trans. Autom. Control **22** (2), 212–222 (1977)
3. Utkin, V.I.: Sliding Modes in Control and Optimization. Springer, Berlin (1992)
4. Itkis, U.: Control Systems of Variable Structures. Wiley, New York (1976)
5. Draženović, B.: The invariance conditions in variable structure systems. Automatica **5**(3), 287–295 (1969)
6. Chang, F.J., Twu, S.H., Chang, S.: Adaptive chattering alleviation of variable structure systems. IEE Proc. D Control Theory Appl. **137**(1), 31–39 (1990)
7. Bartolini, G., Ferrara, A., Usai, E., Utkin, V.I.: On multi-input chattering-free second-order sliding mode control. IEEE Trans. Autom. Control **45**(9), 1711–1717 (2000)
8. Milosavljević, D.: General conditions for the existence of a quasi-sliding mode on the switching hyperplane in discrete variable structure systems. Autom. Remote Control **46**(3), 307–314 (1985)
9. Sarpturk, S.Z., Istefanopulos, Y., Kaynak, O.: On the stability of discrete-time sliding mode control systems. IEEE Trans. Autom. Control **22**(10), 930–932 (1987)
10. Kotta, U., Sarpturk, S.Z., Istefanopulos, Y.: Comments on "On the stability of discrete-time sliding mode control systems". IEEE Trans. Autom. Control **34**(9), 1021–1022 (1989)
11. Utkin, V.I., Drakunov, S.V.: On discrete-time sliding modes. IFAC Proc. Vol. **22**(3), 273–278 (1989)
12. Bartolini, G., Ferrara, A., Utkin, V.I.: Adaptive sliding mode control in discrete-time-systems. Automatica **31**(5), 769–773 (1995)
13. Gao, W.B., Wang, Y., Homaifa, A.: Discrete-time variable structure control systems. IEEE Trans. Industr. Electron. **42**(2), 117–122 (1995)
14. Bartoszewicz, A.: Remarks on 'Discrete-time variable structure control systems'. IEEE Trans. Industr. Electron. **43**, 235–238 (1996)
15. Veselić, B., Perunicic-Draženović, B., Milosavljević, Č.: Improved discrete-time sliding-mode position control using Euler velocity estimation. IEEE Trans. Industr. Electron. **57**(11), 3840–3847 (2010)
16. Bartoszewicz, A., Leśniewski, P.: New switching and nonswitching type reaching laws for SMC of discrete time systems. IEEE Trans. Control Syst. Technol. **24**(2), 670–677 (2016)

17. Ma, H., Li, Y.: Multi-power reaching law based discrete-time sliding-mode control. IEEE Access **7**, 49822–49829 (2019)
18. Bartoszewicz, A., Leśniewski, P.: Reaching law-based sliding mode congestion control for communication networks. IET Proc. Control Theory Appl. **8**(17), 1914–1920 (2014)
19. Golo, G., Milosavljević, Č.: Robust discrete-time chattering free sliding mode control. Syst. Control Lett. **41**, 19–28 (2000)
20. Chakrabarty, S., Bandyopadhyay, B.: A generalized reaching law with different convergence rates. Automatica **63**, 34–37 (2016)
21. Chakrabarty, S., Bartoszewicz, A.: Improved robustness & performance of discrete time sliding mode control systems. ISA Trans. **65**, 143–149 (2016)
22. Bartoszewicz, A., Latosiński, P.: Generalization of Gao's reaching law for higher relative degree sliding variables. IEEE Trans. Autom. Control **63**(9), 3173–3179 (2018)

# Sliding Mode Control with Minimization of IAE and Elastic Constraints on the Input Signal

Mateusz Pietrala(✉)

Institute of Automatic Control, Łodź University of Technology,
18/22 Stefanowskiego Street, 90-924, Łodź, Poland
mateusz.pietrala@edu.p.lodz.pl

**Abstract.** This work considers the second order, continuous-time system affected by unknown, but bounded external disturbances. The main goal is to obtain the robustness with respect to these disturbances from the very beginning of the control process. Moreover, it is essential to drive the representative point to the demand state in finite time. For that purpose, the time-varying, parabolic sliding line is introduced. The performance of our controller is evaluated by minimizing the quality index, which is a sum of Integral Absolute Error and a function, which provides a penalty for the excessive values of the control signal.

**Keywords:** Sliding mode control · Continuous-time systems · Quality index · Integral Absolute Error · Penalty function

## 1 Introduction

Sliding mode control is an effective and popular regulation method due to its computational efficiency and robustness with respect to perturbations and modeling uncertainties [1–8]. Moreover, it was found to be applicable in practice for example in power electronics or network systems [5,9,10]. Firstly, the continuous-time systems were analyzed, and after that many researchers extended this strategy for the sake of controlling discrete-time plants. The sliding mode control can be divided into two main phases: the reaching phase and the sliding phase. In the first one the representative point, staring from the initial position, moves towards the sliding hyperplane, until it reaches its certain neighborhood. In the second one, this point slides along the predefined switching surface to the demand state. However, the insensitivity to perturbations can be guaranteed only in the sliding phase. Therefore, by eliminating the reaching phase, the robustness is obtained from the very beginning of the control process. In order to achieve this desired property, the time-varying sliding hyperplane is introduced [11–20]. However, this hyperplane has to be selected in a specific manner. Our goal is to drive the representative point to the demand state in finite time. Hence, a nonlinear switching hypersurface can be applied [21–24]. In order to guarantee good


R. Szewczyk et al. (Eds.): AUTOMATION 2020, AISC 1140, pp. 14–23, 2020.
https://doi.org/10.1007/978-3-030-40971-5_2

dynamical performance of the plant one can minimize a selected quality index. In this paper, we combine the Integral Absolute Error (IAE) [14–18,20] with a function, which provides a penalty if the control input exceeds its predefined acceptable value [17,25].

This paper is organized as follows. Section 2 contains the second order plant model and the selection of the time-varying sliding line. Moreover, the control signal, which guarantees the stable sliding motion, is introduced. In Sect. 3 the quality index is presented and its value is optimized. We consider two cases: when the sliding line moves for the whole regulation process and when the sliding line moves for the certain period of time, after that it stops and remains fixed to the end of the control process. Furthermore, a more beneficial control strategy is selected. In Sect. 4 the computer simulations are presented and the paper is concluded in Sect. 5.

## 2 Sliding Mode Controller Design

In this paper we will analyze the continuous-time, second order system given by the following state equations

$$\begin{cases} \frac{d}{dt}x_1(t) = x_2(t) \\ \frac{d}{dt}x_2(t) = f(x_1(t), x_2(t), t) + d(t) + bu(t) \end{cases}, \tag{1}$$

where $x_i(t)$ for $i = 1, 2$ are state variables. Moreover $x_1(t)$ can be considered as system's position and $x_2(t)$ as its velocity. The initial conditions are: $x_1(0) \neq 0$ and $x_2(0) = 0$. Function $f$ is any function of system's position, velocity and time and describes modeling uncertainties. External, unknown disturbances are denoted by $d(t)$. However, the absolute value of these disturbances is bounded from above by a predefined, positive value. Furthermore, the following inequality $|f(x_1(t), x_2(t), t) + d(t)| \leq M$, where $M > 0$ has to be satisfied. Parameter $b$ is related to the control input $u$. Our main goal is to eliminate the reaching phase in order to ensure the robustness for the whole regulation process. Moreover, we want to drive the representative point of our system to the predefined demand state selected as $x_1(t_f) = x_2(t_f) = 0$, where $t_f$ denotes the regulation time. In order to achieve this property we introduce the time-varying sliding line $s(t) = 0$, where the sliding variable $s$ is given as follows

$$s(t) = x_2(t) + c(t)\,\mathrm{sgn}[x_1(t)]\sqrt{|x_1(t)|}. \tag{2}$$

Function $c$ describes the movement of the time-varying sliding line and takes a form

$$c(t) = \begin{cases} \alpha t & \text{for} \quad t \leq t_0 \\ \alpha t_0 & \text{for} \quad t > t_0 \end{cases}. \tag{3}$$

Parameter $\alpha$ describes the speed of the evolution of our switching curve. Therefore, it has to be positive in order to ensure the stability of our system and the convergence of the representative point to the demand state in finite time.

Moment $t_0$ denotes the time in which our sliding line stops and further it remains fixed to the end of the control process. Let us observe, that $s(0) = 0$, which results in the elimination of the reaching phase. In order to fulfill the designated goals we propose the following control signal

$$u(t) = -\frac{1}{b}\left\{ \operatorname{sgn}[x_1(t)]\sqrt{|x_1(t)|}\frac{d}{dt}c(t) + \frac{x_2(t)\,c(t)}{2\sqrt{|x_1(t)|}} + M\operatorname{sgn}[s(t)] \right\}. \tag{4}$$

## 3 The Quality Index Minimization

In this section we will introduce and minimize the quality index. One of its summands will be Integral Absolute Error (IAE) and its second summand will be the penalty function, which increases if the control signal exceeds its admissible value $u_{\max}$. Therefore, the most desirable situation is, when the inequality

$$\forall_{t\geq 0}\, |u(t)| \leq u_{\max} \tag{5}$$

is true. Transforming the above inequality and using (4) we obtain

$$\left| \frac{1}{b}\operatorname{sgn}[x_1(t)]\sqrt{|x_1(t)|}\frac{d}{dt}c(t) + \frac{x_2(t)\,c(t)}{2b\sqrt{|x_1(t)|}} \right| \leq u_{\max} - \frac{M}{|b|}. \tag{6}$$

To simplify latter equations we will use the substitution

$$U = u_{\max} - \frac{M}{|b|}. \tag{7}$$

Now we can introduce our quality index as follows

$$Q_{IAE} = \int_0^{t_f} |x_1(t)|\,dt + q\left[\frac{\max\left|u(t) - \frac{M}{b}\operatorname{sgn}[s(t)]\right|}{U}\right]^n, \tag{8}$$

where $q$ is a positive weighting factor and $n \geq 1$ determines how stretchable the constraint is. First summand evaluates the quality of our regulation method and the second one provides a penalty for the excessive values of the control input. In order to calculate the quality index we will consider two cases. First, in order to calculate IAE we have to find the formula for the system's position. From the fact that our control signal guarantees the stable sliding motion along the++++ sliding line and (2) we have that

$$x_2(t) = -c(t)\operatorname{sgn}[x_1(t)]\sqrt{|x_1(t)|}. \tag{9}$$

Substituting (9) into (1) one can get that

$$\frac{d}{dt}x_1(t) + c(t)\operatorname{sgn}[x_1(t)]\sqrt{|x_1(t)|} = 0. \tag{10}$$

Solving the above differential equation we obtain the formula for the square root of the absolute value of the system's position as follows

$$\sqrt{|x_1(t)|} = \begin{cases} C_1 - \frac{1}{4}\alpha t^2 & \text{for} \quad t < t_0 \\ C_2 - \frac{1}{2}\alpha t_0 t & \text{for} \quad t_0 \le t < t_f \\ 0 & \text{for} \quad t \ge t_f \end{cases}, \tag{11}$$

where $C_1$ and $C_2$ are positive constants, obtained during the integration process. However, function (11) has to be continuous. Therefore we substitute $t = t_0$ and $t = t_f$ into (11) obtaining

$$|x_1(t)| = \begin{cases} \left(\sqrt{|x_1(0)|} - \frac{1}{4}\alpha t^2\right)^2 & \text{for} \quad t < t_0 \\ \left(\sqrt{|x_1(0)|} + \frac{1}{4}\alpha t_0^2 - \frac{1}{2}\alpha t_0 t\right)^2 & \text{for} \quad t_0 \le t < t_f \\ 0 & \text{for} \quad t \ge t_f \end{cases}. \tag{12}$$

Moreover, the settling time can be expressed as follows

$$t_f = \frac{1}{2}t_0 + \frac{2\sqrt{|x_1(0)|}}{\alpha t_0}. \tag{13}$$

However, in this case the sliding line moves from the beginning of the control process, after that it stops and remains fixed. In the second case the representative point reaches the demand state along the moving sliding line. Now we have only two time intervals and the absolute value of the system's position is of the form

$$|x_1(t)| = \begin{cases} \left(\sqrt{|x_1(0)|} - \frac{1}{4}\alpha t^2\right)^2 & \text{for } t \le t_f \\ 0 & \text{for } t > t_f \end{cases} \tag{14}$$

and

$$t_f = \frac{2\sqrt[4]{|x_1(0)|}}{\sqrt{\alpha}}. \tag{15}$$

We will consider both cases and calculate the maximum of the control signal by determining optimal values of parameters $\alpha$ and $t_0$.

### 3.1 Convergence to the Demand State Along the Moving Sliding Line

In this subsection we will consider the situation, when $t_0 \ge t_f$. Therefore, we will be using Eq. (14). The Integral Absolute Error is equal to

$$I = \int_0^{t_f} \left(\sqrt{|x_1(0)|} - \frac{1}{4}\alpha t^2\right)^2 dt = |x_1(0)|\, t_f - \frac{\alpha t_f^3}{6}\sqrt{|x_1(0)|} + \frac{\alpha^2 t_f^5}{80}. \tag{16}$$

Substituting (15) into the above equation one can get

$$I = \frac{16\,|x_1(0)|^{\frac{5}{4}}}{15\sqrt{\alpha}}. \tag{17}$$

Now we have to determine the formula for the penalty function. In order to achieve this goal, we will calculate the maximum of the control signal without the switching factor $\frac{M}{b}\operatorname{sgn}\left[s\left(t\right)\right]$. Calculating its derivative we have

$$\frac{d}{dt}\left(u\left(t\right)-\frac{M}{b}\operatorname{sgn}\left[s\left(t\right)\right]\right)=-\frac{\frac{d}{dt}\sqrt{\left|x_1\left(t\right)\right|\frac{d}{dt}c\left(t\right)}}{b}+\frac{d}{dt}\left[\frac{c^2\left(t\right)}{2b}\right]. \tag{18}$$

Taking into account that in this case $c\left(t\right)=\alpha t$ we obtain that

$$\frac{d}{dt}\left(u\left(t\right)-\frac{M}{b}\operatorname{sgn}\left[s\left(t\right)\right]\right)=-\frac{\frac{d}{dt}\sqrt{\left|x_1\left(t\right)\right|\alpha}}{b}+\frac{\alpha^2 t}{b}. \tag{19}$$

After calculating the derivative of the square root of the system's position we get that the above equation takes the form

$$\frac{d}{dt}\left(u\left(t\right)-\frac{M}{b}\operatorname{sgn}\left[s\left(t\right)\right]\right)=\frac{3\alpha^2 t}{2b}. \tag{20}$$

Equating (20) to 0 we can observe that the control signal without the switching factor is a monotonic function. Therefore, its absolute value reaches its extremum for $t=0$ or $t=t_f$. Substituting these values we get

$$\max\left|u\left(0\right)-\frac{M}{b}\operatorname{sgn}\left[s\left(0\right)\right]\right|=\frac{\sqrt{\left|x_1\left(0\right)\right|\alpha}}{\left|b\right|}, \tag{21}$$

$$\max\left|u\left(t_f\right)-\frac{M}{b}\operatorname{sgn}\left[s\left(t_f\right)\right]\right|=\frac{2\sqrt{\left|x_1\left(0\right)\right|\alpha}}{\left|b\right|}. \tag{22}$$

One can notice, that (22) is always greater than (21). Hence, substituting (17) and (22) into (8) we present the quality index as follows

$$Q_{IAE}=\frac{16\left|x_1\left(0\right)\right|^{\frac{5}{4}}}{15\sqrt{\alpha}}+q\left[\frac{2\sqrt{\left|x_1\left(0\right)\right|\alpha}}{\left|b\right|U}\right]^n. \tag{23}$$

We can observe, that in this case the quality index is not affected by the parameter $t_0$. It is caused by the fact that the representative point reaches the demand state along the moving sliding line. In order to obtain the optimal value of parameter $\alpha$ we will calculate the derivative of (23) with respect to $\alpha$ obtaining

$$\frac{d}{d\alpha}Q_{IAE}=-\frac{8\left|x_1\left(0\right)\right|^{\frac{5}{4}}}{15\sqrt{\alpha^3}}+\frac{2^n qn\sqrt{\left|x_1\left(0\right)\right|^n}\alpha^{n-1}}{\left|b\right|^n U^n} \tag{24}$$

Equating (24) to 0 we get that the optimal parameter $\alpha$ for which $Q_{IAE}$ is minimized is given by

$$\alpha=\sqrt[2n+1]{\frac{64\left|b\right|^{2n}U^{2n}}{225\cdot 2^{2n}q^2 n^2\left|x_1\left(0\right)\right|^{\frac{2n-5}{2}}}} \tag{25}$$

Substituting (25) into (23) we obtain that

$$Q_{IAE} = \frac{16\,|x_1(0)|^{\frac{5}{4}}}{15\sqrt{\alpha}}\left(1+\frac{1}{2n}\right). \tag{26}$$

One can observe that if $n \to \infty$, then $Q_{IAE}$ is equal to the value of Integral Absolute Error.

### 3.2 Convergence to the Demand State Along the Fixed Sliding Line

This subsection comprises the case, when the sliding line starts moving, after that it stops at time $t_0$ and remains fixed to the end of the control process. Now the Integral Absolute Error takes the form

$$\begin{aligned} I &= \int_0^{t_0}\left(\sqrt{|x_1(0)|}-\frac{1}{4}\alpha t^2\right)^2 dt + \int_{t_0}^{t_f}\left(\sqrt{|x_1(0)|}+\frac{1}{4}\alpha t_0^2-\frac{1}{2}\alpha t_0 t\right)^2 dt \\ &= \frac{t_0}{2}|x_1(0)| - \frac{\alpha t_0^3}{24}\sqrt{|x_1(0)|} + \frac{\alpha^2 t_0^5}{480} + \frac{2\sqrt{|x_1(0)|^3}}{3\alpha t_0}. \end{aligned} \tag{27}$$

In order to derive the formula for the quality index, we compute the maximum value of the control signal without the switching factor. The only possible maximal values are: $u(0), \lim_{t\to t_0^-} u(t)$ and $u(t)$ for $t \in [t_0, t_f)$. From (4) and (9) we get that first two summands of our control signal have opposite signs. Moreover, at the moment 0 the second summand is equal to 0 and when the sliding line stops, then the first summand is equal to 0. Hence, we conclude that values of the control signal without the switching factor at $u(0)$ and $u(t)$ for $t \in [t_0, t_f)$ have opposite signs. In the case, when the sliding line moves, the value at the moment $t = 0$ is the same as in the previous subsection. Therefore, the quality index is given as follows

$$Q_{IAE} = \frac{t_0}{2}|x_1(0)| - \frac{\alpha t_0^3}{24}\sqrt{|x_1(0)|} + \frac{\alpha^2 t_0^5}{480} + \frac{2\sqrt{|x_1(0)|^3}}{3\alpha t_0} + q\left[\frac{\sqrt{|x_1(0)|}\alpha}{|b|\,U}\right]^n. \tag{28}$$

Now we compute $\lim_{t\to t_0^-} u(t)$ obtaining

$$\lim_{t\to t_0^-}\left|u(t)-\frac{M}{b}\operatorname{sgn}[s(t_f)]\right| = \left|\frac{3\alpha^2 t_0^2}{4|b|} - \frac{\alpha\sqrt{|x(0)|}}{|b|}\right|. \tag{29}$$

Comparing (21) with (29) one can observe that (21) is always greater if (21) and (29) have the same signs. However, if these equations have opposite signs we are not able to settle which one is greater. Hence, we have to consider the second case when the sliding line is fixed. Using (3) we have that $c(t) = \alpha t_0$, so $\frac{d}{dt}c(t) = 0$. Hence, using that fact and (4) one can get that

$$\frac{d}{dt}\left(u(t)-\frac{M}{b}\operatorname{sgn}[s(t)]\right) = 0. \tag{30}$$

Therefore, in this time interval the control signal is constant. It means that we can substitute any $t \in [t_0, t_f)$. We select $t = t_0$ obtaining

$$\max \left| u(t) - \frac{M}{b} \operatorname{sgn}[s(0)] \right| = \frac{\alpha^2 t_0^2}{2|b|}. \tag{31}$$

Hence, in this case the quality index is of the form

$$Q_{IAE} = \frac{t_0}{2} |x_1(0)| - \frac{\alpha t_0^3}{24} \sqrt{|x_1(0)|} + \frac{\alpha^2 t_0^5}{480} + \frac{2\sqrt{|x_1(0)|^3}}{3\alpha t_0} + q \left[ \frac{\alpha^2 t_0^2}{2|b| U} \right]^n. \tag{32}$$

Now, we only have to examine which value: (29) or (31) is greater if these values have the same signs. After calculations we obtain that (31) is greater than (29) if the inequality $\sqrt{|x_1(0)|} - \frac{1}{4}\alpha t^2 > 0$ holds. However, it is always true from the fact that (11) has to be non-negative. Therefore, the maximum of the control signal will never be equal to (29). Hence, the quality index will take the form (28) or (32). In order to derive the minimum value of $Q_{IAE}$ we calculate derivatives of (28) and (32) with respect to $\alpha$ and $t_0$, equate them to 0 obtaining that our index has no stationary points. Therefore, the extremum will be reached on the boundary of the set in which (28) and (32) holds. Hence, we substitute

$$\alpha = \frac{2\sqrt{|x_1(0)|}}{t_0^2} \tag{33}$$

to both (28) and (32) obtaining

$$Q_{IAE} = \frac{91}{120} |x_1(0)| t_0 + \frac{2^n q |x_1(0)|^n}{|b|^n U^n t_0^{2n}}. \tag{34}$$

Searching for the optimal parameter $t_0$ we will calculate the derivative of (34) with respect to $t_0$ getting

$$\frac{\partial}{\partial t_0} Q_{IAE} = \frac{91}{120} |x_1(0)| - \frac{2^{n+1} q n |x_1(0)|^n}{|b|^n U^n t_0^{2n+1}}. \tag{35}$$

Equating (35) to zero we have that the optimal parameter $t_0$ is given as

$$t_0 = \sqrt[2n+1]{\frac{120 \cdot 2^{n+1} q n |x_1(0)|^{n-1}}{91 |b|^n U^n}}. \tag{36}$$

Substituting (36) into (33) we obtain that the optimal parameter $\alpha$ is

$$\alpha = \sqrt[2n+1]{\frac{8281 |b|^{2n} U^{2n}}{28800 q^2 n^2 |x_1(0)|^{\frac{2n-5}{2}}}}. \tag{37}$$

Calculating the minimum value of our quality index $Q_{IAE}$ in the case, when the sliding line stops at the time $t_0$, we obtain that it is equal to

$$Q_{IAE} = \frac{91}{120} |x_1(0)| t_0 \left(1 + \frac{1}{2n}\right). \tag{38}$$

Again, we can see that if $n \to \infty$, then $Q_{IAE}$ is equal to the value of Integral Absolute Error. Moreover, in this case optimal parameters are

$$\begin{cases} \alpha = \frac{|b|u_{\max} - M}{\sqrt{|x_1(0)|}} \\ t_0 = \frac{\sqrt{2}\sqrt{|x_1(0)|}}{\sqrt{|b|u_{\max} - M}} \end{cases} \tag{39}$$

These parameters are equal to the optimal ones in the strategy presented in [26], when the quality index is IAE and the control signal is strictly constrained.

### 3.3 Selecting the Optimal Strategy

In order to choose the most beneficial strategy, which minimizes our quality index we have to compare (26) for $\alpha$ as in (25) with (38) for $t_0$ as in (36). After calculations and simplifications, we obtain that values of these quality indexes differ only in a constant that multiplies a factor containing the same expression of the parameters of our system. That constant in (26) is equal to $\frac{16384\sqrt{2}}{14400}$ and in (38) is equal to $\frac{16562}{14400}$. Therefore, one can observe, that the more favorable strategy, which minimizes our quality index is the one in which the sliding line stops at some moment and remains fixed to the end of the control process.

## 4 Simulation Example

This section contains the simulation example, which verifies the proposed control strategy. We select the following parameters of the system (1): function $f$ is a sine function of the system's position, external disturbances switch between $-1$ and 1 ten times per second. Hence, in order to compensate for these perturbations, $M$ has to be equal to 2. Parameter $b = 2$ and $u_{\max} = 15$. Therefore, $U = u_{\max} - \frac{M}{b} = 14$. The system's initial position is $-100$. The weighting factor $q = 3$ and $n = 20$. From the fact that the more beneficial strategy is the one in which the sliding line stops at some moment and remains fixed for the rest of the control process we substitute these parameters into (36) and (37) obtaining $t_0 \approx 2.6387$ and $\alpha \approx 2.8725$. Calculating the settling time we get $t_f \approx 3.958$. Integral Absolute Error is $I \approx 200.0988$ and the whole quality index is $Q_{IAE} \approx 205.1013$. Fig. 1 presents the state trajectory. One can observe that the system's position increases monotonically until it reaches the demand value 0. Moreover, the finite-time convergence is obtained. In Fig. 2 the control signal without the switching summand decreases monotonically until the sliding line stops moving. From the moment $t_0$ it is constant up to $t_f$. After reaching the demand state, the control signal switches between $-1$ and 1 in order to maintain the representative point in the demand state. We can observe that the control signal slightly exceeds $u_{\max}$ at the very beginning of the control process and when the representative point slides along the fixed switching line.

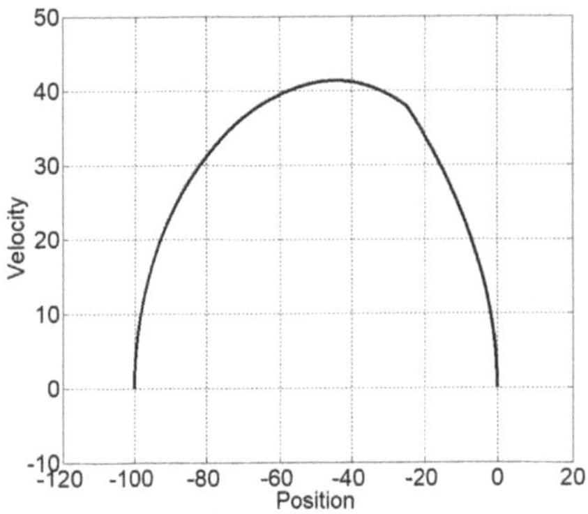


Fig. 1. State Trajectory

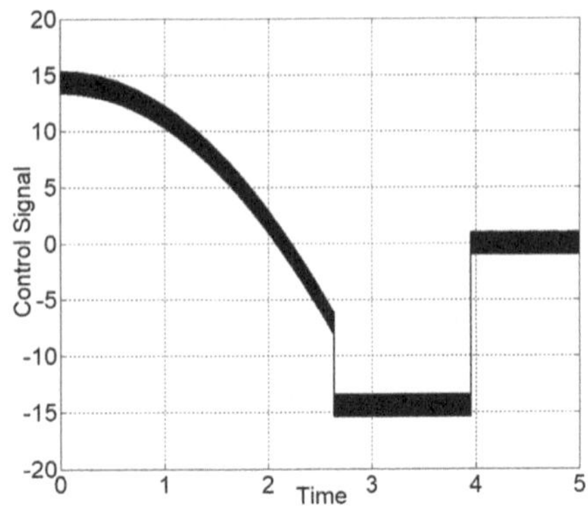


Fig. 2. Control Signal

## 5 Conclusions

In this paper the sliding mode control technique was applied to a second order, continuous-time plant. The time-varying switching line was selected in order to eliminate the reaching phase, which resulted in the robustness to perturbations. Moreover, the parabolic shape of that line ensured the finite regulation time. The quality index was introduced as a sum of the Integral Absolute Error and a penalty function connected with the control signal constraint. Two cases were considered: when the sliding line moves for the whole regulation process and when the sliding line starts moving, then at some time instant stops and remains fixed to the end of the control process. Subsequently, the more beneficial strategy, which minimizes our quality index is selected. Computer simulations verified the theoretical considerations presented in this paper.

## References


1. Bartoszewicz, A., Kaynak, O., Utkin, V.I.: Sliding mode control in industrial appl. IEEE Trans. Ind. Electron. (Special Section) **55**, 3805–4103 (2008)
2. Bartoszewicz, A., Patton, R.: Sliding mode control. Int. J. Adapt. Control Signal Process. (Special Issue) **21**, 635–822 (2007)
3. Basin, M., Fridman, L., Shi, P.: Optimal sliding mode algorithms for dynamic systems. J. Franklin Inst. (Special Issue) **349**, 1317–1616 (2012)
4. Gao, W.: Variable structure control. IEEE Trans. Ind. Electron. (Special Issue) **40**, 1–165 (1993)
5. Kaynak, O., Bartoszewicz, A., Utkin, V.I.: Sliding mode control in industrial applications. IEEE Trans. Ind. Electron. (Special Issue) **56**, 3271–3274 (2009)
6. Misawa, A., Utkin, V.I.: Variable structure systems. Trans. ASME J. Dyn. Syst. Meas. Contr. (Special Issue) **122**, 585–819 (2000)
7. Shtessel, Y., Spurgeon, S., Fridman, L.: Advances in sliding mode observation and estimation. Int. J. Syst. Sci. (Special Issue) **38**, 845–942 (2007)
8. Utkin, V.I.: Sliding mode control. Int. J. Control (Special Issue) **57**, 1003–1259 (1993)

9. Bartoszewicz, A., Leśniewski, P.: Reaching law based sliding mode congestion control for communication networks. IET Control Theory Appl. **8**, 1914–1920 (2014)
10. Zhang, X., Sun, L., Zhao, K., Sun, L.: Nonlinear speed control for PMSM system using sliding-mode control and disturbance compensation techniques. IEEE Trans. Power Electron. **28**, 1358–1365 (2013)
11. Choi, S.B., Park, D.W., Jayasuriya, S.: A time-varying sliding surface for fast and robust tracking control of second-order uncertain systems. Automatica **30**(5), 899–904 (1994)
12. Bartoszewicz, A.: A Comment on A time-varying sliding surface for fast and robust tracking control of second-order uncertain systems. Automatica **31**(12), 1893–1895 (1995)
13. Yilmaz, C., Hurmuzlu, Y.: Eliminating the reaching phase from variable structure control. J. Dyn. Syst. Meas. Contr. **122**(4), 753–757 (2000)
14. Bartoszewicz, A., Nowacka, A.: Switching plane design for the sliding mode control of systems with elastic input constraints. Proc. Inst. Mech. Eng. Part I J. Syst. Control Eng. **219**(6), 393–403 (2005)
15. Bartoszewicz, A., Nowacka, A.: Reaching phase elimination in variable structure control of the third order system with state constraints. Kybernetika **42**(1), 111–126 (2006)
16. Bartoszewicz, A., Nowacka, A.: Sliding mode control of the third-order system subject to velocity, acceleration and input signal constraints. Int. J. Adapt. Control Signal Process. **21**(8–9), 779–794 (2007)
17. Bartoszewicz, A., Nowacka-Leverton, A.: SMC without the reaching phase - the switching plane design for the third-order system. Proc. IET Part D Control Theory Appl. **1**(5), 1461–1470 (2007)
18. Bartoszewicz, A., Nowacka-Leverton, A.: Time-Varying Sliding Modes for Second and Third Order Systems. Springer, Berlin (2009)
19. Bartoszewicz, A., Nowacka-Leverton, A.: ITAE optimal sliding modes for third order systems with input signal and state constraints. IEEE Trans. Autom. Control **55**(8), 1928–1932 (2010)
20. Nowacka-Leverton, A., Michałek, M., Pazderski, D., Bartoszewicz, A.: Experimental verification of SMC with moving switching lines applied to hoisting crane vertical motion control. ISA Trans. **51**(2), 682–693 (2012)
21. Bartoszewicz, A.: Time-varying sliding modes for second-order systems. Proc. IEE Part D Control Theory Appl. **143**(5), 455–462 (1996)
22. Man, Z., Yu, X.H.: Terminal sliding mode control of MIMO linear systems. IEEE Trans. Circuits Syst. I Fundam. Theory Appl. **44**(11), 1065–1070 (1997)
23. Man, Z., Paplinski, A., Wu, H.R.: A robust MIMO terminal sliding mode control scheme for rigid robotic manipulators. IEEE Trans. Autom. Control **39**(12), 2464–2469 (1994)
24. Tang, Y.: Terminal sliding mode control for rigid robots. Automatica **34**(1), 51–56 (1998)
25. Nowacka-Leverton, A., Bartoszewicz, A.: Elastic and conventional constraints in sliding mode control of second order system. In: 2008 34th Annual Conference of IEEE Industrial Electronics, pp. 182–187. IEEE (2008)
26. Pietrala, M., Jaskuła, M., Bartoszewicz, A.: IAE optimal sliding mode control for second order dynamical systems. In: 2018 19th International Carpathian Control Conference (ICCC), pp. 169–174. IEEE (2018)

# Constraining the State and Control Signal Variation Using the Sliding Mode Control

Marek Jaskuła(✉)

Institute of Automatic Control, Łodź University of Technology,
18/22 Stefanowskiego St., 90-924 Łodź, Poland
marek.jaskula@edu.p.lodz.pl

**Abstract.** In this article the discrete-time system violated by unknown, but bounded perturbations was considered. Using the reaching law approach the sliding mode controller was designed. The fastest, finite-time, monotonic convergence of the representative point to the predefined, fixed sliding hyperplane in the presence of state and control input constraints was analyzed. Moreover, the reaching law was improved by considering also the limitation of state and control signal variations. Proposed strategy does not require the fulfillment of the matching conditions.

**Keywords:** Sliding mode control · Discrete-time systems · State variation constraint · Control signal variation constraint · Reaching law

## 1 Introduction

Recently, the sliding mode control technique is being applied to a wide range of uncertain systems. This regulation method is popular mostly due to its effectiveness, robustness and computational efficiency. The origins of this approach can be found in the works of Utkin [19] and Emelyanov [10], where it was implemented only for continuous-time systems. Later, in [2,12,17,20], sliding mode control was successfully extended also to the discrete-time domain. Since then it has found many applications, especially in power electronics and network systems [3,6,11,22].

This control strategy can be divided into two phases. At first, the representative point is driven from the initial state to the predefined switching hyperplane. Then, during the so-called sliding phase, it glides along the given manifold to the desired state.

The process of designing the sliding mode controller begins with the selection of the switching hyperplane parameters. Subsequently, the control input can be calculated using one of two different ways: by proving the stability of the sliding motion or introducing the so-called reaching law. In this paper we consider only the second approach which focuses on the prior determination of the desirable dynamics of the sliding variable.


R. Szewczyk et al. (Eds.): AUTOMATION 2020, AISC 1140, pp. 24–33, 2020.
https://doi.org/10.1007/978-3-030-40971-5_3

Firstly, the reaching law technique was used only in continuous-time domain [13]. Nevertheless, shortly afterwards, Gao, Wang and Homaifa in [14] extended this approach also for discrete-time systems. Over the years the reaching law technique was developed by many researchers [4,5,7–9,15,18,21].

Despite the advantages, the sliding mode control strategy may result in undesirably high absolute values of state variables, control signal and their variations. Referring to our work [16], in this paper we extend our previous approach by constraining changes of both the state and the control input. Moreover, we demand the fastest, finite-time, monotonic convergence of the representative point to the predefined switching manifold.

This paper is organized as follows. Section 2 presents the selection of the sliding hyperplane, the reaching law and the design of the control signal. In Sect. 3 we consider both state and control signal constraints while in Sect. 4 we analyze their variations. Section 5 contains sufficient conditions for the fastest, finite-time, monotonic convergence of the representative point to the demand state in the presence of given limitations. Section 6 comprises the simulation example and finally the conclusions are presented in Sect. 7.

## 2 Sliding Mode Controller Design

To begin with, we consider the $n-$dimensional, discrete-time system under the influence of unknown, bounded perturbations. The plant dynamics is described by the following state equation

$$\boldsymbol{x}(k+1) = \boldsymbol{A}\boldsymbol{x}(k) + \boldsymbol{b}u(k) + \boldsymbol{d}(k), \tag{1}$$

where $\boldsymbol{x}(k) = [x_1(k), \ldots, x_n(k)]^T$ and $\boldsymbol{A}$ are the state vector and the state matrix, respectively. The control input, denoted by $u(k)$, is a scalar value connected with the $n \times 1$ dimensional vector $\boldsymbol{b} = [b_1, \ldots, b_n]^T$. External disturbances $\boldsymbol{d}(k) = [d_1(k), \ldots, d_n(k)]^T$ are bounded in the following sense: $\max |d_i(k)| = D_i$ for each $i \in \{1, \ldots, n\}$. Therefore, we do not require the fulfillment of the matching conditions. At first, in order to design the discrete-time sliding mode controller, we define the following switching hyperplane

$$s(k) = \boldsymbol{c}^T\boldsymbol{x}(k) = 0, \tag{2}$$

where $\boldsymbol{c}^T = [c_1, \ldots, c_{n-1}, 1]$. We select vector $\boldsymbol{c}$ in such a manner that condition $\boldsymbol{c}^T\boldsymbol{b} \neq 0$ is satisfied. Moreover, we do not require crossing the sliding manifold in each consecutive step same as in [1,2]. Using (1), (2) and the following reaching law

$$s(k+1) = s(k) - K\text{sgn}[s(k)] + \boldsymbol{c}^T\boldsymbol{d}(k), \tag{3}$$

where $K$ is a strictly positive constant, we compute the control signal

$$u(k) = \left(\boldsymbol{c}^T\boldsymbol{b}\right)^{-1}\Big\{-\boldsymbol{c}^T\boldsymbol{A}\boldsymbol{x}(k) + \boldsymbol{c}^T\boldsymbol{x}(k) - K\text{sgn}\left[\boldsymbol{c}^T\boldsymbol{x}(k)\right]\Big\}. \tag{4}$$

The fastest, finite-time, monotonic convergence of the representative point to the switching hyperplane is obtained when the condition

$$|s(k)| - |s(k+1)| \geq \lambda > 0, \tag{5}$$

is satisfied for the whole reaching phase. Taking into account the reaching law (3) we rewrite (5) as

$$K - \boldsymbol{c}^T \boldsymbol{d}(k) \operatorname{sgn}[s(k)] \geq \lambda. \tag{6}$$

The greatest possible influence of perturbations on the sliding variable can be calculated in the following manner

$$D_{\max} = |c_1| D_1 + \ldots |c_{n-1}| D_{n-1} + D_n. \tag{7}$$

Thus, (6) is fulfilled when

$$K - D_{\max} \geq \lambda. \tag{8}$$

## 3 Control Signal and State Constraints

In this section we will focus on constraining only the control signal and state variables referring to the results obtained in [16]. In order to constrain the absolute value of the input we need to satisfy

$$-r_u \leq u(k) \leq r_u \tag{9}$$

for every $k \in \mathbb{N}$, where $r_u$ is a positive, real number. Using (4) we can rewrite (9) in the following form

$$-r_u \leq \left(\boldsymbol{c}^T \boldsymbol{b}\right)^{-1} \left[-\boldsymbol{c}^T \boldsymbol{A}\boldsymbol{x}(k) + \boldsymbol{c}^T \boldsymbol{x}(k) - K \operatorname{sgn}\left(\boldsymbol{c}^T \boldsymbol{x}(k)\right)\right] \leq r_u. \tag{10}$$

For the sake of clarity we define $\boldsymbol{g}_u$ as follows

$$\boldsymbol{g}_u = \left(\boldsymbol{c}^T \boldsymbol{b}\right)^{-1} \left(-\boldsymbol{c}^T \boldsymbol{A} + \boldsymbol{c}^T\right). \tag{11}$$

Using the results obtained in [16] one can write the greatest possible value of $K$, for which the absolute value of the control signal is limited by $r_u$, as

$$K_u = \boldsymbol{c}^T \boldsymbol{b} \operatorname{sgn}\left(\boldsymbol{c}^T \boldsymbol{x}(k)\right) \boldsymbol{g}_u \boldsymbol{x}(k) + \left|\boldsymbol{c}^T \boldsymbol{b}\right| r_u. \tag{12}$$

Now we will move to the issue of constraining all of the state variables i.e. satisfying

$$-r_i \leq x_i(k) \leq r_i, \tag{13}$$

for each $i \in \{1, \ldots, n\}$ and $k \in \mathbb{N}$. Assuming that $D_i < \frac{1}{2} r_i$ for every $i \in \{1, \ldots, n\}$ and substituting (4) into (1) we get

$$\begin{aligned} \boldsymbol{x}(k+1) = {} & \boldsymbol{A}\boldsymbol{x}(k) - \boldsymbol{b}\left(\boldsymbol{c}^T \boldsymbol{b}\right)^{-1} \boldsymbol{c}^T \boldsymbol{A}\boldsymbol{x}(k) + \boldsymbol{b}\left(\boldsymbol{c}^T \boldsymbol{b}\right)^{-1} \boldsymbol{c}^T \boldsymbol{x}(k) \\ & - \boldsymbol{b}\left(\boldsymbol{c}^T \boldsymbol{b}\right)^{-1} K \operatorname{sgn}\left[\boldsymbol{c}^T \boldsymbol{x}(k)\right] + \boldsymbol{d}(k). \end{aligned} \tag{14}$$

Multiplying the above equation by the $i$–th versor $\boldsymbol{e}_i$ and using the notation

$$\boldsymbol{G} = \boldsymbol{A} + \boldsymbol{b}\left(\boldsymbol{c}^T\boldsymbol{b}\right)^{-1}\left(-\boldsymbol{c}^T\boldsymbol{A} + \boldsymbol{c}^T\right) = \boldsymbol{A} + \boldsymbol{b}\boldsymbol{g}_u \tag{15}$$

we obtain

$$x_i\left(k+1\right) = \boldsymbol{e}_i\left\{\boldsymbol{G}\boldsymbol{x}\left(k\right) - K\left(\boldsymbol{c}^T\boldsymbol{b}\right)^{-1}\boldsymbol{b}\mathrm{sgn}\left[\boldsymbol{c}^T\boldsymbol{x}\left(k\right)\right] + \boldsymbol{d}\left(k\right)\right\}. \tag{16}$$

Referring to [16] we get the following form of the the greatest possible $K$ for which the condition (13) is fulfilled

$$K_i = \boldsymbol{c}^T\boldsymbol{b}\left(\boldsymbol{e}_i\boldsymbol{b}\right)^{-1}\mathrm{sgn}\left[\boldsymbol{c}^T\boldsymbol{x}\left(k\right)\right]\boldsymbol{g}_i\boldsymbol{x}\left(k\right) + \left|\boldsymbol{c}^T\boldsymbol{b}\left(\boldsymbol{e}_i\boldsymbol{b}\right)^{-1}\right|\left(r_i - D_i\right), \tag{17}$$

where $\boldsymbol{g}_i = \boldsymbol{e}_i\boldsymbol{G}$ and $\left(\boldsymbol{e}_i\boldsymbol{b}\right) \neq 0$.

## 4 Variation of State Constraints

The purpose of our work is to constrain the absolute value of the difference between each two consecutive values of the $i$–th state variable by $\delta_i > 0$ for $i \in \{1, \dots, n\}$, i.e.

$$\left|x_i\left(k+1\right) - x_i\left(k\right)\right| \leq \delta_i. \tag{18}$$

Therefore, our goal is to select the greatest possible parameter $K$, which ensures the fulfillment of inequality (18) for each $k \in \mathbb{N}$. From (16) we get

$$x_i\left(k+1\right) - x_i\left(k\right) = \boldsymbol{e}_i\left\{\boldsymbol{G}\boldsymbol{x}\left(k\right) - K\left(\boldsymbol{c}^T\boldsymbol{b}\right)^{-1}\boldsymbol{b}\mathrm{sgn}\left[\boldsymbol{c}^T\boldsymbol{x}\left(k\right)\right] + \boldsymbol{d}\left(k\right)\right\} - x_i\left(k\right). \tag{19}$$

One can rewrite the above equation as follows

$$x_i\left(k+1\right) - x_i\left(k\right) = \left(\boldsymbol{g}_i - \boldsymbol{e}_i\right)\boldsymbol{x}\left(k\right) - K\left(\boldsymbol{c}^T\boldsymbol{b}\right)^{-1}\boldsymbol{e}_i\boldsymbol{b}\mathrm{sgn}\left[\boldsymbol{c}^T\boldsymbol{x}\left(k\right)\right] + d_i\left(k\right). \tag{20}$$

Substituting (20) into (18) and using the definition of the absolute value we obtain

$$\begin{cases} K\left(\boldsymbol{c}^T\boldsymbol{b}\right)^{-1}\boldsymbol{e}_i\boldsymbol{b}\mathrm{sgn}\left[\boldsymbol{c}^T\boldsymbol{x}\left(k\right)\right] \geq \left(\boldsymbol{g}_i - \boldsymbol{e}_i\right)\boldsymbol{x}\left(k\right) - \delta_i + d_i\left(k\right) \\ K\left(\boldsymbol{c}^T\boldsymbol{b}\right)^{-1}\boldsymbol{e}_i\boldsymbol{b}\mathrm{sgn}\left[\boldsymbol{c}^T\boldsymbol{x}\left(k\right)\right] \leq \left(\boldsymbol{g}_i - \boldsymbol{e}_i\right)\boldsymbol{x}\left(k\right) + \delta_i + d_i\left(k\right). \end{cases} \tag{21}$$

From the fact, that the absolute value of the external disturbance acting on the $i$–th variable is constrained by $D_i$ we get

$$\begin{cases} K\left(\boldsymbol{c}^T\boldsymbol{b}\right)^{-1}\boldsymbol{e}_i\boldsymbol{b}\mathrm{sgn}\left[\boldsymbol{c}^T\boldsymbol{x}\left(k\right)\right] \geq \left(\boldsymbol{g}_i - \boldsymbol{e}_i\right)\boldsymbol{x}\left(k\right) - \delta_i + D_i \\ K\left(\boldsymbol{c}^T\boldsymbol{b}\right)^{-1}\boldsymbol{e}_i\boldsymbol{b}\mathrm{sgn}\left[\boldsymbol{c}^T\boldsymbol{x}\left(k\right)\right] \leq \left(\boldsymbol{g}_i - \boldsymbol{e}_i\right)\boldsymbol{x}\left(k\right) + \delta_i - D_i. \end{cases} \tag{22}$$

After calculations (similar to the ones in the previous section), the greatest admissible $K$ for which the variation of the $i$–th variable is limited is of the form

$$K_{\delta_i} = \boldsymbol{c}^T\boldsymbol{b}\left(\boldsymbol{e}_i\boldsymbol{b}\right)^{-1}\mathrm{sgn}\left[\boldsymbol{c}^T\boldsymbol{x}\left(k\right)\right]\left(\boldsymbol{g}_i - \boldsymbol{e}_i\right)\boldsymbol{x}\left(k\right) + \left|\boldsymbol{c}^T\boldsymbol{b}\left(\boldsymbol{e}_i\boldsymbol{b}\right)^{-1}\right|\left(\delta_i - D_i\right). \tag{23}$$

### 4.1 Variation of the Control Signal

This subsection focuses on constraining the variation of the control signal. Therefore, we require that the condition

$$-\delta_u \le u(k+1) - u(k) \le \delta_u \tag{24}$$

is true for each $k \in \mathbb{N}$. At first, in order to satisfy (24), we add a new state variable which is equal to the control signal in the previous time instant. Hence, the new state vector is of the form

$$\boldsymbol{x}'(k) = \begin{bmatrix} x_1(k) \\ \vdots \\ x_n(k) \\ u(k-1) \end{bmatrix}. \tag{25}$$

The change of the control signal cannot be defined for $k = 0$, hence we assume that $u(-1) = u(0)$. Consequently, we need to expand matrix $\boldsymbol{A}$, vector $\boldsymbol{b}$ and the disturbance vector $\boldsymbol{d}(k)$ in the following manner

$$\boldsymbol{A}' = \left[\begin{array}{ccc|c} & & & 0 \\ & \boldsymbol{A} & & \vdots \\ & & & 0 \\ \hline 0 & \cdots & 0 & 0 \end{array}\right], \quad \boldsymbol{b}' = \left[\begin{array}{c} \boldsymbol{b} \\ \hline 1 \end{array}\right], \quad \boldsymbol{d}'(k) = \left[\begin{array}{c} \boldsymbol{d}(k) \\ \hline 0 \end{array}\right]. \tag{26}$$

We define new vector $(\boldsymbol{c}')^T$ as follows

$$(\boldsymbol{c}')^T = [c_1, \ldots, c_{n-1}, 1, 0]. \tag{27}$$

Applying the proposed strategy and substituting $K_{\delta_u} = K_{\delta_{n+1}}$ we obtain

$$\begin{aligned} K_{\delta_u} = {} & (\boldsymbol{c}')^T \boldsymbol{b}' \left(\boldsymbol{e}'_{n+1}\boldsymbol{b}'\right)^{-1} \operatorname{sgn}\left[(\boldsymbol{c}')^T \boldsymbol{x}'(k)\right] \left(\boldsymbol{g}'_{n+1} - \boldsymbol{e}'_{n+1}\right) \boldsymbol{x}'(k) \\ & + \left|(\boldsymbol{c}')^T \boldsymbol{b}' \left(\boldsymbol{e}'_{n+1}\boldsymbol{b}'\right)^{-1}\right| \left(\delta_u - D_{n+1}\right). \end{aligned} \tag{28}$$

Let us observe that $(\boldsymbol{c}')^T \boldsymbol{b}' = \boldsymbol{c}^T \boldsymbol{b}$, $\boldsymbol{e}'_{n+1}\boldsymbol{b}' = 1$, $(\boldsymbol{c}')^T \boldsymbol{x}'(k) = \boldsymbol{c}^T \boldsymbol{x}(k)$, $D_{n+1} = 0$ and $\boldsymbol{g}'_{n+1} - \boldsymbol{e}'_{n+1} = [\ \boldsymbol{g}_u \ | -1]$. Therefore, we can rewrite (28) as follows

$$K_{\delta_u} = \boldsymbol{c}^T \boldsymbol{b} \operatorname{sgn}\left[\boldsymbol{c}^T \boldsymbol{x}(k)\right] \left[\boldsymbol{g}_u \boldsymbol{x}(k) - u(k-1)\right] + \left|\boldsymbol{c}^T \boldsymbol{b}\right| \delta_u. \tag{29}$$

Our main objective is to satisfy all of the constraints simultaneously and guarantee the fastest, finite-time, monotonic convergence to the sliding hyperplane. In order to achieve this goal we select $K$ in the following manner

$$K = \min\left\{K_1, ..., K_n, K_{\delta_1}, ..., K_{\delta_n}, K_u, K_{\delta_u}\right\}. \tag{30}$$

However, in the case when we demand only the limitation of state variables and their variations we need to eliminate terms $K_u$ and $K_{\delta_u}$ from (30).

## 5 Sufficient Condition for Monotonic, Finite-Time Convergence

This section contains the sufficient condition which guarantees the fastest, monotonic, finite-time convergence of the representative point to the switching manifold while respecting constraints on input, state variables and their variations. Firstly, we define

$$X = \left\{ \boldsymbol{x} \in \Re^n : \forall_{i\in\{1,\ldots,n\}} |x_i| \leq r_i, |\boldsymbol{c}^T \boldsymbol{x}| \leq |\boldsymbol{c}^T \boldsymbol{x}_0| \right\}, \tag{31}$$

and then we refer to the Theorem 1 in [16]:

**Theorem 1.** *In order to fulfill conditions (8), (9) and (13) it is sufficient that the inequalities*

$$\max_{\boldsymbol{x}\in X} |\boldsymbol{g}_i \boldsymbol{x}| < r_i - D_i - \frac{D_{\max}}{\left| \boldsymbol{c}^T \boldsymbol{b} \left( \boldsymbol{e}_i \boldsymbol{b} \right)^{-1} \right|}, \tag{32}$$

$$\max_{\boldsymbol{x}\in X} |\boldsymbol{g}_u \boldsymbol{x}| < r_u - \frac{D_{\max}}{|\boldsymbol{c}^T \boldsymbol{b}|} \tag{33}$$

*are satisfied for every* $i \in \{1, \ldots, n\}$.

The proof of the above theorem can be found in [16]. Next, we will move to the sufficient condition related to the variation of state variables.

**Theorem 2.** *In order to fulfill conditions (8), (9), (13) and (18) it is sufficient that the inequalities (32), (33) and*

$$\max_{\boldsymbol{x}\in X} |(\boldsymbol{g}_i - \boldsymbol{e}_i)\, \boldsymbol{x}(k)| < \delta_i - D_i - \frac{D_{\max}}{\left| \boldsymbol{c}^T \boldsymbol{b} \left( \boldsymbol{e}_i \boldsymbol{b} \right)^{-1} \right|}, \tag{34}$$

*are satisfied for every* $i \in \{1, \ldots, n\}$.

Proving the above theorem is similar to the proof of the previous one with the exception that we replace term $\boldsymbol{g}_i$ by $\boldsymbol{g}_i - \boldsymbol{e}_i$. Extending given control strategy by including also control signal variation constraint is presented in the following theorem:

**Theorem 3.** *In order to fulfill conditions (8), (9), (13), (18) and (24) it is sufficient that the inequalities (32), (33), (34) and*

$$\left(|b_1 g_{i1}| + |g_{u1}|\right) \delta_1 + \cdots + \left(|b_n g_{in}| + |g_{un}|\right) \delta_n \leq \delta_u, \tag{35}$$

$$\left(|b_1 g_{i1}| + |g_{u1}|\right) \delta_1 + \cdots + \left(|b_i (g_{ii} - 1)| + |g_{ui}|\right) \delta_i + \cdots + \left(|b_n g_{in}| + |g_{un}|\right) \delta_n \leq \delta_u \tag{36}$$

*are satisfied for every* $i \in \{1, \ldots, n\}$.

*Proof.* At first we will prove that $K_{\delta_u} \geq K(k-1)$ which ensures fulfillment of condition (8). Substituting (4) for $k-1$ step into (29) and taking into account the fact that in reaching phase $\operatorname{sgn}\left[\boldsymbol{c}^T\boldsymbol{x}(k)\right]$ is constant we get

$$K_{\delta_u} = \boldsymbol{c}^T\boldsymbol{b}\operatorname{sgn}\left[\boldsymbol{c}^T\boldsymbol{x}(k)\right]\boldsymbol{g}_u\left[\boldsymbol{x}(k) - \boldsymbol{x}(k-1)\right] + K(k-1) + \left|\boldsymbol{c}^T\boldsymbol{b}\right|\delta_u. \quad (37)$$

Thus, in order to obtain (8) we need to satisfy the following inequality

$$\boldsymbol{g}_u\left[\boldsymbol{x}(k) - \boldsymbol{x}(k-1)\right] \leq \delta_u. \quad (38)$$

Taking into consideration the fulfillment of conditions (32), (33), (34) and (35) we get

$$\boldsymbol{g}_u\left[\boldsymbol{x}(k) - \boldsymbol{x}(k-1)\right] \leq \left(|b_1 g_{i1}| + |g_{u1}|\right)\delta_1 + \cdots + \left(|b_n g_{in}| + |g_{un}|\right)\delta_n \leq \delta_u. \quad (39)$$

Therefore, (8) is satisfied. Now, we will prove that $K_u \leq K_{\delta_u}^-$ where

$$K_{\delta_u}^- = \boldsymbol{c}^T\boldsymbol{b}\operatorname{sgn}\left[\boldsymbol{c}^T\boldsymbol{x}(k)\right]\left[\boldsymbol{g}_u\boldsymbol{x}(k) - u(k-1)\right] - \left|\boldsymbol{c}^T\boldsymbol{b}\right|\delta_u \quad (40)$$

is the lowest possible $K$ for which the control signal variation is constrained. Using the above equation and (12) we have

$$K_u = K_{\delta_u}^- + \boldsymbol{c}^T\boldsymbol{b}\operatorname{sgn}\left[\boldsymbol{c}^T\boldsymbol{x}(k)\right]u(k-1) + \left|\boldsymbol{c}^T\boldsymbol{b}\right|\left(r_u + \delta_u\right). \quad (41)$$

From (41) and (9) we get that $K_u \leq K_{\delta_u}^-$. Further, we will prove that

$$K_{\delta_u}^- \leq K(k) \leq K_{\delta_u}. \quad (42)$$

Similarly to (37) we can write

$$K_{\delta_u}^- = \boldsymbol{c}^T\boldsymbol{b}\operatorname{sgn}\left[\boldsymbol{c}^T\boldsymbol{x}(k)\right]\boldsymbol{g}_u\left[\boldsymbol{x}(k) - \boldsymbol{x}(k-1)\right] + K(k-1) - \left|\boldsymbol{c}^T\boldsymbol{b}\right|\delta_u \quad (43)$$

Substituting (37), (43) into (42) and subtracting from both sides term

$$\boldsymbol{c}^T\boldsymbol{b}\operatorname{sgn}\left[\boldsymbol{c}^T\boldsymbol{x}(k)\right]\boldsymbol{g}_u\left[\boldsymbol{x}(k) - \boldsymbol{x}(k-1)\right] + K(k-1)$$

we obtain

$$\left|K(k) - K(k-1) - \boldsymbol{c}^T\boldsymbol{b}\operatorname{sgn}\left[\boldsymbol{c}^T\boldsymbol{x}(k)\right]\boldsymbol{g}_u\left[\boldsymbol{x}(k) - \boldsymbol{x}(k-1)\right]\right| \leq \left|\boldsymbol{c}^T\boldsymbol{b}\right|\delta_u. \quad (44)$$

Using the subadditivity of the absolute value the above inequality is true when

$$\left|K(k) - K(k-1)\right| + \left|\boldsymbol{c}^T\boldsymbol{b}\boldsymbol{g}_u\left[\boldsymbol{x}(k) - \boldsymbol{x}(k-1)\right]\right| \leq \left|\boldsymbol{c}^T\boldsymbol{b}\right|\delta_u. \quad (45)$$

From the fact that there exists $i \in \{1, \ldots, n\}$ for which

$$\left|K(k) - K(k-1)\right| \leq \left|K_i(k) - K_i(k-1)\right| = \left|b_i\boldsymbol{g}_i\left[\boldsymbol{x}(k) - \boldsymbol{x}(k-1)\right]\right| \quad (46)$$

or

$$\left|K(k) - K(k-1)\right| \leq \left|K_{\delta_i}(k) - K_{\delta_i}(k-1)\right| = \left|b_i\left(\boldsymbol{g}_i - \boldsymbol{e}_i\right)\left[\boldsymbol{x}(k) - \boldsymbol{x}(k-1)\right]\right| \quad (47)$$

we have that in order to satisfy (24) we need to fulfill the conditions

$$\left(|b_i\boldsymbol{g}_i| + |\boldsymbol{g}_u|\right)\left|\boldsymbol{x}(k) - \boldsymbol{x}(k-1)\right| \leq \delta_u, \quad (48)$$

$$\left(\left|b_i\left(\boldsymbol{g}_i - \boldsymbol{e}_i\right)\right| + |\boldsymbol{g}_u|\right)\left|\boldsymbol{x}(k) - \boldsymbol{x}(k-1)\right| \leq \delta_u \quad (49)$$

for every $i \in \{1, \ldots, n\}$. Taking into account that every state variation is already constrained we finally obtain (35) and (36) which ends the proof.

## 6 Simulation Example

In order to verify the properties of the proposed control strategy computer simulations are carried out. Let us consider the discrete-time plant given by (1) where

$$\boldsymbol{A} = \begin{bmatrix} 2 & 1 \\ 1 & -1 \end{bmatrix}, \qquad \boldsymbol{b} = \begin{bmatrix} 1 \\ 1 \end{bmatrix}, \qquad \boldsymbol{c}^T = \begin{bmatrix} 0.2 \; 1 \end{bmatrix}, \qquad \boldsymbol{d}\,(k) = \begin{bmatrix} d_1(k) \\ d_2(k) \end{bmatrix} \tag{50}$$

and $d_1(k), d_2(k)$ are uniform random number functions which absolute values limited by $1, 0.1$, respectively. Our objective is to monotonically and in finite-time drive the representative point from the initial state $\boldsymbol{x}_0 = [70, -5]^T$ to the sliding line $\boldsymbol{c}^T \boldsymbol{x}\,(k) = 0$. Moreover, we demand fulfilling the following constraints: $|x_1\,(k)\,| \leq 70$, $|x_2\,(k)\,| \leq 5$, $|x_1\,(k) - x_1\,(k-1)\,| \leq 20$, $|x_2\,(k) - x_2\,(k-1)\,| \leq 3$, $|u(k)| \leq 80$ and $|u(k) - u(k-1)| \leq 45$ for every $k \in \mathbb{N}$. To begin with, we calculate matrix $\boldsymbol{G}$, vector $\boldsymbol{g}_u$ and $\frac{D_{\max}}{|\boldsymbol{c}^T\boldsymbol{b}|}$ obtaining

$$\boldsymbol{G} = \begin{bmatrix} 1 & 2.5 \\ 0 & 0.5 \end{bmatrix}, \qquad \boldsymbol{g}_u = \begin{bmatrix} -1 \; 1.5 \end{bmatrix}, \qquad \frac{D_{\max}}{|\boldsymbol{c}^T\boldsymbol{b}|} = 0.3.$$

From the fact that conditions (32–36) are fulfilled we guarantee state, control signal and their variations constraints. Figure 1 presents the state trajectory. We can see that both state variables limits are satisfied for the whole regulation process and the quasi-sliding motion is obtained. From Fig. 2 one can observe that the control signal, starting from minimal admissible value $-80$, rapidly increases and then fluctuates around zero in order to maintain the representative point in demand state. Figures 3 and 4 illustrate the evolution of state variable variations while Fig. 5 shows variation of the control signal. From these figures we can see that all demanded restrictions are satisfied.

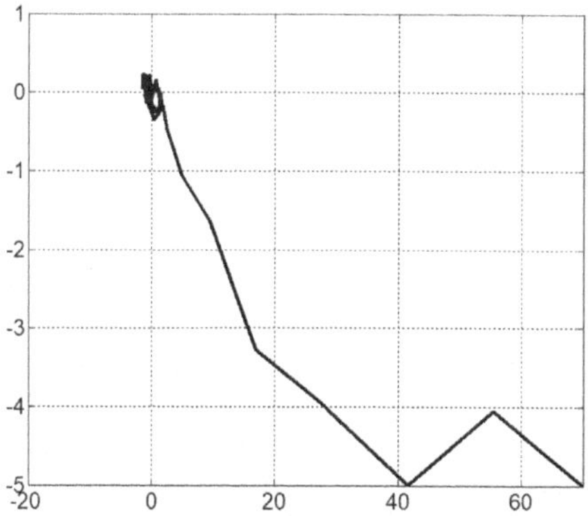


**Fig. 1.** State trajectory

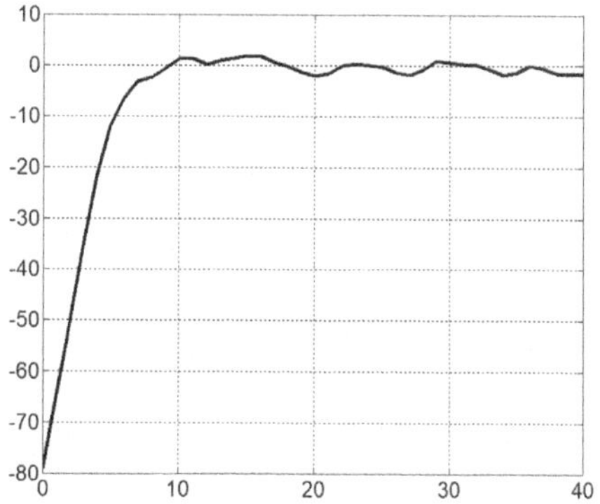


**Fig. 2.** Control signal

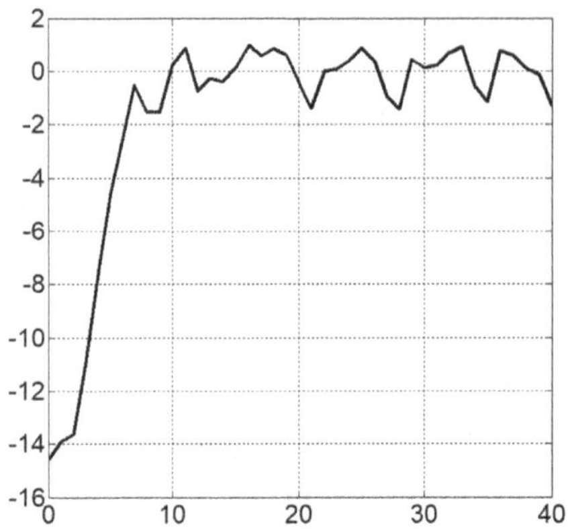


**Fig. 3.** First state variable variation

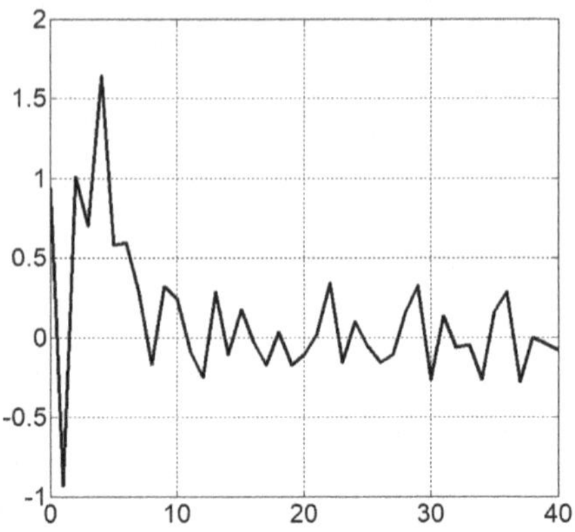


**Fig. 4.** Second state variable variation

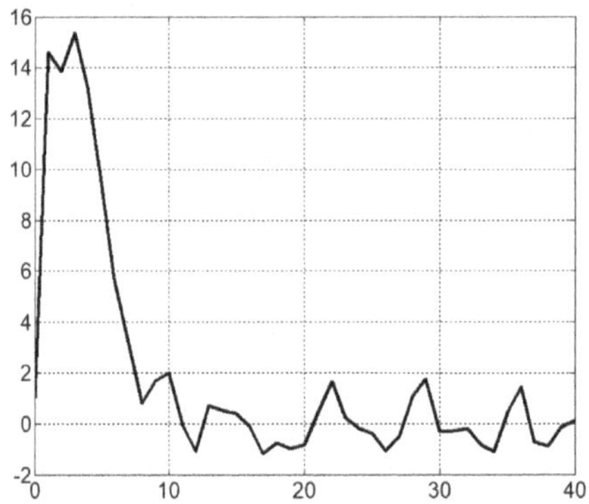


**Fig. 5.** Control signal variation

## 7 Conclusions

In this paper the discrete-time system under the influence of unknown, bounded perturbations has been considered. The problem of constraining not only the state variables and the control signal but simultaneously also their variations has been analyzed. For the purpose of the fastest, finite-time, monotonic convergence of the representative point to the sliding manifold, the new reaching law has been proposed. Computer simulations confirmed theoretical results. Our next aim is to weaken the sufficient condition presented in this work.

## References

1. Bartolini, G., Ferrara, A., Utkin, V.I.: Adaptive sliding mode control in discrete-time systems. Automatica **31**, 769–773 (1995)
2. Bartoszewicz, A.: Discrete-time quasi-sliding-mode control strategies. IEEE Trans. Ind. Electron. **45**, 633–637 (1998)

3. Bartoszewicz, A., Kaynak, O., Utkin, V.I.: Special section: sliding mode control in industrial applications. IEEE Trans. Ind. Electron. **55**, 3805–4103 (2008)
4. Bartoszewicz, A., Latosiński, P.: Discrete time sliding mode control with reduced switching - a new reaching law approach. Int. J. Robust Nonlinear Control **26**, 47–68 (2016)
5. Bartoszewicz, A., Leśniewski, P.: Reaching law approach to the sliding mode control of periodic review inventory systems. IEEE Trans. Autom. Sci. Eng. **11**, 810–817 (2014)
6. Bartoszewicz, A., Leśniewski, P.: Reaching law based sliding mode congestion control for communication networks. IET Control Theor. Appl. **8**, 1914–1920 (2014)
7. Bartoszewicz, A., Leśniewski, P.: New switching and nonswitching type reaching laws for SMC of discrete time systems. IEEE Trans. Control Syst. Technol. **24**, 670–677 (2016)
8. Chakrabarty, S., Bandyopadhyay, B.: A generalized reaching law for discrete sliding mode control. Automatica **52**, 83–86 (2015)
9. Chakrabarty, S., Bandyopadhyay, B.: A generalized reaching law with different convergence rates. Automatica **63**, 34–37 (2016)
10. Emelyanov, S.V.: Variable structure control systems. Nauka, Moscow (1967)
11. Fallaha, C.J., Saad, M., Kanaan, H.Y., Al-Haddad, K.: Sliding-mode robot control with exponential reaching law. IEEE Trans. Ind. Electron. **58**, 600–610 (2011)
12. Furuta, K.: Sliding mode control of a discrete system. Syst. Control Lett. **14**, 145–152 (1990)
13. Gao, W., Hung, J.C.: Variable structure control of nonlinear systems: a new approach. IEEE Trans. Ind. Electron. **40**, 45–55 (1993)
14. Gao, W., Wang, Y., Homaifa, A.: Discrete-time variable structure control systems. IEEE Trans. Ind. Electron. **42**, 117–122 (1995)
15. Golo, G., Milosavljević, C.: Robust discrete-time chattering free sliding mode control. Syst. Control Lett. **41**, 19–28 (2000)
16. Jaskuła, M., Leśniewski, P.: Discrete time sliding mode control in the presence of state and control signal constraints. In: 24th International Conference on Methods and Models in Automation and Robotics, Miedzyzdroje, Poland, pp. 201–204 (2019)
17. Milosavljević, C.: General conditions for the existence of a quasi-sliding mode on the switching hyperplane in discrete variable structure systems. Autom. Rem. Control **46**, 307–314 (1985)
18. Niu, Y., Ho, D.W.C., Wang, Z.: Improved sliding mode control for discrete-time systems via reaching law. IET Control Theor. Appl. **4**, 2245–2251 (2010)
19. Utkin, V.I.: Variable structure systems with sliding modes. IEEE Trans. Autom. Control **22**, 212–222 (1977)
20. Utkin, V.I., Drakunow, S.V.: On discrete-time sliding mode control. In: IFAC Conference Nonlinear Control, pp. 484–489 (1989)
21. Wang, A., Jia, X., Dong, S.: A new exponential reaching law of sliding mode control to improve performance of permanent magnet synchronous motor. IEEE Trans. Magn. **49**, 2409–2412 (2013)
22. Zhang, X., Sun, L., Zhao, K., Sun, L.: Nonlinear speed control for PMSM system using sliding-mode control and disturbance compensation techniques. IEEE Trans. Power Electron. **28**, 1358–1365 (2013)

# Accuracy Estimation of the Discrete, Approximated Atangana-Baleanu Operator

Krzysztof Oprzędkiewicz(✉)

Faculty of Electrical Engineering, Automatics, Computer Science and Biomedical Engineering, Department of Automatic Control and Robotics, AGH University, Kraków, Poland
kop@agh.edu.pl

**Abstract.** In the paper the accuracy analysis of the discrete approximations of the Atangana-Baleanu (AB) operator is addressed. The AB operator is the nonsingular kernel operator proposed by Atangana and Baleanu in the papers [1,2]. It is obtained by replacing the exponential function in the Caputo-Fabrizio operator by the Mittag-Leffler function. The use of the the AB operator at a digital platform (PLC, microcontroller) reuqires to apply the discrete approximation of the factor $s^{\alpha}$. This can be done using discrete ORA or CFE approximations. The step responses of the both approximations are compared to the analytical response of operator. As the cost function the FIT function available in MATLAB is employed. Results of simulations show that the discrete ORA approximation gives better results than the CFE approximation calculated at the same time grid.

**Keywords:** Fractional order systems · Fractional order transfer function · Atangana-Baleanu operator · ORA approximation · CFE approximation

## 1 Introduction

The fractional order (FO) operator using Atangana-Baleanu (AB) derivative has been proposed in the paper [1]. Its properties are analyzed in [2]. This operator is employed to modeling of some physical phenomena, mainly diffusion and heat transfer. Examples of its application cover for example analytical solution of the Hristov diffusion equation [13], a number of interesting recent results is collected in the book [7].

Calculations executed with the use of of AB operator require to employ an approximation of $s^{\alpha}$ element. This can be done using different methods, in time-continuous case the known Oustaloup approximation (ORA) can be used.

This paper deals with the accuracy analysis during use the ORA approximation to express the AB operator. As the reference the analytical formula of the step response is used, analogically, as it was done in: [8].


R. Szewczyk et al. (Eds.): AUTOMATION 2020, AISC 1140, pp. 34–44, 2020.
https://doi.org/10.1007/978-3-030-40971-5_4

The paper is organized as follows: preliminaries give the basic information from fractional calculus necessary to present the proposed results. Next the Atangana-Baleanu operator and its Laplace transform as well as the analytical formula of step and frequency responses are recalled. The CFE based approximation are recalled. Frequency responses are discussed too. Finally the simulations illustrating the dependence between accuracy of the approximation and CFE parameters are given.

## 2 Preliminaries

### 2.1 Basic Ideas

Presentation of elementary ideas is started with a definition of a fractional order, integro-differential operator given for example by [5]:

**Definition 1** *(The elementary fractional order operator). The fractional order integro-differential operator is defined as follows:*

$$ {}_aD_t^{\alpha} f(t) = \begin{cases} \frac{d^{\alpha} f(t)}{dt^{\alpha}} & \alpha > 0 \\ 1 \ \alpha = 0 \\ \int\limits_a^t f(\tau)(d\tau)^{-\alpha} & \alpha < 0 \end{cases}. \tag{1} $$

*where $a$ and $t$ denote time limits of operator calculation, $\alpha \in \mathbb{R}$ denotes the fractional order of the operation.*

Next an idea of Gamma Euler function is recalled (see for example [6]):

**Definition 2.** *The Gamma function*

$$ \Gamma(x) = \int\limits_0^{\infty} t^{x-1} e^{-t} dt. \tag{2} $$

An idea of Mittag-Leffler function needs to be given next. It is a non-integer order generalization of exponential function $e^{\lambda t}$ and it plays crucial role in solution of fractional order state equation. The one parameter Mittag-Leffler function is defined as follows:

**Definition 3.** *The one parameter Mittag-Leffler function*

$$ E_{\alpha}(x) = \sum_{k=0}^{\infty} \frac{x^k}{\Gamma(k\alpha + 1)}. \tag{3} $$

The fractional order operator can be described by different definitions, given by: The Grünwald-Letnikov (the GL definition), Riemann-Liouville (RL definition) and Caputo (C definition). They are recalled below.

**Definition 4.** *The Grünwald-Letnikov definition of the FO operator ([3,10])*

$$ {}_0^{GL}D_t^{\alpha} f(t) = \lim_{h \to 0} h^{-\alpha} \sum_{j=0}^{[\frac{t}{h}]} (-1)^j \binom{\alpha}{j} f(t - jh). \tag{4} $$

In (4) $\binom{\alpha}{j}$ is a generalization of Newton symbol into real numbers:

$$ \binom{\alpha}{j} = \begin{cases} 1, & j = 0 \\ \frac{\alpha(\alpha-1)\ldots(\alpha-j+1)}{j!}, & j > 0 \end{cases} \tag{5} $$

**Definition 5.** *The Riemann - Liouville definition of the FO operator*

$$ {}_0^{RL}D_t^{\alpha} f(t) = \frac{1}{\Gamma(N-\alpha)} \frac{d^N}{dt^N} \int_0^{\infty} (t-\tau)^{N-\alpha-1} f(\tau) d\tau. \tag{6} $$

*where $N - 1 < \alpha < N$ denotes the non integer order of operation, $N$ is the integer order nearest to $\alpha$ and $\Gamma(..)$ is the complete Gamma function expressed by (2).*

The Caputo definition is described as underneath:

**Definition 6.** *The Caputo definition of the FO operator*

$$ {}_0^{C}D_t^{\alpha} f(t) = \frac{1}{\Gamma(N-\alpha)} \int_0^{\infty} \frac{f^{(N)}(\tau)}{(t-\tau)^{\alpha+1-N}} d\tau. \tag{7} $$

### 2.2 The Atangana-Baleanu Operator

The fractional order derivative Atangana-Baleanu operator is obtained by replacing the exponential kernel in the Caputo-Fabrizio (CF) operator by the Mittag-Leffler kernel. It is defined using the C or RL definition of fractional order derivative, expressed by (7) or (6) respectively. Using these definitions we obtain the Atangana-Baleanu-Caputo (ABC) or Atangana-Baleanu-Riemann (ABR)operator respectively [1]:

**Definition 7.** *(The Atangana-Baleanu-Caputo (ABC) operator)*

$$ {}^{ABC}{}_aD_t^{\alpha}(f(t)) = M_{\alpha} \int_a^t f'(x) E_{\alpha}\left(-\alpha \frac{(t-x)^{\alpha}}{1-\alpha}\right) dx. \tag{8} $$

*where $E_{\alpha}(..)$ is the one parameter Mittag-Leffler function, $M_{\alpha}$ is the normalization function equal:*

$$ M_{\alpha} = 1 - \alpha + \frac{\alpha}{\Gamma(\alpha)}. \tag{9} $$

*In (9) $\Gamma(..)$ is the Gamma function.*

**Definition 8.** *(The Atangana-Baleanu-Riemann (ABR) operator)*

$$ {}^{ABR}{}_{a}D_t^{\alpha}(f(t)) = M_{\alpha}\frac{d}{dt}\int_a^t f(x)E_{\alpha}\left(-\alpha\frac{(t-x)^{\alpha}}{1-\alpha}\right)dx. \tag{10} $$

*where* $E_{\alpha}(..)$ *is the one parameter Mittag-Leffler function,* $M_{\alpha}$ *is the normalization function expressed by* (9), $\Gamma(..)$ *is the Gamma function.*

The Laplace transforms for the ABC and ABR derivatives are as follows:

**Definition 9.** *(The Laplace transform of the ABC operator)*

$$ {}^{ABC}G(s) = \mathcal{L}\{{}^{ABC}{}_{a}D_t^{\alpha}(f(t))\}(s) = \frac{M_{\alpha}}{1-\alpha}\frac{s^{\alpha}\{f(t)\}(s) - s^{\alpha-1}f(0)}{s^{\alpha}+\frac{\alpha}{1-\alpha}}. \tag{11} $$

**Definition 10.** *(The Laplace transform of the ABR operator)*

$$ {}^{ABR}G(s) = \mathcal{L}\{{}^{ABR}{}_{a}D_t^{\alpha}(f(t))\}(s) = \frac{M_{\alpha}}{1-\alpha}\frac{s^{\alpha}\{f(t)\}(s)}{s^{\alpha}+\frac{\alpha}{1-\alpha}}. \tag{12} $$

For the homogenous initial condition: $f(0) = 0$ the both Laplace transforms are equal:

$$ \mathcal{L}\{{}^{ABR}{}_{a}D_t^{\alpha}(f(t))\}(s) = \mathcal{L}\{{}^{ABC}{}_{a}D_t^{\alpha}(f(t))\}(s) = G(s). \tag{13} $$

Consequently the Laplace transform of the AB operator takes the following form:

$$ G(s) = \frac{M_{\alpha}}{1-\alpha}\frac{s^{\alpha}}{s^{\alpha}+\frac{\alpha}{1-\alpha}}. \tag{14} $$

where: $\alpha_1 = 1-\alpha$.

### 2.3 The ORA Approximation

The method proposed by Oustaloup (see [9]) allows to approximate an elementary FO transfer function $s^{\alpha}$ by the finite, integer-order transfer function, close to Pade approximation:

$$ s^{\alpha} \approx k_f \prod_{n=1}^{N}\frac{1+\frac{s}{\mu_n}}{1+\frac{s}{\nu_n}} = \frac{L_{ORA}(s)}{D_{ORA}(s)}. \tag{15} $$

In (15) $N$ is the order of the approximation, $k_f$ is the steady state gain, expected to assure the convergence the step response of the approximation to step response of the real plant in a steady state. Coefficients $\mu_n$ and $\nu_n$ are calculated as follows:

$$ \begin{aligned} \mu_1 &= \omega_l\sqrt{\eta} \\ \nu_n &= \mu_n\gamma,\ n = 1,..,N \\ \mu_{n+1} &= \nu_n\eta,\ n = 1,..,N-1 \end{aligned} \tag{16} $$

where:

$$\gamma = \left(\frac{\omega_h}{\omega_l}\right)^{\frac{\alpha}{N}}$$
$$\eta = \left(\frac{\omega_h}{\omega_l}\right)^{\frac{1-\alpha}{N}} \tag{17}$$

In (17) $\omega_l$ and $\omega_h$ describe the bandwidth, for which parameters are calculated. During further considerations it will be expressed in the logarithmic scale:

$$\omega_l = 10^{-W_B},$$
$$\omega_h = 10^{W_B} \tag{18}$$

where $W_B \in \mathbb{Z}$ describes the bandwidth.

The use of ORA approximation (15) to (11) or (12) allows to obtain the approximation of the operator. It is as follows:

$$\mathcal{L}\{{}^{ABR}{}_aD_t^{\alpha}(f(t))\}(s) = \mathcal{L}\{{}^{ABC}{}_aD_t^{\alpha}(f(t))\}(s) \approx \frac{L_{ORA}(s)}{\alpha_1 L_{ORA}(s) + \alpha D_{ORA}(s)}F(s). \tag{19}$$

Consequently the transfer function of the approximation takes the following form:

$$G_{ORA}(s) = \frac{L_{ORA}(s)}{\alpha_1 L_{ORA}(s) + \alpha D_{ORA}(s)}. \tag{20}$$

The discretization of the transfer function (20) can be done using the Matlab function*c2dm*. Consequently we obtain the discrete transfer function $G_{ORA}^{+}(z^{-1})$.

An another way to obtain the discrete fractional operator is to apply the CFE approximation. It is presented in the next subsection.

## 2.4 The CFE Approximation

An implementation of operator (1) at each digital platform (PLC, microcontroller) requires to apply an integer order, finite dimensional, discrete approximant. The most known are PSE (Power Series Expansion) and CFE (Continuous Fraction Expansion). They allow us to estimate a non integer order element with the use of digital FIR or IIR filter. The PSE approximant bases directly on discrete version of GL definition (4) and it has the form of FIR filter containing only zeros. However its digital implementation to keep a good quality requires us to apply long memory buffer (high order of the filter). The CFE approximant has the form of IIR filter containing both poles and zeros. It is faster covergent and easier to implement because its order is relatively low, typically not higher that 5.

The discretization of fractional order element $s^{\alpha}$, $\alpha \in \mathbb{R}$ can be done with the use of the so called generating function $s \approx \omega(z^{-1})$. The new operator raised to power $\alpha$ has the following form (see for example [4,11], p.119):

$$\begin{aligned}\left(\omega(z^{-1})\right)^{\alpha} &= \left(\tfrac{1+a}{h}\right)^{\alpha} CFE\{\left(\tfrac{1-z^{-1}}{1+az^{-1}}\right)^{\alpha}\}_{M,M} \\ &= \frac{W(z^{-1})}{V(z^{-1})} = \left(\tfrac{1+a}{h}\right)^{\alpha} \frac{CFE_N(z^{-1},\alpha)}{CFE_D(z^{-1},\alpha)} = \frac{\sum\limits_{m=0}^{M} w_m z^{-m}}{\sum\limits_{m=0}^{M} v_m z^{-m}}.\end{aligned} \tag{21}$$

In (21) $a$ is the coefficient depending on approximation type (for example: $a = 0$ for Euler approximation, $a = 1$ for Tustin approximation), $h$ denotes the sample time, $M$ is the order of approximation. Numerical values of coefficients $w_m$ and $v_m$ and different values of parameter $a$ can be calculated using MATLAB function given by Petras in [12]. This MATLAB function was applied in experiments described in the next section. If the Tustin approximation is considered ($a$=1) then $CFE_D(z^{-1},\alpha) = CFE_N(z^{-1},-\alpha)$ and the polynomial $CFE_D(z^{-1},\alpha)$ can be given in the direct form (see [4]). Examples of polynomial $CFE_D(z^{-1},\alpha)$ for $M = 1, 3, 5$ are given in Table 1.

The time response of the approximated FO element (21) in $k$-th time moment is expressed as underneath:

$$y^{+}_{CFE}(k) = \frac{1}{v_0}\left[-\sum_{m=1}^{M} v_m y^{+}(k-m) + \sum_{m=0}^{M} w_m u^{+}(k-m)\right]. \tag{22}$$

where $y^{+}_{CFE}(k-m)$ and $u^{+}(k-m)$ denote the output and input signals in $k-m$ - th time moments respectively, $v_m$ and $w_m$ are coefficients of CFE approximation, given for example in the Table 1.

**Table 1.** coefficients of CFE polynomials $CFE_{N,D}(z^{-1},\alpha)$ for Tustin approximation with respect to [4]

| Order $M$ | $w_m$ | $v_m$ |
|---|---|---|
| $M$=1 | $w_1 = -\alpha$ | $v_1 = \alpha$ |
| | $w_0 = 1$ | $v_0 = 1$ |
| $M$=3 | $w_3 = -\frac{\alpha}{3}$ | $v_3 = \frac{\alpha}{3}$ |
| | $w_2 = \frac{\alpha^2}{3}$ | $v_2 = \frac{\alpha^2}{3}$ |
| | $w_1 = -\alpha$ | $v_1 = \alpha$ |
| | $w_0 = 1$ | $v_0 = 1$ |
| $M$=5 | $w_5 = -\frac{\alpha}{5}$ | $v_5 = \frac{\alpha}{5}$ |
| | $w_4 = \frac{\alpha^2}{5}$ | $v_4 = \frac{\alpha^2}{5}$ |
| | $w_3 = -\left(\frac{\alpha}{5} + \frac{2\alpha^3}{35}\right)$ | $v_3 = -\left(\frac{-\alpha}{5} + \frac{-2\alpha^3}{35}\right)$ |
| | $w_2 = \frac{2\alpha^2}{5}$ | $v_2 = \frac{2\alpha^2}{5}$ |
| | $w_1 = -\alpha$ | $v_1 = \alpha$ |
| | $w_0 = 1$ | $v_0 = 1$ |

## 3 The Step Responses of the AB Operator

The time-continuous step response of the Laplace transform (11) or (12) is described by the following remark:

**Remark 1.** *(The step response of the AB operator)*
*Assume that the Laplace transform of the ABC/R operator with initial condition equal to zero is given by* (12). *Its step response is equal:*

$$y(t) = \frac{M_\alpha}{\alpha_1} E_\alpha \left( -\frac{\alpha}{\alpha_1} t^\alpha \right). \tag{23}$$

*where* $E_\alpha(..)$ *is the Mittag-Leffler function* (3), $M_\alpha$ *is expressed by* (9), $\alpha_1 = 1 - \alpha$.

Using (19) or (20) the time continuous and discrete, approximated step response take the following form:

$$y_{ORA}(t) = \mathcal{L}^{-1}\Big\{\frac{1}{s}\frac{L_{ORA}(s)}{\alpha_1 L_{ORA}(s) + \alpha D_{ORA}(s)}\Big\}. \tag{24}$$

$$y^+_{ORA}(k) = \mathcal{Z}^{-1}\Big\{\frac{1}{1-z^{-1}} G^+_{ORA}(z^{-1})\Big\}. \tag{25}$$

The discrete, approximated step response employing the CFE approximation is as follows:

$$y^+_{CFE}(k) = \mathcal{Z}^{-1}\Big\{\frac{1}{1-z^{-1}}\frac{W(z^{-1})}{\alpha_1 W(z^{-1}) + \alpha V(z^{-1})}\Big\}. \tag{26}$$

In (26) $W$ and $V$ are expressed by (21).

## 4 Simulations

The performance of the approximation is estimated using the following cost function, describing the fitting of the approximated result $y_a$ to the analytical one $y$. The results can be time or frequency functions.

$$FIT = \frac{||y - y_a||}{||y - \overline{y_a}||}. \tag{27}$$

In (27) $\overline{y_a}$ is the average value from approximated result $y_a$.The function (27) can be also expressed in %, as it is done in the MATLAB functions *compare* and *tfest*:

$$FIT_{100} = (1 - FIT) \cdot 100\%. \tag{28}$$

Firstly time continuous step response were examined. They are shown in the Fig. 1. The analytical step response $y$ was calculated using (23), the approximation $y_{ORA}(t)$ was calculated using (24) and MATLAB function *step*. The cost function (28) as a function of ORA approximation order $N$ for different non integer orders $\alpha$ is shown in the Fig. 2, the cost function (28) as a function of non integer order $\alpha$ for different bandwidth $W$ is illustrated by Fig. 3.

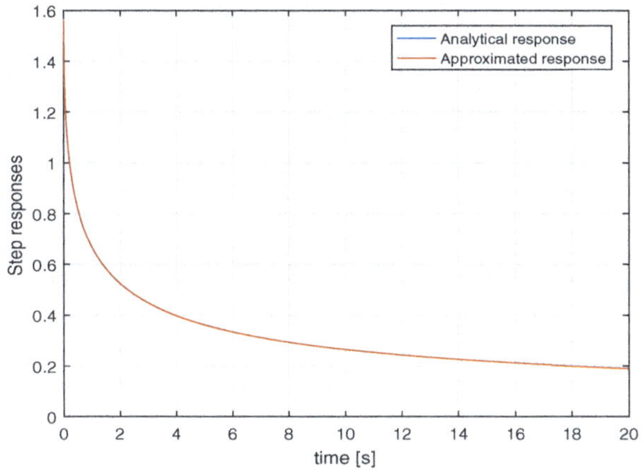


**Fig. 1.** The comparing of approximated (24) to analytical step response (23) for $\alpha = 0.5$, $N = 8$, $W = 4$.

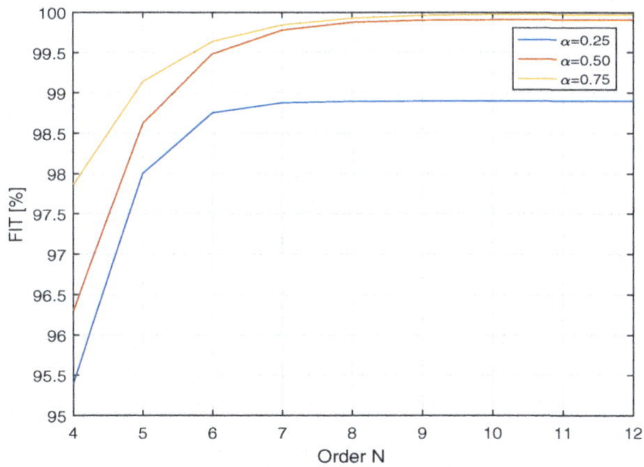


**Fig. 2.** The fitting of approximation (24) to analytical solution (23) as a function of ORA approximation order $N$ for different orders $\alpha$, the bandwidth $W = 4$.

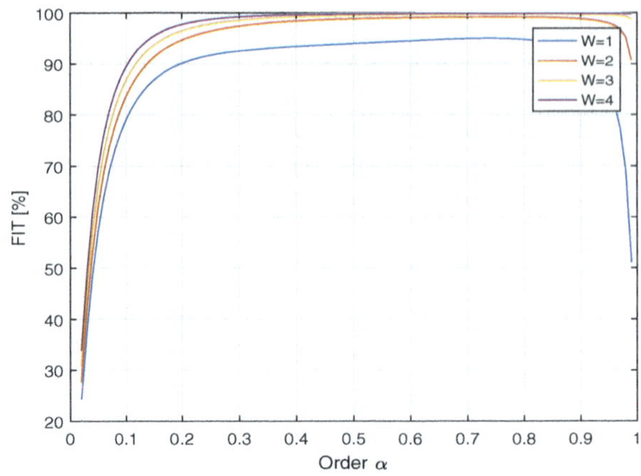


**Fig. 3.** The fitting of approximation (24) to analytical solution (23) as a function of non integer order $\alpha$ for different bandwidth $W$, the ORA approximation order $N = 8$.

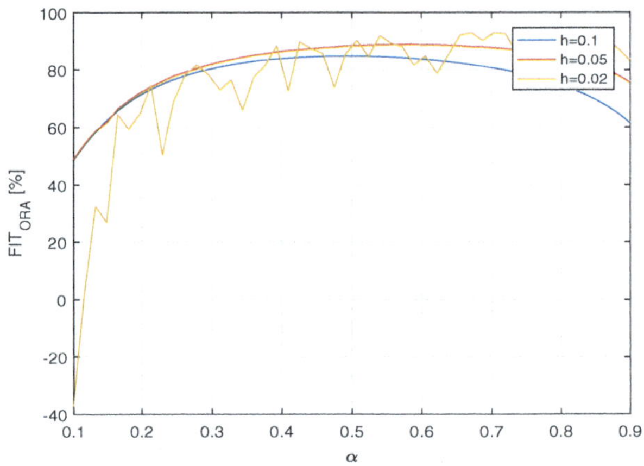


**Fig. 4.** The fitting of discrete ORA approximation (25) to analytical solution (23) as a function of non integer order $\alpha$ for different $h$, the discretization method: Tustin.

Next the discrete step response using discrete ORA approximation was examined. The approximation $y^+_{ORA}(k)$ was calculated using (25) The fitting approximation to analytical solution as a function of order $\alpha$ and for different sample times $h$ is given in the Fig. 4.

Finally discrete time step response using discrete ORA approximation was investigated. The approximated step response $y^+_{CFE}(k)$ was calculated using (26) and MATLAB function *step*. The fitting of approximation to analytical solution as a function of order $\alpha$ for different sample time $h$ is given in the Fig. 5, the fitting of approximation to analytical solution for different discretization methods (determined by the values of coefficient $a$) is given in the Fig. 6.

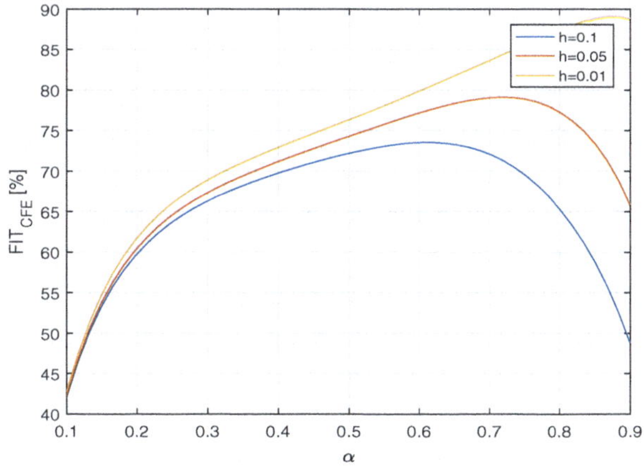


**Fig. 5.** The fitting of CFE approximated step response (26) to analytical solution (23) as a function of non integer order $\alpha$ for different sample times $h$, $a = 0$ (Euler method).

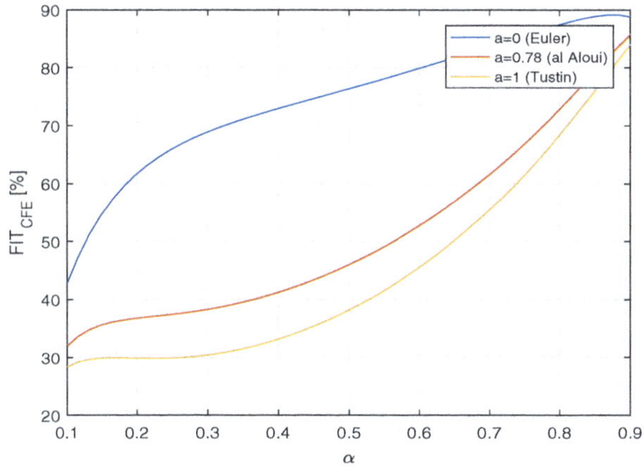


**Fig. 6.** The fitting of CFE approximated step response (26) to analytical solution (23) as a function of non integer order $\alpha$ for different $a$, sample time $h = 0.1$.

## 5 Conclusions

The main conclusion from the paper is that the continuous ORA approximation is more accurate in the sens of FIT function than all considered discrete approximations. Next, the smaller sample time $h$ gives the better fitting of approximation to analytical solution as well as the use of the simpliest Euler discrtization method.

**Acknowledgment.** This paper was sponsored by AGH project no 16.16.120.773.

## References


1. Atangana, A., Baleanu, D.: New fractional derivatives with nonlocal and non-singular kernel: theory and application to heat transfer model (2016)
2. Baleanu, D., Fernandez, A.: On some new properties of fractional derivatives with Mittag-Leffler kernel (2017). arXiv:1712.01762v1 [math.CA]. Accessed 5 Dec 2017
3. Caponetto, R., Dongola, G., Fortuna, I., Petras, I.: Fractional Order Systems. Modeling and Control Applications. World Scientific Series on Nonlinear Science, vol. 72. World Scientific Publishing, Singapore (2010)
4. Chen, Y.Q., Moore, K.L.: Discretization schemes for fractional order differentiators and integrators. IEEE Trans. Circ. Syst. I Fundam. Theor. Appl. **49**(3), 263–269 (2002)
5. Kaczorek, T.: Selected Problems in Fractional Systems Theory. Springer, Heidelberg (2011)
6. Kaczorek, T., Rogowski, K.: Fractional Linear Systems and Electrical Circuits. Springer, Heidelberg (2014). Bialystok University of Technology
7. Gomez, J.F., Torres, L., Escobar, R.F.: Fractional derivatives with mittag-leffler kernel trends and applications in science and engineering. In: Studies in Systems, Decision and Control, vol. 194, Springer, Heidelberg (2019)
8. Oprzedkiewicz, K., Mitkowski, W., Gawin, E.: An estimation of accuracy of Oustaloup approximation. In: Szewczyk, R., Zieliński, C., Kaliczyńska, M. (eds.) Challenges in Automation, Robotics and Measurement Techniques : Proceedings of AUTOMATION-2016, Advances in Intelligent Systems and Computing , Warsaw, Poland, 2–4 March 2016, 440, pp. 299–307 (2016). ISBN: 978-3-319-29356-1, e-ISBN: 978-3-319-29357-8
9. Oustaloup, A., Levron, F., Mathieu, B., Nanot, F.M.: Frequency-band complex noninteger differentiator: characterization and synthesis. IEEE Trans. Circ. Syst. I Fundam. Theor. Appl. **47**(1), 25–39 (2000)
10. Ostalczyk, P.: Discrete Fractional Calculus: Applications in Control and Image Processing. Series in Computer Vision, vol. 4. World Scientific Publishing, Singapore (2016)
11. Petras, I.: Fractional order feedback control of a dc motor. J. Electr. Eng. **60**(3), 117–128 (2009)
12. Petras, I.: http://people.tuke.sk/igor.podlubny/USU/matlab/petras/dfod1.m
13. Sene, N.: Analytical solutions of Hristov diffusion equations with non-singular fractional derivatives. Chaos **29**, 023112 (2019). https://doi.org/10.1063/1.5082645

# Linearization of Fractional Nonlinear Systems by State-Feedbacks

Tadeusz Kaczorek

Bialystok University of Technology, Bialystok, Poland
kaczorek@ee.pw.edu.pl

**Abstract.** Using the conformable fractional derivative the notions of Lie derivative and full relative degree are extended to the fractional nonlinear systems. The canonical form of the fractional nonlinear systems is introduced and sufficient conditions for the existence of the canonical form for the systems are established. A method for finding nonlinear state-feedbacks linearizing the fractional nonlinear system is proposed.

**Keywords:** Fractional · Nonlinear · System · Lie derivative · Linearization · State-feedback

## 1 Introduction

Mathematical fundamentals of the fractional calculus are given in the monographs [1–3]. The positive fractional linear systems have been investigated in [4, 5]. Stability of fractional linear 1D discrete-time and continuous-time systems has been investigated in the papers [6–8]. The notion of practical stability of positive fractional discrete-time linear systems has been introduced in [9]. Some recent interesting results in fractional systems theory and its applications can be found in [10–13]. The minimum energy control problem for standard linear systems has been formulated and solved by J. Klamka [14–17] and for 2D linear systems with variable coefficients in [18]. The controllability and minimum energy control problem of fractional discrete-time linear systems has been investigated by Klamka in [16]. The minimum energy control of fractional positive discrete-time linear systems with bounded inputs has been addressed in [19]. The theory of geometrical approach to analysis of nonlinear systems based on the Lie algebra has been given in the monographs [20–22]. The problem of linearization of nonlinear systems by nonlinear state-feedbacks has been investigated in many papers and books [21, 23–30].

In this paper the method for linearization of fractional nonlinear systems of full relative order by nonlinear state-feedbacks will be proposed. The paper is organized as follows. In Sect. 2 the conformable fractional derivative and its properties are given. The notion of the Lie derivative is extended to fractional nonlinear systems in Sect. 3. The concept of the relative full degree systems is introduced in Sect. 4. The existence of canonical form for fractional nonlinear systems is addressed in Sect. 5. The linearization problem of fractional nonlinear systems by nonlinear state-feedbacks is considered in Sect. 6. Concluding remarks are given in Sect. 7.


R. Szewczyk et al. (Eds.): AUTOMATION 2020, AISC 1140, pp. 45–56, 2020.
https://doi.org/10.1007/978-3-030-40971-5_5

The following notation will be used: $\Re$ – the set of real numbers, $\Re^{n \times m}$ – the set of $n \times m$ real matrices and $\Re^n = \Re^{n \times 1}$.

## 2 Preliminaries

The following conformable fractional derivative of the function $f(t)$ of order $\alpha$, $0 < \alpha < 1$ will be used [31]

$$\frac{d^\alpha f(t)}{dt^\alpha} = \lim_{\varepsilon \to 0} \frac{f(t + \varepsilon t^{1-\alpha}) - f(t)}{\varepsilon}, t > 0. \tag{1}$$

If (1) exists then the function $f(t)$ is called $\alpha$-differentiable [19].

**Theorem 1 [31].** If the functions $f(t)$ and $g(t)$ are $\alpha$-differentiable, $0 < \alpha < 1$ then

$$\frac{d^\alpha}{dt^\alpha}[af(t) + by(t)] = a\frac{d^\alpha f(t)}{dt^\alpha} + b\frac{d^\alpha g(t)}{dt^\alpha}, \tag{2}$$

where $a, b \in \Re$ are some real numbers.

**Theorem 2 [31].** If the functions $f(t)$ and $g(t)$ are $\alpha$-differentiable, then

$$\frac{d^\alpha}{dt^\alpha}[f(t)g(t)] = f(t)\frac{d^\alpha g(t)}{dt^\alpha} + g(t)\frac{d^\alpha f(t)}{dt^\alpha}, \tag{3}$$

$$\frac{d^\alpha}{dt^\alpha}\left[\frac{f(t)}{g(t)}\right] = \frac{g(t)\frac{d^\alpha f(t)}{dt^\alpha} - f(t)\frac{d^\alpha g(t)}{dt^\alpha}}{[g(t)]^2}, \tag{4}$$

$$\frac{d^\alpha f(t)}{dt^\alpha} = t^{1-\alpha}\frac{df(t)}{dt}. \tag{5}$$

**Theorem 3**

$$\frac{d^\alpha t^q}{dt^\alpha} = qt^{q-\alpha} \text{ for all } q \in \Re, \tag{6}$$

$$\frac{d^\alpha e^{qt}}{dt^\alpha} = qt^{1-\alpha}e^{qt} \text{ for all } q \in \Re. \tag{7}$$

**Proof.** Using (1) for $t^q$ we have

$$\frac{d^\alpha t^q}{dt^\alpha} = \lim_{\varepsilon \to 0} \frac{(t + \varepsilon t^{1-\alpha})^q - t^q}{\varepsilon} \tag{8}$$

and taking info account that

$$(t+\varepsilon t^{1-\alpha})^q = \sum_{k=0}^{q} \frac{q!}{(q-k)!k!} t^{q-k} (\varepsilon t^{1-\alpha})^k \tag{9}$$

we obtain

$$\frac{d^\alpha t^q}{dt^\alpha} = \lim_{\varepsilon \to 0} \frac{\sum_{k=1}^{q} \frac{q!}{(q-k)!k!} t^{q-k} (\varepsilon t^{1-\alpha})^k}{\varepsilon} = qt^{q-\alpha} \tag{10}$$

Similarly, using (1) for $e^{qt}$ we have

$$\frac{d^\alpha e^{qt}}{dt^\alpha} = \lim_{\varepsilon \to 0} \frac{e^{q(t+\varepsilon t^{1-\alpha})} - e^{qt}}{\varepsilon} = \lim_{\varepsilon \to 0} \frac{e^{qt}(e^{q\varepsilon t^{1-\alpha}} - 1)}{\varepsilon} \tag{11}$$

and taking into account that

$$e^{q\varepsilon t^{1-\alpha}} = \sum_{k=0}^{\infty} \frac{(q\varepsilon t^{1-\alpha})^k}{k!} \tag{12}$$

we obtain

$$\frac{d^\alpha e^{qt}}{dt^\alpha} = qt^{1-\alpha} e^{qt}. \square \tag{13}$$

**Theorem 4.** Let $f[g(t)] = f \circ g$ be the composition of the α-differentiable functions $f$ and $g$. Then

$$\frac{d^\alpha [f \circ g]}{dt^\alpha} = \frac{d^\alpha f}{dg^\alpha} \frac{d^\alpha g}{dt^\alpha}, 0 < \alpha < 1 \tag{14}$$

**Proof.** Using (1) we may write

$$\begin{aligned} \frac{d^\alpha [f \circ g]}{dt^\alpha} &= \lim_{\varepsilon \to 0} \frac{f[g(t) + \varepsilon g(t)^{1-\alpha}] - f[g(t)]}{g(t+\varepsilon^{t1-\alpha}) - g(t)} \frac{g(t+\varepsilon t^{1-\alpha}) - g(t)}{\varepsilon} \\ &= \lim_{\varepsilon \to 0} \frac{f[g(t) + \varepsilon g(t)^{1-\alpha}] - f[g(t)]}{g(t+\varepsilon^{t1-\alpha}) - g(t)} \lim_{\varepsilon \to 0} \frac{g(t+\varepsilon t^{1-\alpha}) - g(t)}{\varepsilon} = \frac{d^\alpha f}{dg^\alpha} \frac{d^\alpha g}{dt^\alpha} \end{aligned} \tag{15}$$

since $g(t+\varepsilon^{t1-\alpha}) - g(t) \to 0$ for $\varepsilon \to 0$. □

## 3 Definition of the Fractional Lie Derivative

**Definition 1.** Let $h(x)$ be the $\alpha$-differentiable scalar function of the vector $x = x(t) \in \Re^n$ and $f = f(x) \in \Re^n$ be the vector function of $x$. The scalar function defined by the formula

$$L_f^\alpha h = \frac{\partial^\alpha h}{\partial x^\alpha} f = \begin{bmatrix} \frac{\partial^\alpha h}{\partial x_1^\alpha} & \cdots & \frac{\partial^\alpha h}{\partial x_n^\alpha} \end{bmatrix} \begin{bmatrix} f_1(x) \\ \vdots \\ f_n(x) \end{bmatrix} = \sum_{k=1}^{n} \frac{\partial^\alpha h}{\partial x_k^\alpha} f_k(x) \tag{16}$$

is called the fractional Lie derivative of the order $\alpha$ $(0 < \alpha < 1)$ of the scalar function $h = h(x)$ along the vector function $f(x)$.

Using the formula

$$L_f^{k\alpha} h = L_f(L_f^{(k-1)\alpha} h) \text{ for } k = 1, 2, \ldots (L_f^0 h = h) \tag{17}$$

we can define the fractional Lie derivative of the order $k\alpha$ for $k = 1, 2, \ldots$ of the scalar function $h$ along the vector function $f = f(x)$.

From Definition 1 it follows that for two vector function $f(x) \in \Re^n$, $g(x) \in \Re^n$ and a scalar function $h(x)$ we have

$$L_g^\beta L_f^\alpha h = \frac{\partial^\beta (L_f^\alpha h)}{\partial x^\beta} g(x), 0 < \alpha, \beta < 1 \tag{18}$$

where $\alpha$ and $\beta$ are the fractional order of the Lie derivative.

It is easy to prove that

$$L_{af_1 + bf_2}^\alpha h = a L_{f_1}^\alpha h + b L_{f_2}^\alpha h, 0 < \alpha < 1 \tag{19}$$

for any vector functions $f_1, f_2$ and a scalar function $h$.

**Example 1.** Consider the vector function

$$f = f(x) = \begin{bmatrix} f_1(x) \\ f_2(x) \end{bmatrix}, \quad x = \begin{bmatrix} x_1 \\ x_2 \end{bmatrix} \tag{20}$$

and the scalar function $h(x) = x_1 x_2$.

By Definition 1 the fractional Lie derivative of the order $\alpha$, $0 < \alpha < 1$, of $h(x)$ along the vector (20) has the form

$$L_f^\alpha h(x) = \frac{\partial^\alpha h}{\partial x^\alpha} f = \begin{bmatrix} \frac{\partial^\alpha h}{\partial x_1^\alpha} & \frac{\partial^\alpha h}{\partial x_2^\alpha} \end{bmatrix} \begin{bmatrix} f_1(x) \\ f_2(x) \end{bmatrix} = f_1(x) x_1^{1-\alpha} x_2 + f_2(x) x_1 x_2^{1-\alpha} \tag{21}$$

since by (6) $\frac{\partial^\alpha x_k}{\partial x_k^\alpha} = x_k^{1-\alpha}$, $k = 1, 2$.

## 4 Relative Degree of Fractional Nonlinear Systems

Consider the fractional nonlinear continuous-time system described by the equations

$$\begin{aligned} \frac{d^\alpha x(t)}{dt^\alpha} &= f(x) + g(x)u, \ 0<\alpha<1, \\ y &= h(x), \end{aligned} \tag{22}$$

where $x = x(t) \in \Re^n$, $u = u(t) \in \Re$, $y = y(t) \in \Re$ are the state vector, input and output of the system, $f(x) \in \Re^n$ and $g(x) \in \Re^n$ are analytical vector functions of $x$ and $h(x) \in \Re$ is an analytical scalar function of $x$.

**Definition 2.** A positive integer $r$ is called the relative degree of fractional nonlinear system (22) if

$$L_g^\alpha L_f^{k\alpha} h \begin{cases} = 0 & \text{for } k = 0, 1, \ldots, r-2 \\ \neq 0 & \text{for } \quad k = r-1 \end{cases} \tag{23}$$

for all $x \in \Re^n$ in the neighborhood of $x_0 \in \Re^n$.

**Example 2.** It will be shown that the fractional nonlinear system (22) with $0<\alpha<1$ and

$$x = \begin{bmatrix} x_1 \\ x_2 \end{bmatrix}, \ f(x) = \begin{bmatrix} x_2 \\ f_2(x_1, x_2) \end{bmatrix}, \ g(x) = \begin{bmatrix} 0 \\ 1 \end{bmatrix}, \ h(x) = x_1 \tag{24}$$

has the relative degree $r = 2$ for any $x \in \Re^2$ and arbitrary function $f_2(x_1, x_2)$.

Using (16), (17) and (24) we obtain

$$L_g^\alpha h(x) = \frac{\partial^\alpha h(x)}{\partial x^\alpha} g(x) = \begin{bmatrix} \frac{\partial^\alpha h(x)}{\partial x_1^\alpha} & 0 \end{bmatrix} \begin{bmatrix} 0 \\ 1 \end{bmatrix} = [x_1^{1-\alpha} \quad 0] \begin{bmatrix} 0 \\ 1 \end{bmatrix} = 0, \tag{25}$$

$$L_f^\alpha h(x) = \frac{\partial^\alpha h(x)}{\partial x^\alpha} f(x) = \begin{bmatrix} \frac{\partial^\alpha h(x)}{\partial x_1^\alpha} & 0 \end{bmatrix} \begin{bmatrix} x_2 \\ f_2(x_1, x_2) \end{bmatrix} = x_2 \frac{\partial^\alpha h(x)}{\partial x_1^\alpha} = x_1^{1-\alpha} x_2, \tag{26}$$

and

$$L_g^\alpha L_f^\alpha h(x) = \frac{\partial^\alpha \left( x_2 \frac{\partial^\alpha h(x)}{\partial x_1^\alpha} \right)}{\partial x^\alpha} g(x) = x_1^{1-\alpha} x_2^{1-\alpha} \neq 0. \tag{27}$$

Therefore, by Definition 2 the relative degree of fractional nonlinear system (22) with (24) is equal to $r = 2$.

**Definition 3.** The fractional nonlinear system (22) is called of full relative degree if $r = n$.

## 5 Reduction of the Fractional Nonlinear System to Its Canonical Form

Consider the fractional full degree ($r = n$) nonlinear system (22). For the system we introduce the following new state vector

$$z = \begin{bmatrix} z_1 \\ \vdots \\ z_n \end{bmatrix} = \begin{bmatrix} h(x) \\ L_f^\alpha h(x) \\ \vdots \\ L_f^{\alpha(n-1)} h(x) \end{bmatrix} = \phi(x). \tag{28}$$

It will be shown that using (28) it is possible to reduce the nonlinear system (22) to the form

$$\frac{d^\alpha z}{dt^\alpha} = \begin{bmatrix} \frac{d^\alpha z_1}{dt^\alpha} \\ \vdots \\ \frac{d^\alpha z_n}{dt^\alpha} \end{bmatrix} = \begin{bmatrix} z_2 \\ \vdots \\ z_n \\ \bar{f}(z_1, \ldots, z_n) \end{bmatrix} + \begin{bmatrix} 0 \\ \vdots \\ 0 \\ \bar{g}(z_1, \ldots, z_n) \end{bmatrix} u \tag{29}$$

where

$$\bar{f}(z) = L_f^{n\alpha} h(x)\Big|_{x=\phi^{-1}(z)}, \tag{30}$$

$$\bar{g}(z) = L_g^\alpha L_f^{(n-1)\alpha} h(x)\Big|_{x=\phi^{-1}(z)}. \tag{31}$$

Using (28), (14) and (23), (22) we obtain

$$\frac{d^\alpha z_1}{dt^\alpha} = \frac{\partial^\alpha h(x)}{\partial x^\alpha}[f(x) + g(x)u] = \frac{\partial^\alpha h(x)}{\partial x^\alpha} f(x) = L_f^\alpha h(x) = z_2 \tag{32}$$

since by (23) $\frac{\partial^\alpha h(x)}{\partial x^\alpha} g(x) = L_g^\alpha h(x) = 0$ and

$$\frac{d^\alpha z_2}{dt^\alpha} = \frac{\partial^\alpha (L_f^\alpha h(x))}{\partial x^\alpha}[f(x) + g(x)u] = \frac{\partial^\alpha (L_f^\alpha h(x))}{\partial x^\alpha} f(x) = L_f^{2\alpha} h(x) = z_3 \tag{33}$$

since by (23) $\frac{\partial^\alpha (L_f^\alpha h(x))}{\partial x^\alpha} g(x) = L_g^\alpha L_f^\alpha h(x) = 0$.

Continuing this procedure we obtain

$$\begin{aligned} \frac{d^\alpha z_{n-1}}{dt^\alpha} &= \frac{\partial^\alpha (L_f^{(n-2)\alpha} h(x))}{\partial x^\alpha}[f(x) + g(x)u] \\ &= \frac{\partial^\alpha (L_f^{(n-2)\alpha} h(x))}{\partial x^\alpha} f(x) = L_f^{(n-1)\alpha} h(x) = z_n \end{aligned} \tag{34}$$

but

$$\frac{d^\alpha z_n}{dt^\alpha} = \frac{\partial^\alpha (L_f^{(n-1)\alpha} h(x))}{\partial x^\alpha}[f(x) + g(x)u] = L_f^{n\alpha} h(x) + L_g^\alpha L_f^{(n-1)\alpha} h(x)u \tag{35}$$

since by (23)

$$L_g^\alpha L_f^{(n-1)\alpha} h(x) \neq 0. \tag{36}$$

From (29) it follows that the fractional nonlinear full order system (22) can be reduced to canonical form (29) if there exists the inverse matrix $x = \phi^{-1}(z)$ of the matrix (28).

Therefore, the following theorem has been proofed.

**Theorem 5.** The fractional nonlinear full order system (22) can be reduced to its canonical form (29) if there exists the inverse matrix of the matrix (28).

**Example 3.** Consider the fractional nonlinear system (22) with

$$f(x) = \begin{bmatrix} e^{\beta x_1} + x_1 x_2 \\ x_1 + x_2 \\ x_2 + x_3 \end{bmatrix}, \quad g(x) = \begin{bmatrix} 1 \\ 0 \\ 0 \end{bmatrix}, \quad h(x) = x_3, \ 0 < \alpha < 1, \tag{37}$$

where $\beta$ is given parameter.

First we shall show that the system is of full degree, i.e. $r = n = 3$.

Using (1) and (37) we obtain

$$\begin{aligned} L_g^\alpha h(x) &= \frac{\partial^\alpha h(x)}{\partial x^\alpha} g(x) = \begin{bmatrix} \frac{\partial^\alpha h(x)}{\partial x_1^\alpha} & \frac{\partial^\alpha h(x)}{\partial x_2^\alpha} & \frac{\partial^\alpha h(x)}{\partial x_3^\alpha} \end{bmatrix} g(x) \\ &= \begin{bmatrix} 0 & 0 & \frac{\partial^\alpha x_3}{\partial x_3^\alpha} \end{bmatrix} \begin{bmatrix} 1 \\ 0 \\ 0 \end{bmatrix} = [0 \quad 0 \quad x_3^{1-\alpha}] \begin{bmatrix} 1 \\ 0 \\ 0 \end{bmatrix} = 0, \end{aligned} \tag{38}$$

$$\begin{aligned} L_f^\alpha h(x) &= \frac{\partial^\alpha h(x)}{\partial x^\alpha} f(x) = \begin{bmatrix} 0 & 0 & \frac{\partial^\alpha x_3}{\partial x_3^\alpha} \end{bmatrix} f(x) \\ &= [0 \quad 0 \quad x_3^{1-\alpha}] \begin{bmatrix} e^{\beta x_1} + x_1 x_2 \\ x_1 + x_2 \\ x_2 + x_3 \end{bmatrix} = x_3^{1-\alpha}(x_2 + x_3) \end{aligned} \tag{39}$$

$$L_g^\alpha L_f^\alpha h(x) = L_g^\alpha [x_3^{1-\alpha}(x_2 + x_3)] = \begin{bmatrix} 0 & x_3^{1-\alpha} \frac{\partial^\alpha x_2}{\partial x_2^\alpha} & x_3^{1-\alpha} \frac{\partial^\alpha x_3}{\partial x_3^\alpha} \end{bmatrix} \begin{bmatrix} 1 \\ 0 \\ 0 \end{bmatrix} = 0 \tag{40}$$

and

$$L_f^{2\alpha}h(x) = L_f^{\alpha}(L_f^{\alpha}h(x)) = \begin{bmatrix} 0 & x_2^{1-\alpha}x_3^{1-\alpha} & (2-\alpha)x_3^{2(1-\alpha)} \end{bmatrix} \begin{bmatrix} e^{\beta x_1} + x_1x_2 \\ x_1 + x_2 \\ x_2 + x_3 \end{bmatrix}, \quad (41)$$
$$= x_2^{1-\alpha}x_3^{1-\alpha}(x_1 + x_2) + (2-\alpha)x_3^{2(1-\alpha)}(x_2 + x_3) \neq 0$$

$$L_g^{\alpha}L_f^{2\alpha}h(x) = \begin{bmatrix} \frac{\partial^{\alpha}L_f^{2\alpha}h(x)}{\partial x_1^{\alpha}} & \frac{\partial^{\alpha}L_f^{2\alpha}h(x)}{\partial x_2^{\alpha}} & \frac{\partial^{\alpha}L_f^{2\alpha}h(x)}{\partial x_3^{\alpha}} \end{bmatrix} g(x) = (x_1x_2x_3)^{1-\alpha} \neq 0. \quad (42)$$

Therefore, the condition (23) is satisfied for $r = 3$ and by Definition 3 the fractional nonlinear system (22) with (37) is of full degree.

To find the new state vector (28) we compute its components

$$\begin{aligned} z_1 &= h(x) = x_3, \\ z_2 &= L_f^{\alpha}h(x) = c_1(x_2 + x_3), \; c_1 = x_3^{1-\alpha} \\ z_3 &= L_f^{2\alpha}h(x) = c_2(x_1 + x_3) + c_3(x_2 + x_3), \; c_2 = (x_2x_3)^{1-\alpha}, \; c_3 = (2-\alpha)x_3^{2(1-\alpha)}. \end{aligned} \quad (43)$$

Therefore

$$z = \begin{bmatrix} z_1 \\ z_2 \\ z_3 \end{bmatrix} = \begin{bmatrix} 0 & 0 & 1 \\ 0 & c_1 & c_1 \\ c_2 & c_3 & c_2 + c_3 \end{bmatrix} \begin{bmatrix} x_1 \\ x_2 \\ x_3 \end{bmatrix} = \phi(x) \quad (44)$$

and

$$\begin{bmatrix} x_1 \\ x_2 \\ x_3 \end{bmatrix} = \begin{bmatrix} 0 & 0 & 1 \\ 0 & c_1 & c_1 \\ c_2 & c_3 & c_2 + c_3 \end{bmatrix}^{-1} \begin{bmatrix} z_1 \\ z_2 \\ z_3 \end{bmatrix} = \begin{bmatrix} -z_1 - \frac{c_3}{c_1c_2}z_2 + \frac{1}{c_2}z_3 \\ -z_1 + \frac{1}{c_1}z_2 \\ z_1 \end{bmatrix} = \phi^{-1}(z) \quad (45)$$

since $c_1 \neq 0$ and $c_2 \neq 0$.

The desired canonical form of the system is given by

$$\frac{d^{\alpha}z}{dt^{\alpha}} = \begin{bmatrix} \frac{d^{\alpha}z_1}{dt^{\alpha}} \\ \frac{d^{\alpha}z_2}{dt^{\alpha}} \\ \frac{d^{\alpha}z_3}{dt^{\alpha}} \end{bmatrix} = \begin{bmatrix} z_2 \\ z_3 \\ \bar{f}(z_1, z_2, z_3) \end{bmatrix} + \begin{bmatrix} 0 \\ 0 \\ \bar{g}(z_1, z_2, z_3) \end{bmatrix} u, \quad (46)$$

where

$$\bar{f}(z) = L_f^{3\alpha}h(x)\Big|_{x=\phi^{-1}(z)} = \left[c_2\frac{\partial^\alpha x_1}{\partial x_1^\alpha} \quad c_3\frac{\partial^\alpha x_2}{\partial x_2^\alpha} \quad (c_2+c_3)\frac{\partial^\alpha x_3}{\partial x_3^\alpha}\right]f(x)\Big|_{x=\phi^{-1}(z)}, \tag{47}$$

$$\bar{g}(z) = L_g^\alpha L_f^{2\alpha}h(x)\Big|_{x=\phi^{-1}(z)} = c_2\frac{\partial^\alpha x_1}{\partial x_1^\alpha}\Big|_{x=\phi^{-1}(z)} \tag{48}$$

and $\phi^{-1}(z)$ is defined by (45).

## 6 Linearization by State-Feedback

Consider the fractional nonlinear system in the canonical form (29).

**Theorem 6.** The fractional nonlinear system in the canonical form (29) can be linearized by the use of the state feedback

$$u = \frac{1}{\bar{g}(z)}(v - \bar{f}(z)) \tag{49}$$

to the form

$$\frac{d^\alpha z}{dt^\alpha} = Az + bv, \tag{50a}$$

$$y = cz, \tag{50b}$$

where

$$A = \begin{bmatrix} 0 & 1 & 0 & \dots & 0 \\ 0 & 0 & 1 & \dots & 0 \\ \vdots & \vdots & \vdots & \ddots & \vdots \\ 0 & 0 & 0 & \dots & 1 \\ 0 & 0 & 0 & \dots & 0 \end{bmatrix} \in \Re^{n\times n}, \; b = \begin{bmatrix} 0 \\ \vdots \\ 0 \\ 1 \end{bmatrix} \in \Re^n, \; c = [1 \quad 0 \quad \dots \quad 0] \in \Re^{1\times n} \tag{51}$$

and $v$ is a new input of the system (Fig. 1).

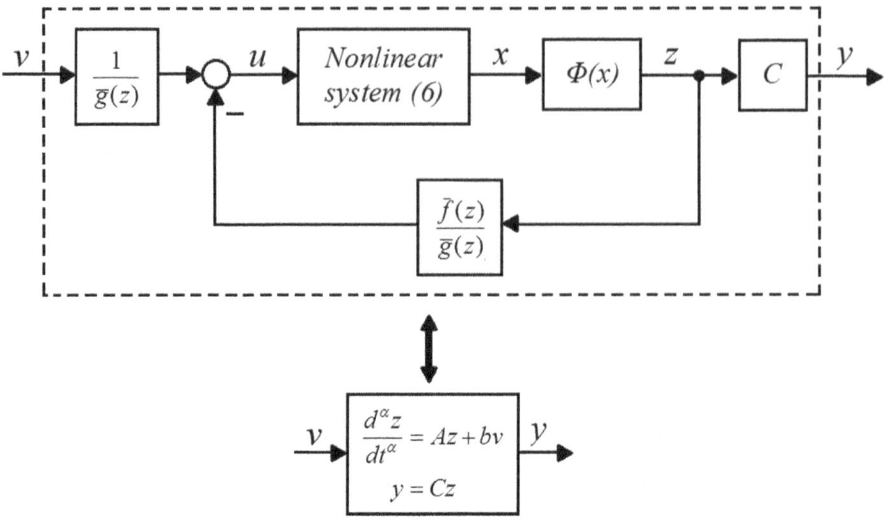


**Fig. 1.** Illustration of Theorem 6

**Proof.** Substituting (49) into (29) we obtain

$$\frac{d^\alpha z}{dt^\alpha} = \begin{bmatrix} z_2 \\ \vdots \\ z_n \\ \bar{f}(z) \end{bmatrix} + \begin{bmatrix} 0 \\ \vdots \\ 0 \\ \bar{g}(z) \end{bmatrix} \frac{1}{\bar{g}(z)}(v - \bar{f}(z)) = Az + bv \tag{52a}$$

and

$$y = h(x) = z_1 = cz, \tag{52b}$$

where the matrices $A$, $b$, $c$ are defined by (50a, 50b). □

**Example 4.** (Continuation of Example 3). Consider the fractional nonlinear system (22) with (37) and its canonical form (46).

By Theorem 6 there exists the canonical form (46) since the matrix (44) is invertible. Therefore, there exists the state-feedback (49) which linearize the fractional nonlinear system (22) with (37).

Applying to the system (46) the state-feedback (49) with $\bar{g}(z)$ and $\bar{f}(z)$ defined by (47) and (48) we obtain the linear system (50a, 50b) with the matrices

$$A = \begin{bmatrix} 0 & 1 & 0 \\ 0 & 0 & 1 \\ 0 & 0 & 0 \end{bmatrix}, \; b = \begin{bmatrix} 0 \\ 0 \\ 1 \end{bmatrix}, \; c = [1 \quad 0 \quad 0]. \tag{53}$$

The considerations can be easily extended to multi-input multi-output fractional nonlinear systems.

## 7 Concluding Remarks

Using conformable fractional derivative the notion of Lie derivative and full relative degree have been extended to the fractional nonlinear systems. The canonical form of nonlinear systems has been extended to the fractional nonlinear systems. Sufficient conditions for the existence of the canonical form for the systems have been established. A method for finding nonlinear state-feedbacks linearizing the fractional nonlinear system have been proposed. The considerations have been illustrated by examples of the fractional nonlinear systems. The presented considerations for single-input single-output fractional nonlinear systems can be easily extended to multi-input multi-output fractional nonlinear systems. An open problem is an extension of these considerations to positive fractional nonlinear systems.

**Acknowledgment.** This work was supported by National Science Centre in Poland under work No. 2017/27/B/ST7/02443.

## References

1. Oldham, K.B., Spanier, J.: The Fractional Calculus. Academic Press, New York (1974)
2. Ostalczyk, P.: Epitome of the Fractional Calculus: Theory and Its Applications in Automatics. Wydawnictwo Politechniki Łódzkiej, Łódź (2008). (in Polish)
3. Podlubny, I.: Fractional Differential Equations. Academic Press, San Diego (1999)
4. Kaczorek, T.: Fractional positive continuous-time systems and their reachability. Int. J. Appl. Comput. Sci. **18**(2), 223–228 (2008)
5. Kaczorek, T.: Positive linear systems consisting of n subsystems with different fractional orders. IEEE Trans. Circuits Syst. **58**(6), 1203–1210 (2011)
6. Busłowicz, M.: Stability of linear continuous time fractional order systems with delays of the retarded type. Bull. Pol. Acad. Sci. Tech. **56**(4), 319–324 (2008)
7. Dzieliński, A., Sierociuk, D.: Stability of discrete fractional order state-space systems. J. Vib. Control **14**(9/10), 1543–1556 (2008)
8. Kaczorek, T.: Selected Problems of Fractional Systems Theory. Springer, Berlin (2012)
9. Kaczorek, T.: Practical stability of positive fractional discrete-time linear systems. Bull. Pol. Acad. Sci. Tech. **56**(4), 313–317 (2008)
10. Dzieliński, A., Sierociuk, D., Sarwas, G.: Ultracapacitor parameters identification based on fractional order model. In: Proceedings of ECC 2009, Budapest (2009)
11. Jumerie, G.: The Leibniz rule for fractional derivatives holds with non-differentiable functions. Math. Stat. **1**(2), 50–52 (2013)
12. Radwan, A.G., Soliman, A.M., Elwakil, A.S., Sedeek, A.: On the stability of linear systems with fractional-order elements. Chaos Solitones Fractals **40**(5), 2317–2328 (2009)
13. Kaczorek, T.: An extension of Klamka's method of minimum energy control to fractional positive discrete-time linear systems with bounded inputs. Bull. Pol. Acad. Tech. **62**(2), 227–231 (2014)
14. Klamka, J.: Controllability of Dynamical Systems. Kluwer Academic Press, Dordrecht (1991)
15. Klamka, J.: Minimum energy control of 2D systems in Hilbert spaces. Syst. Sci. **9**(1–2), 33–42 (1983)

16. Klamka, J.: Controllability and minimum energy control problem of fractional discrete-time systems. In: Baleanu, D., Guvenc, Z.B., Tenreiro Machado, J.A. (eds.) New Trends in Nanotechnology and Fractional Calculus Applications, pp. 503–509. Springer, Dordrecht (2010)
17. Klamka, J.: Relative controllability and minimum energy control of linear systems with distributed delays in control. IEEE Trans. Autom. Contr. **21**(4), 594–595 (1976)
18. Kaczorek, T., Klamka, J.: Minimum energy control of 2D linear systems with variable coefficients. Int. J. Control **44**(3), 645–650 (1986)
19. Solteiro Pires, E.J., Tenreiro Machado, J.A., Moura Oliveira, P.B.: Functional dynamics in genetic algorithms. In: Workshop on Fractional Differentiation and Its Application, vol. 2, pp. 414–419 (2006)
20. Brockett, R.W.: Nonlinear systems and differential geometry. Proc. IEEE **64**(1), 61–71 (1976)
21. Isidori, A.: Nonlinear Control Systems. Springer, Berlin (1989)
22. Marino, R., Tomei, P.: Nonlinear Control Design – Geometric, Adaptive, and Robust. Prentice Hall, London (1995)
23. Aguiller, J.L.M., Garcia, R.A., D'Attellis, C.E.: Exact linearization of nonlinear systems: trajectory tracking with bounded control and state constrains. In: Proceedings of the 38th Midwest Symposium on Circuits and Systems, Rio de Janeiro, pp. 620–622 (1995)
24. Charlet, B., Levine, J., Marino, R.: Sufficient conditions for dynamic state feedback linearization. SIAM J. Contr. Optim. **29**(1), 38–57 (1991)
25. Daizhan, C., Tzyh-Jong, T., Isidori, A.: Global external linearization of nonlinear systems via feedback. IEEE Trans. Autom. Control **30**, 808–811 (1985)
26. Fang, B., Kelkar, A.G.: Exact linearization of nonlinear systems by time scale transformation. In: Proceedings of the American Control Conference, Denver-Colorado, pp. 3555–3560 (2003)
27. Jakubczyk, B.: Introduction to Geometric Nonlinear Control; Controllability and Lie Bracket. Summer Schools on Mathematical Control Theory, Trieste (2001)
28. Jakubczyk, B., Respondek, W.: On linearization of control systems. Bull. Pol. Acad. Sci. Tech. **28**, 517–521 (1980)
29. Melham, K., Saad, M., Abou, S.C.: Linearization by redundancy and stabilization of nonlinear dynamical systems: a state transformation approach. In: IEEE International Symposium on Industrial Electronics, pp. 61–68 (2009)
30. Taylor, J.H., Antoniotti, A.J.: Linearization algorithms for computer-aided control engineering. Control Syst. Mag. **13**, 58–64 (1993)
31. Khalil, R., Horani, M., Yousef, A., Sababheh, M.: A new definition of fractional derivative. J. Comput. Appl. Math. **264**, 65–70 (2014)

# Control of the Operation Mode of the Production Facility Based on the Relevant Characteristics of the Technological Process

Igor Korobiichuk[1], Liudmyla Davydenko[2(✉)], Nina Davydenko[3(✉)], and Volodymyr Davydenko[3]

[1] ŁUKASIEWICZ Research Network – Industrial Research Institute for Automation and Measurements PIAP, Warsaw, Poland
ikorobiichuk@piap.pl

[2] Lutsk National Technical University, Lutsk, Ukraine
l.davydenko033@gmail.com

[3] National University of Water and Environmental Engineering, Rivne, Ukraine
ninadavydenko1992@gmail.com,
v.a.davydenko@nuwm.edu.ua

**Abstract.** The article deals with the control of the operation mode of a production facility taking into consideration its actual operation conditions. Control procedure, which is based on the application of Shewhart charts for technological process characteristics, has been proposed. This procedure provides not only identification of the discrepancies in the process, but also the reasons that caused them. Construction of control charts is based on data obtained from the monitoring system and organized taking into consideration cyclic changes in the production process. This allows taking into consideration the actual operation conditions of the production facility and setting the correct standards for controlled parameters. The principles of interpretation of control charts and set of alarms about exceeding the standards, which take into consideration the nature of the influence of the controlled parameters on the production process efficiency, have been proposed. The proposed procedure has been applied to control the operation mode of the water supply pumping station. Standards for controlled technological parameters have been set taking into consideration cyclic changes in the water supply process. The description of alarms, which have been used to notify the operator of the control results, has been performed. The analysis of control charts of technological parameters enabled to determine moments of the discrepancy between the operation mode of the pumping station and the actual needs of the technological process, and to identify its reasons. The obtained results confirm the expediency of using the proposed procedure to take a decision on the need to improve the production process efficiency.

**Keywords:** Shewhart charts · Standards · Control limits · Alarms


R. Szewczyk et al. (Eds.): AUTOMATION 2020, AISC 1140, pp. 57–66, 2020.
https://doi.org/10.1007/978-3-030-40971-5_6

# 1 Introduction

The current worldwide awareness of reducing global warming requires energy-efficient production processes. Efficiency is a set of properties that characterize the quality of the system's functioning, which is evaluated as an accordance of the results: needed and achieved [1]. Production processes are characterized by a certain amount of variability that can affect the accordance of planned and obtained result [2]. This causes additional energy resources costs. Therefore, the accomplishment of the energy efficiency requirement requires ensuring and improving the efficiency of the production processes and production systems that implement them. It is possible to ensure and improve the production efficiency via operative management, which includes all operational methods and activities focusing upon monitoring processes and removing causes of discrepancy [3]. The operation mode of the production facility (PF) is determined by the needs of the technological process that characterized by a certain cyclic variability due to the influence of external factors. An important interpretation of the improvement is the reduction of variability in processes [4]. Based on the early detection of significant deviations in the process, it is possible to implement interventions in the process and improve it [2]. There are many examples of detection and prevention of the process variability in different areas [2, 3, 5–9]. However, usually operative management is focused upon the process itself. Herewith, the issues of conformity of the PF operation mode with the actual needs of the technological process remain unaddressed. At the same time, such discrepancy reduces the efficiency of the PF operation mode.

Therefore, solving the issue of increasing the PF efficiency requires monitoring of technological parameters, actual needs of the technological process and external factors that determine the change in these needs. Complex monitoring of the production process is one of the elements of the generalized structure of energy efficiency control of the PF proposed by the authors in [10, 11]. The introduction of monitoring system provides a possibility to create a large databases that contain information on the PF operation modes, as well as external factors that determine the technological process parameters [10, 11]. Therefore, information on all input and output aspects of the production process is available. The analysis of this information allows taking decisions about necessity to adjust PF operation mode in response to the variable demand in the final product of the technological process in order to ensure its energy efficiency [5]. However, it is necessary to solve next issues. First, to identify the moments when the demand for the final product of the production process changes. The principles for identifying these changes have been set out in detail by the authors in [12]. Second, to check the conformity of the actual characteristics of the technological process to the planned values and to determine the moments of their discrepancy, i.e., to perform the control of the PF operation mode, as well as to notify the operator of the control results. The solution of the second issue is the subject of this research. Therefore, the purpose of the article is to develop a procedure for controlling the PF operation mode, which will provide information on the efficiency of the PF operation mode and the need to adjust it according to the actual needs of the technological process.

The article is organized as follows. Section 2 describes the substantiation for selecting the control tool, the principles for implementing the control procedure, organizing the notification of control results and their application. Section 3 contains experimental research and discussion of obtained results. The final Section concludes the paper and presents further researches.

## 2 Research Methodology

Control of the technological parameters and characteristics of the efficiency of the organization of the operation mode is one way to control the accordance of the actual PF operation mode with the planned. The main stages of control are: (1) obtaining of primary information on the actual state of the control object, its controlled indicators; (2) obtaining of secondary information – deviations from the given parameters by comparison of the primary information with the established standards; (3) preparation of information for the development of appropriate control effects on the object.

The current values of the controlled parameters is determined and their deviation from the given (or determined) normative value is evaluated to control the operation modes. Controlled parameters should satisfy the limits:

$$\underline{x}_i \leq x_i \leq \bar{x}_i, \qquad i = \overline{1,N} \tag{1}$$

Otherwise, it is considered that the object does not function properly. Therefore, it is necessary to determine the moments of the discrepancy of the process parameters with the planned, and to correct anomalies and discrepancies and improve process efficiency [5]. It is advisable to use Statistical Process Control (SPC) to solve the issue [4]. SPC can be applied to any process for which the output data can be measured. The SPC is a set of tools used to understand and control a production process in order to achieve its stability [13] and improve efficiency [4].

The SPC tools allow identifying the unnatural variability in processes that cause to low efficiency [3]. These tools indicate when corrective and improvement actions should be taken [5].

Control charts is a key tool of the SPC, through which process characteristics can be visualized with statistically determined limits. They are used to differentiate between common and special variations in the process in order to establish the need for corrections or changes in the process [3]. The control charts method provides information about the process during its development, allows understanding when the observed values are beyond acceptable limits. Shewhart charts [14] and Hotelling charts [16] are used for controlling.

The Shewhart chart has: a central line (CL) (it corresponds to the average arithmetic value of the controlled indicator [14, 15]); two control limits: upper $U_{CL}$ and lower $L_{CL}$. Control limits are determined by the rule of “three sigma” ($\pm 3\sigma$, where $\sigma$ – standard deviation or its evaluation). The range between the $U_{CL}$ and $L_{CL}$ indicates variability caused by natural reasons.

The Hotelling chart allows controlling groups of technological parameters. This reduces the number of charts that need simultaneous analysis. However, it does not allow determining what parameter is the reason of the process instability. In addition, it has only one control limit. This allows establishing the fact of discrepancy of the standard, but does not allow establishing its nature (there is a tendency to increase the process efficiency or to deteriorate it).

Therefore, it is expedient to use the Shewhart charts to control the technological parameters.

Application of individuals control charts (X-charts) and moving range control charts ($R_m$-charts) is appropriate for technological parameters control. They do not require the definition of rational groups to ensure the assessment of the process variability. The control procedure requires the following stages.

Stage 1 – the verification of the accordance of the data sample to the normal distribution law. A normal probability distribution law is assumed for control charts that use quantitative data. Deviation from this assumption affects the charts efficiency. Mean $\bar{x}$, mode $Mo$, and median $Me$ are the same for a normal distribution law. The hypothesis of the normality of the distribution law is checked using the values of skewness and kurtosis coefficients [17]:

$$A_s = \frac{\mu_3}{(\mu_2)^{3/2}} = \frac{\frac{1}{n}\sum_{i=1}^{n}(x_i - \bar{x})^3}{\left[\frac{1}{n}\sum_{i=1}^{n}(x_i - \bar{x})^2\right]^{3/2}}, \tag{2}$$

$$E_k = \frac{\mu_4}{(\mu_2)^2} - 3, \tag{3}$$

where $\mu_2$, $\mu_3$, $\mu_4$ – the second, third and fourth central moment.

Unbiased estimation of coefficients [17]:

$$A_{ns} = \frac{\sqrt{n(n-1)}}{n-2} A_s, \tag{4}$$

$$E_{nk} = \frac{n-1}{(n-2)(n-3)}\left[(n+1)E_k + 6\right]. \tag{5}$$

The standard deviation of the skewness and kurtosis coefficients:

$$s_{A_s} = \sqrt{\frac{6}{n+3}}, \tag{6}$$

$$s_{E_k} = 2 \cdot s_{A_s} = 2 \cdot \sqrt{\frac{6}{n+3}}. \tag{7}$$

For normal distribution, $A_s = 0$ and $E_k = 0$. Hypothesis $H_0$ is checked: random variable distribution is different from normal. Hypothesis $H_0$ is rejected if:

$$|A_{ns}| < \frac{s_{A_s}}{1.5 \div 2} = (0.5 \div 0.67)\sqrt{\frac{6}{n+3}}, \tag{8}$$

$$E_{nk} < \frac{s_{E_k}}{1.5 \div 2} = (1 \div 1.33)\sqrt{\frac{6}{n+3}}. \tag{9}$$

Stage 2 – the definition of standards. Correct setting of purpose-oriented variables is an important stage for the further application of the control procedure [11]. It is necessary to take into consideration the real conditions of the control object functioning to ensure that the standards would not be overestimated or understated. Therefore, they are expedient to be established based on the accumulated statistics on the technological parameters of the production process and efficiency indicators of the operation mode of the technological equipment. The monitoring system ensures the creation of large databases that should be used for the correct determination of standards. The control limits of the Shewhart charts should be taken as the standards for controlled technological parameters and efficiency indicators of the PF operation mode. Selective mathematical expectation of a controlled parameter is taken as its average value:

$$\bar{x} = \frac{1}{n}\sum_{i=1}^{n} x_i, \tag{10}$$

where $n$ – the number of sample values; $x_i$ – the observed parameter value.

Confidence interval for selective mathematical expectation [17]:

$$\bar{x} - \frac{s}{\sqrt{n}} t_{\left[1-\frac{\alpha}{2},(n-1)\right]} < \bar{x} < \bar{x} + \frac{s}{\sqrt{n}} t_{\left[1-\frac{\alpha}{2},(n-1)\right]}, \tag{11}$$

where $t_{\left[1-\frac{\alpha}{2},(n-1)\right]}$ – quantile of Student's t-distribution for given confidence probability $(1 - \alpha)$ with $(n - 1)$ degrees of freedom; $\alpha$ – significance level, that determines the probability of making mistakes $\alpha = 1 - p$; $s$ – selective standard deviation:

$$s^2 = \frac{1}{n-1}\sum_{i=1}^{n} (x_{i.} - \bar{x})^2. \tag{12}$$

In the case of a normal distribution law, the confidence interval, constructed with a confidence probability p = 0.997, is accepted as the admissible limits of parameter change $[x_{\min}, x_{\max}]$. These limits are in line with the "three sigma" rule. The limits of $\pm 3\sigma$ from the central line indicate that approximately 99.7% of the statistical values will fall within these limits, if the process is in a statistically controlled state [14]. The limits on a distance of $\pm 2\sigma$ (warning) are additionally determined on the control chart to warn of approaching the process to exit from the state of statistical control.

Stage 3 – the identification of "alarms" and notification of exceeding the established standards. Alarms are used for this. They can be specified in different combinations [18]: High and High High; Low and Low Low; Deviation etc. Herewith, it is necessary to group the alarms, depending on the purpose and type of controlled parameters, set the values and priorities of the control limits, as well as the type of signal. The operation mode control is performed to ensure its effective management. Therefore, the controlled parameters should be grouped by their influence on the efficiency level of the operation mode. Energy consumption should be considered as efficiency criterion of the operation mode [19]. There are two types of parameters: (1) parameter increase provides energy consumption decrease (energy efficiency increase); (2) parameter increase provides energy consumption increase. It is advisable to apply correlation analysis to identify the nature of the influence of controlled technological process parameters on energy consumption. The parameter belongs to the first type if the correlation coefficient is negative, otherwise – to the second. Such distribution changes the interpretation of the control charts. For the parameters of the first type, the zone of unsatisfactory state of the PF operation mode corresponds to the range of the control chart below the lower control limit, and for the parameters of the second type – above the upper one. The analysis of control charts will allow identifying unplanned deviations of technological parameters and the reasons for the deterioration of the PF operation mode efficiency. The control results provide information for decision-making on the need to adjust the PF operation mode.

It should be noticed that the control procedure is preceded by the procedure for identifying cyclic changes in the production process [12]. It will allow determining the time ranges that correspond to the different operation conditions of the PF. Forming samples of the required data on the technological parameters and performing of the control stages should be realized for each of these ranges. This will allow taking into consideration the actual operation conditions of the PF and ensuring the correctness of the control results.

## 3 Results

The proposed principles have been applied to control the operation mode of the pumping station (PS) of water supply. Previously analysis of the factors that determine the efficiency of the PS operation mode has been performed [20]. It is established that the volume of water supplied to the water supply network ($Q$) and excessive pressure ($\Delta H$) are the main controlled parameters of the technological process of water supply. The coefficient of energy efficiency of the water supply mode is an additional parameter that reflects the efficiency of the technological process of water supply ($K_{ef}$) under conditions of variable water consumption [20]. The actual PS operation conditions depend on the water consumption, which is variable and depends on a number of seasonal and social factors. The control of the efficiency of the water supply process should provide for taking into consideration the actual PS operation conditions [10, 11]. Therefore, first, the procedure for identifying cyclic changes in the water supply process has been applied [12] and data samples have been formed for typical days of typical seasons. This also allowed excluding anomalous data from the samples. The application of the control procedure of the PS operation mode is demonstrated for workdays of the season “winter”.

Statistical analysis of data samples for technological parameters has been performed. The results of the calculation are presented in Table 1. Its results do not give grounds for rejecting the hypothesis about the normality of the law of distribution of technological parameters.

**Table 1.** The results of statistical analysis of data samples

| Variable | $\bar{x}$ | $Me$ | $Mo$ | $A_s$ | $s_{A_s}$ | $E_k$ | $s_{E_k}$ |
|---|---|---|---|---|---|---|---|
| $Q$ | 47.971 | 47.919 | 47.374 | −0.184 | 0.242 | −0.867 | 0.481 |
| $K_{ef}$ | 0.838 | 0.837 | 0.827 | −0.184 | 0.242 | −0.867 | 0.481 |
| $\Delta H_\Sigma$ | 2.592 | 2.422 | 2.412 | 0.385 | 0.242 | −0.661 | 0.481 |

Correlation analysis has been performed to identify the nature of the influence of the controlled parameters on the power consumption ($W$) of the PS (Table 2).

**Table 2.** The results of the calculation of the correlation coefficients

| | $W$ | $Q$ | $K_{ef}$ | $\Delta H_\Sigma$ |
|---|---|---|---|---|
| $W$ | 1.00 | 0.81 | −0.19 | 0.89 |
| $Q$ | 0.81 | 1.00 | 0.95 | 0.19 |
| $K_{ef}$ | −0.19 | 0.95 | 1.00 | 0.19 |
| $\Delta H_\Sigma$ | 0.89 | 0.19 | 0.19 | 1.00 |

It is established that $Q$ and $\Delta H$ are parameters of the second type; $K_{ef}$ is a parameter of the first type. Description of the alarms has been performed for them (Table 3).

**Table 3.** Description of the alarms

| Parameter | Signal type | Signal value | Notification | Type of alarm |
|---|---|---|---|---|
| $Q$, | «Hi-Hi» | $\bar{x}_i + 3\sigma_i$ | Discrepancy to the planned mode | **Alarm** |
| | «Hi» | $\bar{x}_i + 2\sigma_i$ | Exceeding of the admissible limit may occur. Increase of power consumption | *Warning* |
| | Deviation | $\bar{x}_i \pm \sigma_i$ | Deviation from the norm | Notification |
| | «Lo» | $\bar{x}_i - 2\sigma_i$ | Exceeding of the admissible limit may occur | *Warning* |
| | «Lo-Lo» | $\bar{x}_i - 3\sigma_i$ | Discrepancy to the planned mode | **Alarm** |
| $\Delta H$ | «Hi-Hi» | $\bar{x}_i + 3\sigma_i$ | Exceeding of the admissible limit. Decrease of efficiency | **Alarm** |
| | «Hi» | $\bar{x}_i + 2\sigma_i$ | Exceeding of the admissible limit and decrease of efficiency may occur | *Warning* |
| | Deviation | $\bar{x}_i \pm \sigma_i$ | Deviation from the norm | Notification |
| | «Lo» | $\bar{x}_i - 2\sigma_i$ | Increase of efficiency may occur | Notification |
| | «Lo-Lo» | $\bar{x}_i - 3\sigma_i$ | Increase of efficiency | Notification |
| $K_{ef}$ | «Hi-Hi» | $\bar{x}_i + 3\sigma_i$ | Increase of efficiency | Notification |
| | «Hi» | $\bar{x}_i + 2\sigma_i$ | Increase of efficiency may occur | Notification |
| | Deviation | $\bar{x}_i \pm \sigma_i$ | Deviation from the norm | Notification |
| | «Lo» | $\bar{x}_i - 2\sigma_i$ | Exceeding of the admissible limit and deterioration of efficiency may occur | *Warning* |
| | «Lo-Lo» | $\bar{x}_i - 3\sigma_i$ | Discrepancy to the planned mode. Deterioration of efficiency | **Alarm** |

The limits of the confidence interval to the average value of the corresponding technological parameter, which are determined with the confidence probability p = 0.997, are taken as the limiting control limits.

The control procedure of the PS operation mode provides for the analysis of accordance of the actual water consumption chart with the planned water supply mode for a given day type of the appropriate season. Otherwise, control of the water supply mode efficiency is incorrect. A fragment of the control results is presented in Table 4 (the "+" sign is above the control value; the "−" sign is below the control value).

**Table 4.** The results of the control of the PS technological parameters

| Date | Deviation of technological parameters from control/warning limits | | Identification of the reason | Notification |
|---|---|---|---|---|
| | $Q$, $m^3$ | $\Delta H$, m | | |
| 01.02. | 0/0 | 0/–0.02 | | Parameters are within normal limits |
| 04.02. | +951/+1152.7 | +0.21/+0.30 | Increase in $Q$ | **Discrepancy to the planned mode** |
| 05.02. | 0/−141.67 | –0.13/–0.22 | | **Discrepancy to the planned mode** |
| … | | | | |
| 11.02. | 0/+1.7 | +0.26/+0.35 | Increase in $\Delta H$ | **Decrease of efficiency** |
| 12.02. | 0/0 | 0/+0.05 | | Parameters are within normal limits |
| … | | | | |
| 28.02. | 0/−106.67 | 0/+0.04 | Increase in $\Delta H$ with decreasing $Q$ | **Decrease of efficiency** |

Network damage, which led to significant increase in Q, may be the reason of the water supply mode discrepancy. The insufficient consistency of the water consumption chart and the water supply chart is the reason for the increase in excessive pressure in the network. Therefore, it is necessary to adjust the water supply chart.

The obtained results of the control of the PS operation mode confirm the efficiency of the proposed procedure. Its application provides the PS operator with information to decide on the necessity of adjusting the PS operation mode and improving the efficiency of the water supply process.

## 4 Conclusions

The PF operation mode should correspond to the actual needs of the technological process that it carries out in order to ensure the efficiency of the production process. These needs change under the influence of external factors and determine the actual operation conditions of the facility. Therefore, control of the technological process characteristics that determine the PF operation mode is one way to control the efficiency of the organization of production process. The use of Shewhart charts is a convenient tool for such control. The proposed control procedure provides that the cyclic changes of the production process, due to the influence of external factors, will be predetermined. It will allow establishing correct values of the standards of the controlled parameters depending on the actual PF operation conditions and obtaining the correct control results. The application of the proposed procedure to control the PS operation mode confirms its effectiveness. Analysis of control charts of the characteristics of the PS operation mode allowed objectively determining the times of the discrepancy between the PF operation mode and the actual needs of the process, the reasons for its occurrence and making conclusions about the adjustment of the operation mode.

The obtained results are a continuation of a number of previous researches by the authors [10, 11, 20] and are an element of information support for the energy management system of an industrial enterprise. The application of the proposed control procedure provides information on the efficiency of the PF operation mode and the reasons for its decrease or increase. This makes it easier to take the decision on the need to adjust the production process. In addition, the analysis of this information allows identifying ways of improvement the PF operation and increase its energy efficiency. Developing principles for integrating the results of the proposed control procedure into the decision-making support system of the PF operator will allow increasing the efficiency of its practical application.

## References


1. Volkova, V., Yemelyanov, A.: System Theory and System Analysis in the Management of Organizations. Finance and Statistics, Moscow (2006). (in Russian)
2. Šatanová, A., Figuli, L., Sedliačiková, M.: Optimization of production process through selected statistical methods. Procedia Econ. Financ. **23**, 959–963 (2015). https://doi.org/10.1016/S2212-5671(15)00494-3
3. Gejdoš, P.: Continuous quality improvement by statistical process control. Procedia Econ. Financ. **34**, 565–572 (2015). https://doi.org/10.1016/S2212-5671(15)01669-X
4. Montgomery, D.C.: Introduction to Statistical Quality Control. Wiley, New York (2009)
5. Godina, R., Pimentel, C., Silva, F.J.G., Matias, João C.O.: Improvement of the statistical process control certainty in an automotive manufacturing unit. Procedia Manuf. **17**, 729–736 (2018). https://doi.org/10.1016/j.promfg.2018.10.123
6. Azizi, A.: Evaluation improvement of production productivity performance using statistical process control, overall equipment efficiency, and autonomous maintenance. Procedia Manuf. **2**, 186–190 (2015). https://doi.org/10.1016/j.promfg.2015.07.032
7. Ning, X., Shang, Y., Tsung, F.: Statistical process control techniques for service processes: a review. In: Proceedings - 2009 6th International Conference on Service Systems and Service Management, pp. 927–931. https://doi.org/10.1109/ICSSSM.2009.5175015
8. Škulj, G., Vrabič, R., Butala, P., Sluga, A.: Statistical process control as a service: an industrial case study. Procedia CIRP **7**, 401–406 (2013). https://doi.org/10.1016/j.procir.2013.06.006

9. Sancho, J., Pastor, J.J., Martínez, J., García, M.A.: Evaluation of harmonic variability in electrical power systems through statistical control of quality and functional data analysis. Procedia Eng. **63**, 295–302 (2013). https://doi.org/10.1016/j.proeng.2013.08.224
10. Korobiichuk, I., Davydenko, L., Davydenko, V., Davydenko, N.: Information support the operative control procedures of energy efficiency of operation modes of municipal water supply system facilities. In: Advances in Intelligent Systems and Computing, vol. 920, pp. 571–582 (2020). https://doi.org/10.1007/978-3-030-13273-6_53
11. Davydenko, L., Rozen, V., Davydenko, V., Davydenko, N.: Formalization of energy efficiency control procedures of public water-supply facilities. Adv. Intell. Syst. Comput. **543**, 196–202 (2017). https://doi.org/10.1007/978-3-319-48923-0_24
12. Davydenko, N., Korobiichuk, I., Davydenko, L., Nowicki, M., Davydenko, V.: Identification of cyclic changes in the operation mode of the production facility based on the monitoring data. In: Advances in Intelligent Systems and Computing, vol. 1044, pp. 189–197 (2020). https://doi.org/10.1007/978-3-030-29993-4_24
13. Woodall, W.: Controversies and contradictions in statistical process control (with discussion). J. Qual. Technol. **32**, 341–378 (2000). https://doi.org/10.1080/00224065.2000.11980013
14. ISO 7870-2:2013 «Control charts—Part 2: Shewhart control charts». Switzerland (2013)
15. Wheeler, D.J.: Advanced Topics in Statistical Process Control the Power of Shewharts Charts. SPC Press, Knoxville (2004)
16. Hotelling, H.: Multivariate quality control, illustrated by the testing of sample bombsights. In: Techniques of Statistical Analysis, pp. 113–184. McGraw-Hill, New York (1947)
17. Shashkov, V.B.: Processing of Experimental Data and the Construction of Empirical Formulas. GOU OGU, Orenburg (2005). (in Russian)
18. Alarm and event guidance in InTouch HMI. ZAO Klinkmann Spb, Wonderware, Russia (2008). http://old.intouch.ru/support/pub/ITAlarmsAndEvents_ru_10_300408.pdf
19. Davydenko, L., Rozen, V., Davydenko, V., Davydenko, N.: Control of operation modes efficiency of complex technological facilities based on the energy efficiency monitoring. In: Lecture Notes in Mechanical Engineering, Advances in Design, Simulation and Manufacturing II, pp. 531–540 (2020). https://doi.org/10.1007/978-3-030-22365-6_53
20. Davydenko, N.V.: Formation of the relevant variables set of the energy baseline of the water supply pumping station. Tech. Sci. **187**, 23–25 (2017). Transactions of Kharkiv Petro Vasylenko National Technical University of Agriculture. (in Ukrainian)

# Component-Based Simulation Tool for Virtual Commissioning of Control Systems for Heat Exchange and Distribution Processes

Michał Frątczak[1], Paweł Nowak[1], Tomasz Kłopot[1], Jacek Czeczot[1], Szymon Bysko[2(✉)], and Sara Bysko[2]

[1] Faculty of Automatic Control, Electronics and Computer Science, Institute of Automatic Control, Silesian University of Technology, Gliwice, Poland
Michal.Fratczak@polsl.pl

[2] Simulation Department, ProPoint Sp. z o.o. Sp. K., ul., Bojkowska 37R, Gliwice, Poland
Szymon.Bysko@propoint.pl

**Abstract.** This paper deals with the validation of the concept of the component-based simulation tool and its application for virtual commissioning of the industrial control systems. It is shown how to decompose the heat exchange and distribution system into the general-purpose components and then how to use these components for designing the component-based simulator of the process, which is suitable to use in virtual commissioning environment. Simulator is based on SIMULINK and WinMOD. General hardware configuration and logical layers of the experimental setup, which was used during the experiments, are shown and described in details. Then, this simulator is tested for its simulation accuracy and the hardware experimental setup is tested for its applicability for virtual commissioning of the control systems for continuous dynamical processes of medium and high complexity. Communications lags between examined control system and simulator are analyzed due to checking whether the experiment results obtained depend of simulator's time scaling factor.

**Keywords:** Virtual commissioning · Modeling and simulation · Library components · Heat exchange and distribution systems

---

This work supported by ProPoint Sp. z o.o. Sp. K. under grant POIR.01.01.01-00-0047/15-00.


R. Szewczyk et al. (Eds.): AUTOMATION 2020, AISC 1140, pp. 67–77, 2020.
https://doi.org/10.1007/978-3-030-40971-5_7

# 1 Introduction

Industrial process require significant improvement in their efficiency due to increasing ecological and economical requirements [1–3]. This goal can be obtained by e.g. successive reduction of energy consumption and preserving high safety and environmental requirements [4–7]. At the same time, within a wide class of different industrial processes, heat exchange and distribution systems can be considered as the subject of intensive studies. Their efficient control and design surely allows for very significant reduction of waste heat and for overall improvement in energy efficiency [5–7].

This paper discusses the possibility of component-based modelling of the heat exchange and distribution systems of low and medium complexity. Such processes are widely applied in different manufacturing systems as well as in domestic heating systems. The concept itself is not new but the novelty of this paper results from its application for deriving reconfigurable simulator that can be directly used for virtual commissioning of the design and verification of the dedicated (even advanced) control systems.

The concept of virtual commissioning is similar to more general approach called Hardware in the Loop (HIL) that involves a virtual plant and a real PLC controller. The idea has already been well established method for supporting the design and verification of the control systems and it is more and more widely used in the manufacturing systems – see e.g. [8–11]. In [12], it is reported that the significant reduction of the real commissioning time by about 75% is possible when the virtual commissioning is involved, comparing to the conventional approach when the real commissioning is carried out directly at the plant. Consequently, it is possible to simplify the commissioning stage and to reduce its costs, both for the vendors of the control systems and for their industrial partners (manufacturers). However, this data has been obtained for the application of the virtual commissioning for the example discrete event systems where the control is determined as the logical function based on the binary inputs. Such applications are more and more popular and they are also very effective because the modelling stage requires relatively small amount of time.

In this paper, it is investigated how the component-based modeling can be applied for virtual commissioning of the control systems for the heat and distribution processes. Such processes are not the discrete event systems but they belong to the group of the continuous industrial processes. In fact, the application of the idea of virtual commissioning for such systems is very promising but its application is still very limited. The preliminary discussion on this problem can be found in [13] where the requirements and limitations of the virtual commissioning for the continuous processes are considered based on the very simple example of the pneumatic laboratory setup. In this paper, this concept is further developed by its application to more complex and realistic heat exchange and distribution process. Thus, the major contribution of this work is to present the experiences from developing the component-based model of the example process for the virtual commissioning purposes. However, the results and conclusions have much more general impact because they can be helpful for readers who are planning to apply the procedure of the virtual commissioning for the design and validation of the control systems for the continuous industrial processes.

## 2 Description of the Example Heat Exchange and Distribution Process

The example heat exchange and distribution process is schematically presented in Fig. 1. It consists of the single heat source (electric flow heater, EFH) and the single heat receiver (heat exchanger, HE) with varying heat consumption adjusted by flow rate $F_2$. In this setup, two heat exchangers of very different type can be operated in terms and the channels of the operated heat exchanger are respectively supplied by hot water (left circuit) and cold water (right circuit). One unit is the plate heat exchanger (PHE) LM25-6 made by TAU Energy and it consists of six plates asymmetrically forming two channels: for the hot water of the volume of 0.19 [L] and for the cold water of the volume of 0.29 [L]. The second unit is the shell-tube heat exchanger (STHE) JAD-3.18 made by SECESPOL. It consists of a shell containing a bundle of 18 tubes, which take the form of a coil. In this heat exchanger, cold water is introduced into a shell of the volume 4.2 [L] while hot water goes into tubes of the volume 5 [L]. The outer surface of the tubes form an overall heat exchange surface of 2.12 [m$^2$]. At the same time, two different electric flow heaters can be also operated in terms. They have different nominal powers and volumes. One (EFH1) has the nominal power $P_{nom}$ = 12 [kW] and the nominal volume $V$ = 30 [L]. The other (EFH2) has $P_{nom}$ = 4 [kW] and $V$ = 0.5 [L].

For the whole setup, temperature measurements ($T_{hin}$, $T_{hout}$, $T_{in1}$, $T_{in2}$, $T_{out1}$, $T_{out2}$) are provided by platinum-wire resistance temperature devices (RTD) directly-coupled to 0–10 [V] transmitters. Flow rates $F_1$, $F_2$ are measured by magnetic flow meters and can be adjusted respectively by operating the control valves V1, V2. Power of the electric flow heaters are adjusted by the power supply $P_h$ within the percentage range 0–100 [%] of the respective nominal power $P_{nom}$ by thyristor-based unit controlled by PWM (Pulse Width Modulation) algorithm.

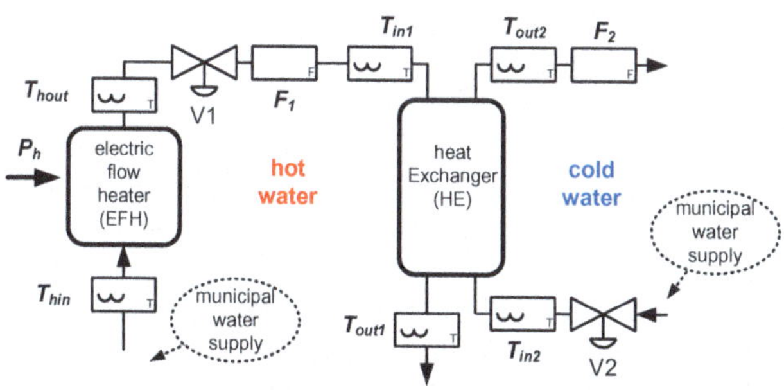


**Fig. 1.** Schematic diagram of laboratory heat exchange and distribution system.

Readers should note that the example setup represents the simple heat exchange and distribution systems with one heat source and one heat receiver with the adjustable heat consumption. However, even such a simple case can represent very realistic processes, such as the domestic heating system, cooling system for the nonisothermical biochemical processes, etc.

# 3 Component-Based Modeling of the Example Heat Exchange and Distribution Process

This section presents the concept of the component-based simulator of the example heat exchange and distribution process. Potentially, in this system, three different general-purpose components can be distinguished and modeled separately: electric flow heater (EFH), heat exchanger (HE) and control valve V. Each component can be encapsulated and they can form a library for further simulation of more complex heating/cooling systems. It was assumed that the suggested models for all components should have the first principle origin. This approach preserves most important properties (dynamics, nonlinearities, etc.) and allows for potential inclusion of the highest numbers of the component inputs. Consequently, very high functionality of each component can be ensured jointly with its relatively highest generality.

## 3.1 Modeling of HE

Figure 2 shows the input-output specification of the component representing HE. For reliable application, it requires four input signals for inlet temperatures and flow rates and values of four model parameters. Its outputs are two outlet temperatures.

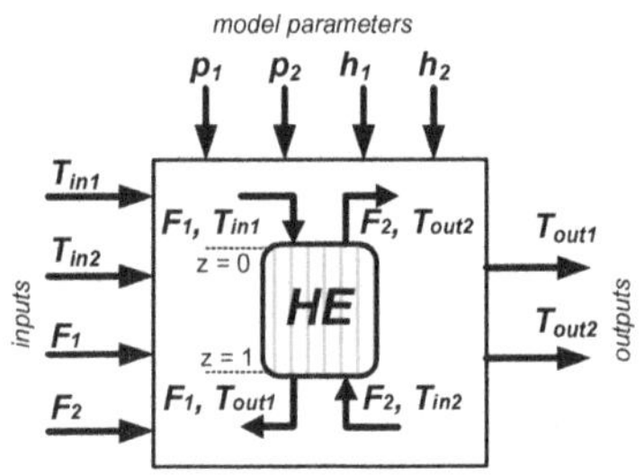


**Fig. 2.** Input-output specification of the component encapsulating model of HE.

Modeling of HE is based on general framework described in details in [14, 15] for plate heat exchanger and it is based on heat conservation considerations suggested for double-pipe counter current heat exchangers. For normalized space variable $z = x/L$, the simplified HE model is suggested in the form of coupled partial differential equations (PDE) describing dynamical evolution of temperatures $T_1(z, t)$, $T_2(z, t)$ that respectively denote the distribution of temperatures of hot and of cold water in both channels.

$$\frac{\partial T_1(z,t)}{\partial t} = -p_1 F_1(t) \frac{\partial T_1(z,t)}{\partial z} - h_1(T_1(z,t) - T_2(z,t)), \tag{1a}$$

$$\frac{\partial T_2(z,t)}{\partial t} = p_2 F_2(t) \frac{\partial T_2(z,t)}{\partial z} + h_2(T_1(z,t) - T_2(z,t)). \tag{1b}$$

Boundary conditions are defined by inlet temperatures $T_1(0,t) = T_{in1}(t)$, $T_2(1,t) = T_{in2}(t)$ and initial profiles are defined as $T_1(z,0) = T_{1,0}(z), T_2(z,0) = T_{2,0}(z)$. The HE component is always initialized with the uniformly distributed initial profiles. The outlet temperatures are denoted as $T_1(1,t) = T_{out1}(t), T_2(0,t) = T_{out2}(t)$. In the model (1a, 1b), all unknown or uncertain physical and construction parameters are lumped and represented by substitute positive model parameters $p_1$ [1/m$^3$], $p_2$ [1/m$^3$], $h_1$ [1/s], $h_2$ [1/s]. For numerical simulation, the orthogonal collocation method is applied for model spatial discretization.

### 3.2 Modelling of EFH

Figure 3 shows the input-output specification of the component encapsulating the model of EFH. It requires three input signals for inlet temperature, flow rate and power supply and values of four model parameters. Its output is the outlet temperature.

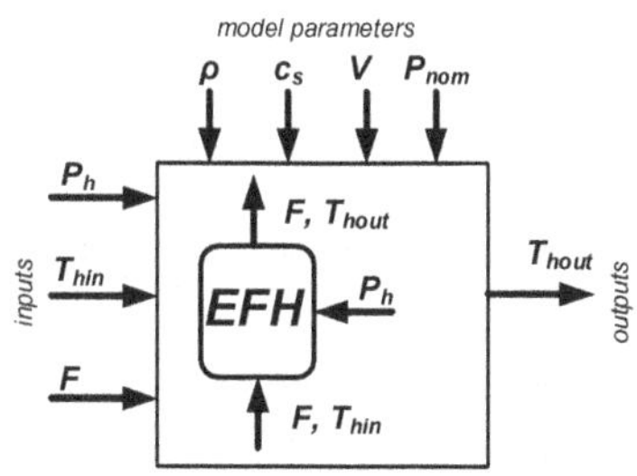


**Fig. 3.** Input-output specification of the component encapsulating model of EFH.

The general concept of first principle-based modelling of EFH is presented in [16–18]. The model consists of the nonlinear dynamical equation describing the outlet temperature $T_{hout}$:

$$\frac{dT_{hout}}{dt} = \frac{F_1}{V}(T_{hin} - T_{hout}) + \frac{P_h \cdot P_{nom}}{100 \cdot c_s \cdot \rho \cdot V}, \qquad (2)$$

where $V$ [L] denotes the volume of EFH, $c_s$ [J/kg °C], is water specific heat and $\rho$ [kg/L] is water density. However, Eq. (2) represents very general and nonlinear dynamics but only of the first order and thus, its accuracy is limited to the cases when very accurate modeling of the heater is not required.

### 3.3 Modelling of Control Valve

The input-output specification of the component that represents the general-purpose control valve is presented in Fig. 4. For this component, two inputs have been defined: $u$ representing percentage opening of the valve (adjustable within the range 0–100 [%]) and the pressure drop over the valve $\Delta p$ [Pa]. The model parameter $k_v$ represents the valve constant (flow coefficient) and the valve dynamics (in the forward path between valve opening $u$ and the flow rate $F$) is represented by the time constant $T$ [s]. The output is the flow rate through the valve.

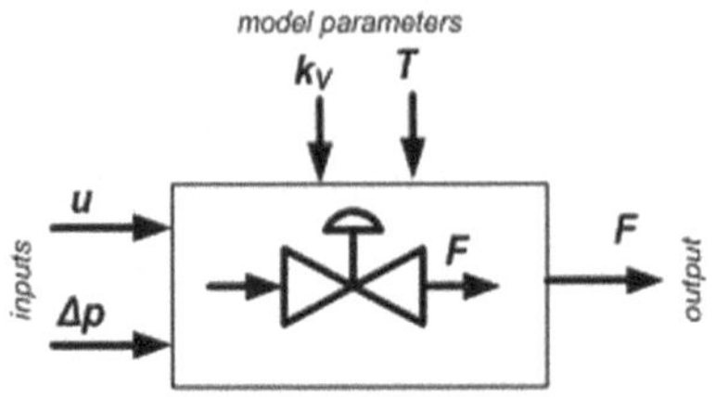


**Fig. 4.** Input-output specification of the component encapsulating model of control valve.

The general model of this component has the form of Hammerstein model described by the nonstationary valve opening characteristics representing the valve nonlinearity:

$$k_v = f(u) \cdot \sqrt{\Delta p}, \tag{3a}$$

and the following transfer function describing valve dynamics:

$$F(s) = \frac{k_v}{(1 + s \cdot T)}. \tag{3b}$$

The generality of the valve component can be preserved only if the function $f(u)$ in the formula (3a) that represents the valve opening characteristic is implemented separately outside the encapsulated component block. It allows for modelling different types of the control valves subject to different required opening characteristics.

## 4 Experimental Setup

The components of the hardware configuration of the experimental setup are presented in Fig. 5. It consists of the single PLC (Programmable Logic Controller) SIEMENS SIMATIC S7 unit and the PC-based platform for the SCADA system. The process simulator is running in the open loop, without any control systems and such a configuration allows for simulating the realistic industrial conditions where the process is directly connected to PLC level. This connection is represented by SimbaPRO device that simulates the real distributed I/O modules. All these components are connected by Ethernet to the PC where the WinMOD environment is running. This programming environment from Mewes & Partner GmbH supports virtual commissioning by integrating the control system with the virtual (simulated) plant. In particular, it simulates the sensors/actuators system with the scaling abilities. The real process is represented by the process simulator running on the separated PC and written in SIMULINK environment. This approach allows for preserving the desired hardware configuration that is to be applied in the practical implementation. After virtual commissioning stage, the whole control system can be reconstructed at the plant and the real commissioning can be completed to ensure its reliability.

**Fig. 5.** General hardware configuration and logical layers of the experimental setup.

## 5 Simulation Results

The experimental setup described in previous section was used for testing the potential possibilities of using the component-based simulator of the heat exchange and distribution system for virtual commissioning of industrial control systems. The tests were divided into two stages.

At the first stage, the modeling accuracy of the process simulator was validated based on the real measurement data collected from the operating setup. The tests were carried out for two configurations of the setup: (C1) including EFH2 with PHE and (C2) that included EFH1 with STHE. In both cases, the laboratory setup was excited by realistic variations of the power supply $P_h$ and of the flow rates $F_1$ and $F_2$. Then, the same variations of the input signals were applied to the simulators implemented in the experimental setup described in Sect. 4. The tests were carried out for different time scales as the experimental setup allows for scaling the simulation time. The scaling of the simulation time allows for speed up or for slow down the simulation by the adjusted ratio [13]. The tests were carried out for the simulation time speeded up 2–4 times comparing to the real time simulation. It was found that the simulation accuracy does not depend on this speed up factor.

The highest possible modelling accuracy is ensured when the parameters of each component are identified based on input-output measurement data collected from operating process. Such a procedure was applied for each component separately and it allows for compensating potential modelling inaccuracies resulting from model simplifications.

- Two different valves V1 and V2 are used and they have different characteristics. V1 is the equal percentage valve while V2 is the linear one. For both valves, the same time constant $T = 5$ [s] was adjusted to represent their dynamics. In the considered case, the pressure drops over the valves are not measurable due to the lack of pressure sensors. Thus, for both valves, the input $\Delta p = 1$ is adjusted and their opening characteristics $f(u)$ in Eq. (3a) are substituted by the corresponding flow characteristics determined experimentally.
- EFH1 was simulated by the component presented in Fig. 3. Based on identification, its parameters were adjusted as $V = 28.8$ [L] and $P_{nom} = 13$ [kW]. For EFH2, apart

from adjusting $V = 0.25$ [L], the additional dynamics must be added and Eq. (2) must be coupled with the second-order-plus-dead-time dynamics of unitary gain, of two time constants depending on the flow rate:

$$\tau_1(F_1) = 38.73 \cdot e^{(-1.209 \cdot F_1)} + 11.97, \tag{4a}$$

$$\tau_2(F_1) = 8.395 \cdot e^{(-0.6665 \cdot F_1)} + 4.852, \tag{4b}$$

and of dead time $\tau_0$ calculated as:

$$\tau_0(F_1) = 16.07 \cdot e^{(-1.133 \cdot F_1)} + 2.85. \tag{4c}$$

The other parameters were adjusted as $c_s = 4200$, $\rho = 1$.

- For both HE, the same identification procedure was applied in the unified form described in details in [15]. For PHE, the highest modelling accuracy can be obtained for $p_1 = 0.902$, $p_2 = 0.827$, $h_1 = 0.09$, $h_2 = 0.08$. For STHE, the parameters are to be adjusted as $p_1 = 1.747$, $p_2 = 1.40$, $h_1 = 0.291$, $h_2 = 0.233$.

Figure 6 shows the example accuracy of simulation obtained for both configurations of the experimental setup and for the speed up factor 4. The simulation accuracy shown in Fig. 6 is very high, both in terms of steady state and of process dynamics representation. These results show the stage of normal operating of the experimental setups. However, in each case, the corresponding SIMULINK-based process simulator must be run to reach the steady state representing the desired operating point. Readers should note that for the real laboratory heat exchange and distribution process, the same situation takes place. The real process also has to be driven to the operating point before it starts normal operation.

At the second stage, the experimental setup was tested for existing the communication lags between its layers shown in Fig. 5. When the virtual commissioning of the control system for the dynamical process is considered, the significant communication lag should be treated as the additional dead time in the process dynamics. When the

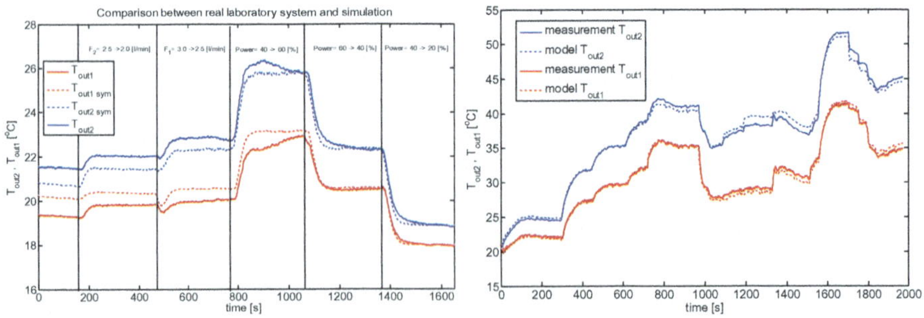


**Fig. 6.** Example simulation accuracy for the experimental setup. Upper diagram corresponds to the configuration (C1) and lower – for (C2). Solid lines represent real measurements and dashed lines – corresponding simulation results.

PID controllers are to be tuned during virtual commissioning, the presence of this additional dead time can influence the tuning stage and consequently, the retuning could be required for real commissioning.

Figure 7 shows the distribution of the total communication lag between PLC and the process simulator implemented in SIMULINK. The results were obtained for transferring the step changes of the single signal (flow rate $F_1$) from PLC to SIMULINK and backward. The period of these changes was 100 [ms] and the data were stored with the period of 2 [ms]. The latter period defines the accuracy for determining the communication lag. During the whole experiment, the process simulator was running. For each change of $F_1$, the communication lag was determined and the histogram was derived. It represents the normal distribution with mean value between 85 and 90 [ms].

During the same experiment, the distribution of the communication lag between PLC and WinMOD environment was also derived and it is shown in Fig. 7. It also represents normal distribution with mean value about 36 [ms], which is half of the mean total communication lag in the experimental setup.

It can be seen that the communication lag is evenly distributed between the logical layers of the experimental setup shown in Fig. 5. At the same time, the existing communication lag does not depend on the time scaling factor, which was tested.

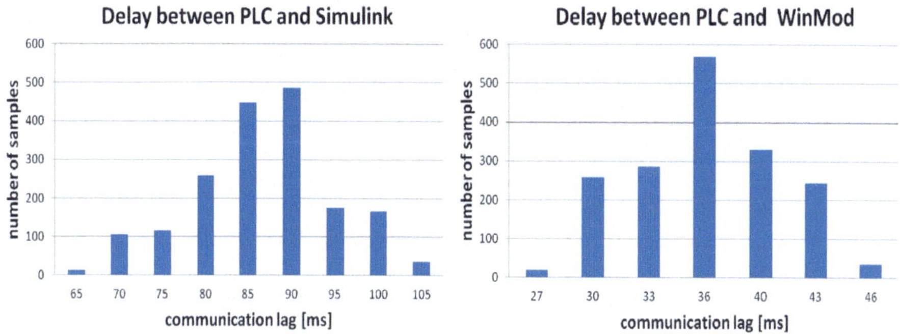


**Fig. 7.** Total communication lag distribution between PLC and SIMULINK-based process simulator and between PLC and WinMOD environment.

## 6 Conclusions

The most significant limitation of the application of the virtual commissioning of the control systems for the continuous industrial processes is the necessity of their accurate modeling. This modeling should be based on first principle approach because it is very important to include the influence of the external process disturbances into the simulator. It provides reliable simulation and allows for testing the disturbance rejection properties of the control system. At the same time, it makes the simulation procedure case-dependent and time-consuming.

In this paper, it is shown that this limitation can be partially removed by designing the dedicated library of general-purpose components. Proper design ensures their flexibility and they can be used for fast and reliable designing of the simulator of the processes of different complexity. It is also shown that the communication lag for the experimental setup is stochastic and its mean value is approximately equal to 90 [ms]. This value is not significant comparing to the dynamics of the industrial continuous processes. In such cases, the cycle time in PLC devices is usually adjusted as $\geq$100 [ms] so the configuration of the setup introduces additional dead time of a single sample. However, it must be noted that this dead time can become more significant when the simulation is speeded up. It is advised to consider its impact for each case.

## References


1. Bauer, M., Craig, I.K.: Economic assessment of advanced process control – a survey and framework. J. Process Control **18**, 2–18 (2008)
2. Rhinehart, R.R., Darby, M.L., Wade, H.L.: Editorial - choosing advanced control. ISA Trans. **50**, 2–10 (2011)
3. Yuan, Z., Chen, B., Sin, G., Gani, R.: State-of-the-art of progress in the optimization-based simultaneous design and control for chemical processes. AIChE J. **58**(6), 1640–1659 (2012)
4. Kemp, I.C.: Pinch Analysis and Process Integration: A User's Guide to Process Integration for the Efficient Use of Energy. Butterworth-Heinemann, Oxford (2007)
5. Bonhivers, J.C., Srinivasan, B., Stuart, P.R.: New analysis method to reduce the industrial energy requirements by heat-exchanger network retrofit: part 1 – concepts. Appl. Therm. Eng. (2014). https://doi.org/10.1016/j.applthermaleng.2014.04.078
6. Semkov, K., Mooney, E., Connolly, M., Adley, C.: Efficiency improvement through waste heat reduction. Appl. Therm. Eng. **70**, 716–722 (2014)
7. Jie, P., Zhu, N., Li, D.: Operation optimization of existing district heating systems. Appl. Therm. Eng. **78**, 278–288 (2015)
8. Hoffmann, P., Shumann, R., Maksoud, T.M.A., Premier, G.C.: Virtual commissioning of manufactoring systems – a review and new approaches for simplifications. In: Proceedings of 24th European Conference on Modelling and Simulation (2010)
9. Lee, C.G., Park, S.C.: Survey on the virtual commissioning of manufacturing systems. J. Comput. Design Eng. **1**(3), 213–222 (2014)
10. Alszer, S., Krystek, J., Bysko, S.: Simulation and virtual commissioning for intelligent buffer control system – case study. In: Proceedings of 2017 2nd International Conference on Robotics and Automation Engineering, China, pp. 207–211 (2017)
11. Bysko, S., Krystek, J., Bysko, S.: Automotive paint shop 4.0. Comput. Ind. Eng. (in press). https://doi.org/10.1016/j.cie.2018.11.056
12. Koo, L.J., Park, C.M., Lee, C.H., Park, S.C., Wang, G.N.: Simulation framework for the verification of PLC programs in automobile industries. Int. J. Prod. Res. **49**(16), 4925–4943 (2011)
13. Fratczak, M., Nowak, P., Klopot, T., Czeczot, J., Bysko, S., Opilski, B.: Virtual commissioning for the control of the continuous industrial processes – case study. In: 20th International Conference on Methods and Models in Automation and Robotics (MMAR), Międzyzdroje, Poland, pp. 1032–1037 (2015)

14. Fratczak, M., Czeczot, J., Nowak, P.: Simplified modeling of plate heat exchangers. In: 19th International Conference on Methods and Models in Automation and Robotics (MMAR), Międzyzdroje, Poland, pp. 578–583 (2014)
15. Fratczak, M., Nowak, P., Czeczot, J., Metzger, M.: Simplified dynamical input–output modeling of plate heat exchangers – case study. Appl. Therm. Eng. **98**, 880–893 (2016)
16. Laszczyk, P.: Simulation based evaluation of PI, PFC and GMC control for electric heater. In: Simulation in industry 2001: 13th European Simulation Symposium, pp. 202–205 (2001)
17. Czeczot, J.: Modelling for the effective control of the electric flow heaters – simulation validation. Simul. Model. Pract. Theory **16**(4), 429–444 (2008)
18. Fratczak, M., Czubasiewicz, R.: Cascade balance-based adaptive control of heating system – simulation validation. In: 21th International Conference on Methods and Models in Automation and Robotics (MMAR), Międzyzdroje, Poland, pp. 1170–1175 (2011)

# The System Optimization of Technical, Technological and Construction Parameters of Polder Systems

Anatoliy Rokochinskiy[1], Igor Korobiichuk[2](✉), Lyudmyla Kuzmych[3], Pavlo Volk[1], and Anna Kuzmych[1]

[1] The National University of Water and Environmental Engineering, Rivne, Ukraine
{a.m.rokochindkiy, p.p.volk}@nuwm.edu.ua, annekuzmych@gmail.com

[2] ŁUKASIEWICZ Research Network – Industrial Research Institute for Automation and Measurements PIAP, Warsaw, Poland
ikorobiichuk@piap.pl

[3] National Aviation University, Kyiv, Ukraine
klv@nau.edu.ua

**Abstract.** The article substantiates the necessity and considers possible approaches to the system optimization of technological and constructive parameters of polder systems as a necessary condition for the creation and functioning of water management and reclamation facilities on ecological and economic bases under changing climatic conditions.

**Keywords:** System optimization · Technological and constructive parameters · Polder systems

## 1 Introduction

Sustainable development of drainage reclamation as one of the most important factors in the intensification of agricultural production and an integral part of the food and resource provision of the population of Polissya requires rational and efficient use of water, land and energy resources and, at the same time, ensuring the ecological stability of drained territories.

The urgency of this approach is determined by the challenges of the nowadays concerning the energy, food and water problems, which is greatly aggravated due to changes in climatic conditions [1, 2].

According to existing experience and practice, the design or operation of drainage systems, including polder systems, requires the transition from their consideration not merely as technical but as complex natural and technical ecological-economic systems. Such systems should have appropriate of change in the whole methodology, technical and technological the strategies for their creation and functioning [3–5], as well as the direct account of time-varying and uncertain climatic conditions, as they, together with


R. Szewczyk et al. (Eds.): AUTOMATION 2020, AISC 1140, pp. 78–86, 2020.
https://doi.org/10.1007/978-3-030-40971-5_8

reclamation factors, have a defining the effect on the overall ecological and economic effects [6–9].

At present, in the area of drainage reclamation, the most technically most advanced are drainage systems of polder type with mechanical water lifting. Such systems have undeniable advantages over gravity, both in terms of the speed of regulation of water regime and ecology. The main disadvantage of the polder system is the relatively high cost of its construction and operation. However, the future of technical upgrading of drainage systems is the future of systems with mechanical water absorption, and in the conditions of periodic flooding of territories with flood and floodwaters, there are no alternatives to polder drainage, which has already been developed well, as in Ukraine in Eastern and Western Europe [8].

The need for additional humidification of drained lands from the zone of western Polissya was proved by studies [9–14], other researchers.

At the same time, as noted by [11], and others, as regards the two-way regulation of water regime, they are quite convenient hydraulic engineering complexes. At them, by changing the mode of pumping water by the pumping station and timely closure of the shields of the gateway regulators, it is possible at certain times to achieve the required position of groundwater levels in the fields.

## 2 Methods and Techniques

The question of regulating the water regime of drained lands on polder systems in different natural climatic zones was developed only on separate sites, in particular, Leonardo Alfonso, Arnold Lobbrecht, Roland Price, Van Overloop P.J. considered the optimization of water levels on the system [12, 13].

Therefore, today it is extremely important to raise the overall technological, ecological and economic efficiency of the functioning of drainage systems, especially polders.

Thanks to the solution of this issue, the following results can be achieved:

- to improve the technological and technical state of the polder system;
- to improve the regimes and technologies of water regulation, the work of the polder pumping station;
- to increase the fire safety of drained peat soils;
- to make a forecast of changes in their ecological and land reclamation status for the nearest and long-term future of the system, taking into account climate changes conditions [14].

The polder systems are an extremely complex and specific object of research due to the conditions of their creation and functioning, in comparison with other traditional drainage systems. As the methodological basis for solving this problem, we have adopted a systematic approach and systematic analysis from its integral components – the method of optimization and modeling of complex objects and systems. With such complexity and complexity of the problem, there is a need for the development and application of modern optimization methods aimed at a more complete reproduction of the interrelations and interactions of production and natural processes in the functioning of the polder systems – a solution to a complex optimization problem with heterogeneous elements.

Nowadays we have many examples of the application of the optimization approach for determining the parameters of drainage systems and schemes of rational water use during the developing of methods for the justification of technical solutions in projects for the construction and reconstruction of water and land reclamation facilities. Generally, such objects were intensively developed in the 70–80 s of the last century as for the irrigation reclamation zone, as well as areas of drainage reclaim lands [15–19].

## 3 Results and Discussion

Consideration of the polder system as a complex natural and technical ecological-economic system determines the necessary condition for finding a general optimum in such a system as a whole based on system optimization. The essence of the system optimization is to find intermediate and local optimums for all of its main components of water-regulating elements (pump station, main channel, gateways regulators, drainage, etc.) and their regimes of operation (pumping units, water supply, etc.), for all the main factors in space and time. These factors influence the efficiency of water regulators (climate, relief, cultivation of crops, soils, soil-reclamation differences, schemes and technologies of water regulation, etc.) [18, 19].

The same goes for finding the optimum for all components of the system *effect-mode-technology-design* and implementation of the corresponding optimization model in their interconnection.

Thus, the optimization of the parameters of heterogeneous performance indicators of priority reclamation measures on operating the polder system can be presented as: *the given* parameters *of economic and ecological effect* ↔ *optimum pumping module* ↔ *optimal parameters of the pump station.*

In this case, the pumping module serves as the key link of such a subsystem, as the main providing factor for water regulation. Then, the functional relationship between its components in the general case determines the need for system optimization when sequentially different decision-making levels are considered in time according to different optimization criteria for optimizing the parameters of the pump modules, which substantiate the design parameters of the pump station and other elements of the polder system in their interconnection the trick.

At the stage of operation of the operating polder system PS, by analogy with [8], the optimum parameters of the pump module can be justified by such a complex optimization model:

$$\begin{cases} D_0 = \max\limits_{\{i\}} \sum\limits_{n=1}^{n_p} D_{ip} \cdot \alpha_p, i = \overline{l, n_i}; \\ q_0 = \min\limits_{\{i\}} \sum\limits_{n=1}^{n_p} |q_s - \hat{q}_{ecol}| \cdot \alpha_p, i = \overline{l, n_i} \end{cases}, \tag{1}$$

where

$D_0$ – the optimal value of the income criterion $D$ for the $i$ – option of the aggregate $\{i\}$, $i = \overline{l, n_i}$;
$D_{ip}$ – the income criterion for the $i$ – projective option of the accounting year;
$\alpha_p$ – value (specified or given) of repetition or particle of the possible state of typical meteorological regimes in the calculated periods of population growth $\{p\}$, $p = \overline{l, n_p}$, within the project life of the object,

$$\sum_{p=1}^{n_p} \alpha_p = l;$$

$q_0$ – the optimum estimated value of the drainage module for the $i$ – option;
$q_s$ – the weighted average value of the drainage module within the system and the design life of the object for the $i$ – option;
$\hat{q}_{ecol}$ – the limit value of the drainage module corresponding to the ecological level of drainage efficiency;
$i$ – option of the plenty $\{i\}$, $i = \overline{l, n_i}$ concerning the type, structure, and parameters of drainage.

According to the economic criterion of optimization the *income indicator D*, which is achieved at the expense of obtaining a certain volume of cultivated agricultural products on the reclaimed land with the application of various options of technological solutions of the plenty $\{i\}$. $i = \overline{1, n_i}$ – the plenty of possible methods and schemes of water regulation on the system due to its type, design, water availability, etc.:

$$D_i = V_i - C_i, \rightarrow i = \overline{l, n_i}, \tag{2}$$

where

$V_i$ – volume (value) of the received products *by i*– option of the plenty $\{i\}$, $i = \overline{l, n_i}$;
$C_i$ – current expenses for production according to variants of technical decisions.

In this case, the optimization criterion is the maximization of the net income indicator:

$$D_i \rightarrow max, \rightarrow i = \overline{l, n_i}, \tag{3}$$

and the function of the goal will be:

$$D_0 = \max_{\{i\}} D_i = \max_{\{i\}} [V_i - (A_i + C_i^{cz} + C_i^{м} + C_i^{в})], i = \overline{l, n_i}, \tag{4}$$

At the stage of reconstruction, the optimal parameters of the pump station and other elements of the polder system are based on such an integrated optimization model [16]:

$$\begin{cases} ZP_0 = \min\limits_{\{i\}} \sum\limits_{n=1}^{n_p} ZP_{ip} \cdot \alpha_p, i = \overline{l, n_i} \\ q_0 = \min\limits_{\{i\}} \sum\limits_{n=1}^{n_p} |q_s - \hat{q}_{ecol}| \cdot \alpha_p, i = \overline{l, n_i} \end{cases}, \tag{5}$$

where

$ZP_0$ – the optimal value of the criterion by $i$ – option of the plenty $\{i\}$, $i = \overline{l, n_i}$;

As a general economic criterion of optimization will be expenses $ZP_i$ by volume (value) of the received products $V$ in the plenty $\{i\}$, $i = \overline{l, n_i}$ – option of technical decisions of the aggregate, types of polder systems, caused by accepted methods and schemes of water regulation [18, 19]:

$$ZP_i = Z_i \cdot k_{Z_i}^V = (C_i + E_n K_i)/V_i i = \overline{l, n_i}, \tag{6}$$

where:

$k_{Z_i}^V$ – the coefficient of the expenses $Z_i$ by volume (value) of the received products $V_i$ by $i$ – option of the plenty $\{i\}$, $i = \overline{l, n_i}$, which is determined by $l/V_i$;
$C_i$ – current expenses for production according to the option of technical decisions;
$E_n$ – normative coefficient of economic efficiency of capital investments $K_i$ according to the option of technical decisions.

The condition of optimization, in this case, is the minimization of the indicator of expenses, will be:

$$ZP_i \rightarrow min, \ i = \overline{l, n_i}, \tag{7}$$

and a function of the goal will be:

$$ZP_0 = \min\limits_{\{i\}} \ ZP_i, \rightarrow i = \overline{l, n_i}. \tag{8}$$

In the formula (2), current expenses $C_i$ consist of agricultural expenses $C_i^{agr}$ and operational expenses $C_i^o$. The operational expenses include deductions for depreciation and repairs $A_i$, reclaimed expenses $C_i^{recl}$ for system maintenance.

Then the formula (5) takes the form:

$$ZP_0 = \min\limits_{\{i\}} \left\{ \frac{\left(A_i + C_i^{agr} + C_i^{recl} + C_i^o\right) + E_n K_i}{V_i} \right\}, i = \overline{l, n_i}. \tag{9}$$

The drainage operation at the polder systems in the regime of drainage leads to a strengthening of the washing water regime in different conditions of the humidity of the vegetation periods. Consequently, we can see the reduction of soil fertility due to the washing of nutrients and the violation of the structure of the soil, etc.

As a projective criterion of environmental optimality can be the deviation of the weighted average value the drainage module within the system $q_s$ and the project lifetime of the object from its limit value $\hat{q}_{ecol}$, which corresponds to the ecological level of efficiency of the work of the object, that is $q_s \to \hat{q}_{ecol}$. Minimization of the drainage module leads to an increase in the total moisture content, which is extremely important for drained peat soils in conditions of climate change.

Experimentally determined that, depending on the granulometric composition of drained soils, the ecologically optimal drainage modules make up an average of $\hat{q}_{ecol} = 0.2, \ldots, 0.4$ (l/s ha).

The solution to these issues was carried out on the example of the drainage system "Birky" (Rivne region, Volodymyrets district). This system is typical of the region and put into operation in 1978. The area of the system is 544.9 (ha). The pottery drainage is laid on an area of the 444 hectares, the area of the peat bogs is 89% of the whole area of the system, two-way regulation is possible in the area of 177.9 hectares. The polder area with mechanical drainage is 470 (ha) (Fig. 1).

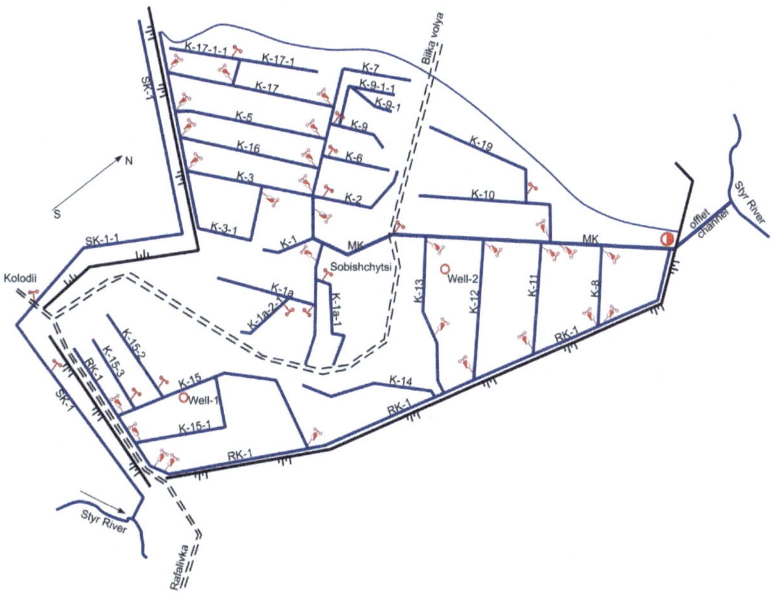


**Fig. 1.** Scheme of the drainage system "Birky".

The water receiver of the system is the Styr River, which in this area is a typical plain river. The dam is a peatland (partially flooded depending on the water content of the year). They protect drained lands from flooding throughout the year. The open channels of the lateral network collect water and take it to the leading channels of the senior order, which, in turn, feed it into the main channel to the pumping station. In arid periods, water from the system does not pump out but remains for the sublimation of moisture. The removal of excess water is carried out by an electrified pumping station and gravity.

A characteristic feature of the polder system "Birky" is that the design of the construction here is the possibility of realization of practically all the basic technologies of water regeneration of drained lands: drainage with mechanical pumping and systematic drainage, preventive gliding and subsoil moisture, can also be applied by irrigation by spraying.

The justification of the optimal technological (pumping module) and design parameters (pumping unit) of the drainage system "Birky" is carried out based on machine experiments on the corresponding complex of forecast-optimization models. The use of these models is regulated by industry standards developed at the Department of Water Engineering and Water Technology of the National University of Water and Environmental Engineering [17].

The sequence of experiment execution is represented by the following scheme:

1. The reasoning of the optimum parameters of the pump module for the operating drainage system (slot in the dam for self-discharge of pumped water) operates in the range from 0.0 to 2.0 (l/s ha);
2. The reasoning of the optimal design and parameters of the jet pump unit of the operating drainage system, which includes an additional projected siphon for unloading the pump station.
3. Cultivation of the project crop rotation: winter wheat with yield (4.0 t/ha), crop rotation 0.2%; potatoes (40.0 t/ha), crop rotation 0.2%; vegetables (30.0 t/ha), crop rotation 0.1%; grass (40.0 t/ha), crop rotation 0.5%. Soils – Peat of medium power of well-laid-out average ash content on alluvial soups.

According to the results of the forecast-optimization calculations, the optimum pump-out module of the drainage system "Birky" is estimated at 0.95 (l/s ha) with the net income indicator determined by the optimization model (1) $D_0 = 15.24 \cdot 10^6$ (\$/ha).

For the optimal pump pumping module due to the reconstruction, the optimal design of the site for pumping out of the polder system, determined by the optimization model (6) at $ZP_0 = 1.32 \cdot 10^6$ (\$).

The determined economically optimal technological and constructive solutions for the considered object are environmentally acceptable in the given conditions since the average weighted value of the drainage module within the system and its design lifetime $q_s = 0.38$ (l/s ha), which corresponds to the ecological level of drainage efficiency $q_s \rightarrow \hat{q}_{ecol}$.

## 4 Conclusions

Thus, the application of system optimization makes it possible at the stage of operation or reconstruction of the operating polder system to substantiate its technological and constructive parameters in the creation and operation of water management and reclamation facilities on the ecological and economic bases under changing climatic conditions.

## References

1. Shumakov, B.B.: Reclamation in the 21st century. Reclam. Water Manag. **3**, 4–6 (1996)
2. Korobiichuk, I., Lysenko, V., Reshetiuk, V., Lendiel, T., Kamiński, M.: Energy-efficient electrotechnical complex of greenhouses with regard to quality of vegetable production. In: Recent Advances in Systems, Control and Information Technology. Advances in Intelligent Systems and Computing, vol. 543, pp. 243–251 (2017). https://doi.org/10.1007/978-3-319-48923-0_30
3. Aydarov, I.P., Golovanov, A.I., Nikolsky, Y.N.: Optimization of meliorative regimes of irrigated and drained agricultural land (recommendations). Agrometeoizdat, Moskva (1990). 60 p. [in Russian]
4. Kovalenko, P.I.: Actual problems of using water resources and reclaimed land at the present stage. J. Reclam. Water Manag. Issue **99**, 2–16 (2011)
5. Korobiichuk, I., Ladanyuk, A., Zaiets, N., Vlasenko, L.: Modern development technologies and investigation of food production technological complex automated systems. In: ACM International Conference Proceeding Series. 2nd International Conference on Mechatronics Systems and Control Engineering ICMSCE 2018, pp. 52–57 (2018). https://doi.org/10.1145/3185066.3185075
6. Shang, S.H.: A general multi-objective programming model for minimum ecological flow or water level of inland water bodies. J. Arid Land **7**(2), 166–176 (2014). https://doi.org/10.1007/s40333-014-0077-6
7. Korobiichuk, I., Lysenko, V., Opryshko, O., Komarchyk, D., Pasichnyk, N., Juś, A.: Crop monitoring for nitrogen nutrition level by digital camera. In: Szewczyk, R., Zieliński, C., Kaliczyńska, M. (eds.) Automation 2018. AUTOMATION 2018. Advances in Intelligent Systems and Computing, vol. 743, pp. 595–603. Springer (2018). https://doi.org/10.1007/978-3-319-77179-3_56
8. Schultz, B.: Water management and flood protection of the polders in the Netherlands under the impact of climate change and man-induced changes in land use. J. Water Land Dev. **12**, 71–94 (2008)
9. Kuzmych, L., Kobylianskyi, O., Duk, M.: The current state of tools and methods of control of deformations and mechanical stresses of complex technical systems. In: Proceedings of SPIE 10808, Photonics Applications in Astronomy, Communications, Industry, and High-Energy Physics Experiments, p. 108085J (2018). https://doi.org/10.1117/12.2501661
10. Kuzmych, L., Kvasnikov, V.: Study of the durability of reinforced concrete structures of engineering buildings. In: Advances in Intelligent Systems and Computing, vol. 543, pp. 659–663 (2017)
11. Dolid, M.A.: Optimal length conducting network of polder systems of Polesye Ukraine. Hydromelioration and hydro technical construction: Lvov, pp.17–27 (1990)

12. Alfonso, L., Lobbrecht, A., Price, R.: Optimization of water level monitoring network in polder systems using information theory. Water Resour. Res. **46**(12), 1–13 (2010). https://doi.org/10.1029/2009WR008953
13. Van Overloop, P.J.: Drainage control in water management of polders in the Netherlands. Irrigat. Drain. Syst. **20**(1), 99–109 (2006)
14. Kovalenko, P., Rokochinskiy, A., Jeznach, J., Volk, P., Koptyuk, R., Prykhodko, N.: Evaluation of climate change in Polissia region and ways of adaptation to it. J. Water Land Dev. **41**(1), 72–82 (2019). https://doi.org/10.2478/jwld-2019-0030
15. Babich, V., Dovbenko, V., Kuzmych, L., Dovbenko, T.: Estimation of flexures of the reinforced concrete elements according to the National Ukrainian & European standards. In: MATEC Web of Conferences, Volume 116, Article number 02005 (2017). https://doi.org/10.1051/matecconf/201711602005
16. Korobiichuk, I., Kuzmych L., Kvasnikov, V., Nowak, P.: The use of remote ground sensing data for assessment of environmental and crop conditions of the reclaimed land. Advances in Intelligent Systems and Computing (AISC), ICA 2017: Automation 2017, vol. 550, pp. 418–424 (2017). https://doi.org/10.1007/978-3-319-54042-9_39
17. Su, X., Chiang, P., Pan, S., Chen, G., Tao, Y., Wu, G., Wang, F., Cao, W.: Systematic approach to evaluating environmental and ecological technologies for wastewater treatment. Chemosphere **218**, 778–792 (2019). https://doi.org/10.1016/j.chemosphere.2018.11.108
18. Rokochinsky, A.M.: System optimization of water regulation as a necessary condition for the creation and operation of water management and reclamation facilities on the ecological and economic basis. Reclam. Water Manag. **104**, 67–71 (2016)
19. Rokochinskiy, A., Volk, P., Pinchuk, O., Turcheniuk, V., Frolenkova, N., Gerasimov, Y.: Forecasted estimation of the efficiency of agricultural drainage on drained lands. J. Water Land Dev. **40**(1), 149–153 (2019). https://doi.org/10.2478/jwld-2019-0016

# The Method of Multi-criteria Parametric Optimization

Igor Korobiichuk[1(✉)], Volodymyr Drevetsky[2], Lyudmyla Kuzmych[3], and Ivan Kovela[4]

[1] ŁUKASIEWICZ Research Network – Industrial Research Institute for Automation and Measurements PIAP, Warsaw, Poland
ikorobiichuk@piap.pl
[2] National University of Water and Environmental Engineering, Rivne, Ukraine
westra.rv@ukr.net
[3] National Aviation University, Kyiv, Ukraine
klv@nau.edu.ua
[4] Lviv Polytechnic National University, Lviv, Ukraine
kim@ukr.net

**Abstract.** The theoretical bases of the method of multi-criteria parametric optimization are developed, algorithms of calculation of continuous systems with typical linear regulators are given. Algorithms of the calculation of systems with digital controllers are given and shown on the example of the calculation of system with digital PI- regulator and objects with self-alignment and delay at a given value of the period of discreteness.

**Keywords:** Multi-criteria parametric optimization · Automatic control system · Proportional integrable differentiating regulator · Amplitude-frequency characteristic

## 1 Introduction

Resource savings and production efficiency are typically associated with minimizing the dispersion of established processes in automated process control systems [1–3]. Classical methods for the synthesis of such systems (methods of linear optimal filtration by N. Wiener and R. Kalman), in spite of mathematical excellence, in many cases have more methodical than practical significance [1–8]. This is due to the need for a priori information on the statistical characteristics of perturbations affecting the system (spectral densities, correlation functions, or other equivalent initial data) to solve an optimization problem [4–7]. Obtaining such information – the task is quite laborious, which requires a long experiment and it cannot always be fulfilled in the conditions of industrial production [7, 8].

There is also another requirement that must satisfy automatic control systems, namely: reliable control is possible only with the provision of intensive extinction of the free components of the controlled signals and the low sensitivity of the quality criteria to the variation of the parameters of the system elements [9–14]. Thus, the


R. Szewczyk et al. (Eds.): AUTOMATION 2020, AISC 1140, pp. 87–97, 2020.
https://doi.org/10.1007/978-3-030-40971-5_9

efficiency of control of the technological process cannot be fully characterized by only one partial criterion.

Consequently, during the solving of the problems of optimization of the automatic control system (ACS) it is necessary to focus on technologically and economically grounded management criteria. The contradiction between the various requirements of the system requires the application of many criteria. The known attempts to reduce them to one generalized indicator, subjectively attributing to each an individual criterion included in it, some weight. Ultimately, it leads to the arbitrary "optimal" solution. Thus, the problem of optimal control of a given technological process should be considered and solved as multi-criteria [15], so it is desirable to provide the simultaneous optimization of the most important characteristics of the system in the steady-state and transition processes.

## 2 Methods and Techniques

The method of multi-criteria parametric optimization (MMPO) is suitable for the synthesis of industrial ACS, was proposed in [16, 17] for continuous systems and developed in the direction of its propagation to systems with digital controllers. This method implements the principle of modal control with mods which are considered to be poles of the transfer function of a closed system. It uses the apparatus of complex – frequency characteristics (CFC) and allows simultaneously maximizing the speed, degree of stability, filtering properties of the system concerning the harmonic components of the spectrum of perturbations, as well as criteria of economic efficiency.

The choice of these functions is based on the economic substantiation of the requirements relating to the process of a particular process. In this case, as the criteria for speed, the geometric pole $\Omega_0$ and degree of stability $\eta$ are considered:

$$\Omega_0 \equiv \sqrt[n]{\prod_{k=1}^{n} |s_k|} \equiv \sqrt[n]{a_0},\ \eta \equiv \inf(-\beta_k) \geq \eta_\partial, \tag{1}$$

where $a_0$ – a free member of the characteristic equation of a closed ACS; $\eta_\partial$ – an acceptable value of $\eta$; $\beta_k$ – the real part of the root of the characteristic equation of a closed system, which can also be written in the form:

$$\Delta(s) = a_0 s^n + a_1 s^{n-1} + \ldots + a_{n-1}s + a_n = 0. \tag{2}$$

Then the geometric pole will be described by the expression:

$$\Omega_0 \equiv \sqrt[n]{a_n/a_0} \equiv \sqrt[n]{|s_1 s_2 \cdots s_n|}. \tag{3}$$

Using the geometric pole, the characteristic equation takes on a form:

$$\Delta(s) = s^n + C_1\Omega_0 s^{n-1} + C_2\Omega_0^2 s^{n-2} + \ldots + C_{n-1}\Omega_0^{n-1} s + \Omega_0^n = 0, \tag{4}$$

where $C_k = a_k \Omega_0^{n-k} k = 1, \cdots, n-1$.

The nature of the transition process depends on the coefficients $C_k$ and does not depend on the coefficients $\Omega_0$. The change $\Omega_0$ with the selected coefficients $C_k$ changes only the duration of the process, without changing its nature, for example, overregulation $\sigma$. Therefore, the geometric root is a measure of system performance.

In a compact record, the characteristic equation of a closed system acquires the form:

$$\Delta(s) \equiv \sum_{k=0}^{n} a_k s^k \equiv \prod_{k=1}^{n} (s - s_k) = 0. \tag{5}$$

The filtering properties of the system are estimated by the value of the amplitude-frequency characteristic (AFC) of the closed system $|\Phi(j\omega)|$.

The optimization of the criteria (1) and $|\Phi(j\omega)|$ is ensured by the appropriate placement of the roots of Eq. (5) by choosing the settings of the regulator (for digital systems – by the period of discretion also).

In practice, feedback on all variables often fails. The reasons for this may be the incomplete observation of the object, lack of necessary technical means, economic considerations, etc. Under these circumstances, not all poles of the system can be located in any predetermined position. However, far from all of them, to the same extent determine the quality of management.

Since the optimization of the system is associated with the change of $\eta$, $a_0$, $|\Phi(j\omega)|$, then the dominant ones are those poles whose change of location most influences on the change of these criteria. It was shown in [13] that in the systems without delay the indicated criteria, as well as the degree of extinction of oscillations with restrictions on power and the rate of change of control actions, simultaneously acquire values which close to the maximum, with the placement of dominant poles which meet the requirements:

$$s_k = (C\beta)/n\,(-1 + j\psi_k)\,, \quad k = \overline{1, n};$$
$$\psi_k = \begin{cases} (-1)^k/m; & 2N \geq k; \\ 0, & 2N < k, \end{cases} \tag{6}$$

where: $N$ – the whole part of $n/2$, $n$ – the number of dominant roots.

The characteristic equation of delayed systems is transcendental and has an infinite number of roots that cannot be placed in a predetermined position with the help of a finite dimension regulator. However, for such systems, the dominant poles may be allocated, which would ensure the approximation to the multicriterial optimum. The corresponding requirements are also given in [16, 17].

The maximum of the geometric pole as one of the criteria of the system's performance, with given restrictions, is achieved if the poles $s_k$, $k = \overline{1,\ n}$ are agreed on the requirements (2.134) and (2.179) [16]. It also shows that for linear autonomous systems, the requirement $\eta$ = max in the appropriate restrictions is provided when the poles of the system are located in such configuration:

$$\beta_k = -C_\beta / n; \quad k = \overline{1,\ n}, C_\beta = -\sum_{k=1}^{n} \beta_k, \tag{7}$$

so their real parts are the same.

The placement of the system's poles following the requirements set provides it with good filtering properties if the bulk of the spectral density of disturbances is concentrated in the low-frequency field.

Thus, the placement of the system's poles following the specified requirements, in addition to the optimum filtering properties, provides maximum performance ratings $\eta$ $\Omega_0$, and a zero parametric sensitivity of these criteria.

It is also important that fact when these requirements are met, the multicriteria optimum is achieved even under the conditions of statistical uncertainty of perturbations that act on the object.

## 3 Results and Discussion

***Algorithms for Calculation of Continuous Systems with Typical Linear Regulators***
The application of this method for parametric optimization of systems with typical regulators is associated with some features of such ACS.

Systems with typical regulators are systems with incomplete feedback on the state. Usually, this connection is carried out only on the scalar output signal of the object, that is, on the available direct control of the variable state – an adjustable value. Consequently, the fulfillment of requirements [16] for all poles of such a system is usually not possible. Besides, many industrial sites are characterized by a delay in the regulatory channel. Therefore, the calculation of the parameter settings of the standard regulators is reduced to the choice of their parameters based on the requirements [16] only for the dominant poles.

The field of admissible placement of the poles for systems with typical regulators, which meets the specified requirements, is shown at Fig. 1.

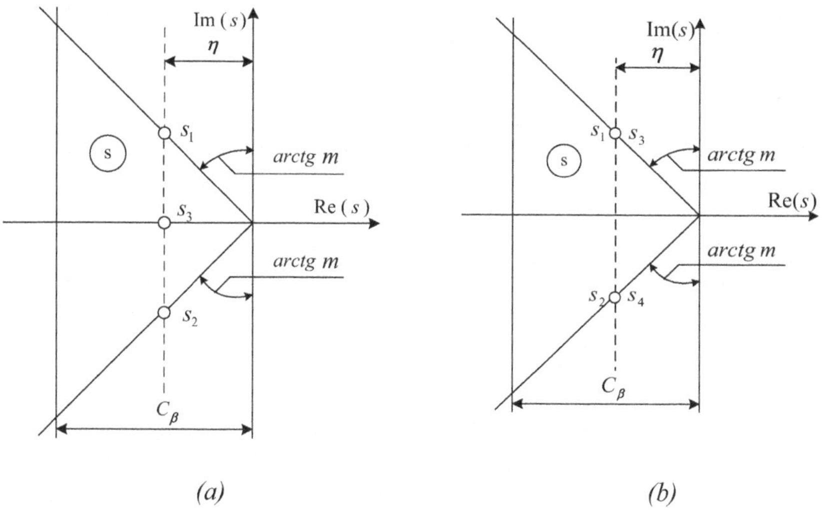


**Fig. 1.** Field of admissible placement of poles for systems with PI (proportional integrable) regulators and PD (proportional differentiating) regulators (a) and for systems with PID (proportional integrable differentiating) regulator (b).

Thus, the optimization of systems with typical regulators by the method of the multicriterial optimum is based on the provision of a given placement of dominant roots of the characteristic equation of a closed system:

$$\Delta(s) = 1 + W_p(s)\ W_o(s) = 0, \tag{8}$$

Based on the requirements [16] in systems with P- and I-regulators, to achieve the multicriterial optimum it is necessary to ensure the existence of a pair of complex conjugate roots of the Eq. (8).

$$s_{1,2} = -m\omega \pm j\omega, \tag{9}$$

Then equality must be fulfilled:

$$\Delta(s)|_{s=s_1} = 0. \tag{10}$$

For systems with PI- and PD- regulators the condition of a multicriterial optimum is to provide an appropriate choice of parameters for their adjustment of the three dominant roots ($n_\partial = 3$) of the characteristic equation of a closed ACS that have the form:

$$s_{1,2} = -m\omega \pm j\omega,\ s_3 = -m\omega. \tag{11}$$

In this case, equality must be fulfilled:

$$\Delta(s)|_{s=s_1} = 0, \ \Delta(s)|_{s=s_3} = 0. \tag{12}$$

We can obtain systems of equations describing the conditions for achieving the multicriterial optimum by substituting (11) into (12), and separating the real and imaginary parts.

The requirements for the achievement of the multicriterial optimum for systems with PID- regulators are to provide two identical pairs of complex conjugate roots of the characteristic equation ($n_\partial = 4$) which have the form:

$$s_{1,2} = -m\omega \pm j\omega; \ s_{3,4} = -m\omega \pm j\omega. \tag{13}$$

Since the pole $s_1$ is the root of the characteristic of the equation $\Delta(s) = 0$ of the system, equality must be fulfilled:

$$\Delta(s)|_{s=s_1} = 0. \tag{14}$$

In addition, given that $s_1 = s_2$ the expressions for $\Delta(s)$ and $d\Delta(s)/ds$ can be presented in the following form:

$$\begin{aligned} &\Delta(s) = (s - s_1)^2 \tilde{\Delta}(s); \\ &d\Delta(s)/ds = 2(s - s_1)\tilde{\Delta}(s) + (s - s_1)^2 \tilde{\Delta}(s)/ds, \end{aligned} \tag{15}$$

where $\tilde{\Delta}(s)$ – the value that does not depend on $s_1$.

Then, taking into account (15), the condition must be fulfilled:

$$\left.\frac{d\Delta(s)}{ds}\right|_{s=s_1} = 0, \tag{16}$$

The validity of this expression is also confirmed by the well-known mathematical notation that $k$ – the multiple roots of the polynomial is also the root of all its derivatives up to and including the ($k$ – 1) – degree.

Thus, substituting (13) in expressions (14), (16) and equating their real and imaginary parts to zero, we obtain a system of four equations that are nonlinear concerning the parameters of the configuration. By solving this system of equations, you can define the parameters of the adjustment $K_p$, $T_i$, $T_d$, and the own frequency of the ACS with the PID- regulator. The resulting solution thus also minimizes the variance of the regulated value at random low-frequency perturbations [9].

One disadvantage of the method is its computational complexity, but with the use of modern information technologies, such as MAPLE [15], these difficulties are quite easy to overcome. Besides, in [16] the requirements for choosing a value of the indicator $m$ are not clearly defined, which leads to the need to use some additional quality criterion.

An important advantage of the method is its highest among known methods of universality. This is since, in spite of the necessity of numerical solution of the system of equations, it is essentially analytical. This property of the method makes it possible to implement new approaches to the solution of numerous problems of synthesis of industrial ACS of different structures and that is why it is accepted in this article as a basis.

Since the MMPO is based on the provision of a given placement of the dominant roots of the characteristic equation of a closed system, the optimal criteria adopted in this method for the values of the parameters of the regulator do not depend either on the type or on the place of application of the input actions, and only depend on the structure of the system. However, the quality of the transient processes has placed the application of input actions strongly and that this factor is not taken into account, it should be considered a disadvantage of the method.

Taking into account the above considerations and expressions, algorithms of synthesis of systems with different algorithms of regulators functioning can be specified.

For example, for systems with P- and I-regulators, the requirements (6) can be performed for $n = 2$ dominant poles. After the substitution of the complex roots in (10) we obtain a complex expression. Equating the real and imaginary parts of this expression to zero, we obtain a system of real equations:

$$\left.\begin{aligned} eq_1 &:= \operatorname{Re}\left[\Delta(s)\right]_{s=s_1} = 0, \\ eq_2 &:= \operatorname{Im}\left[\Delta(s)\right]_{s=s_1} = 0. \end{aligned}\right\} \quad (17)$$

For solving the system (17) by one of the numerical methods, we can determine the coefficient $K_p$ of the proportional regulator or $K_i = 1/T_i$ for the integral regulator, as well as the frequency of the proper oscillations $\omega$, which indirectly characterizes the performance of a one-circuit system.

Similarly, for systems with PI- and PD- regulators we obtain the system of Eq. (18) based on expressions (11), (12). With its numerical solution, you can define the adjustment parameters $K_p$, $T_i$ and the own frequency $\omega$ of the ACS with the PI-regulator, or $K_p$, $T_d$ and $\omega$ for the system with the PD- regulator:

$$\left.\begin{aligned} eq_1 &:= \operatorname{Re}\left[\Delta(s)\right]_{s=s_1} = 0, \\ eq_2 &:= \operatorname{Im}\left[\Delta(s)\right]_{s=s_1} = 0, \\ eq_3 &:= \left[\Delta(s)\right]_{s=s_3} = 0. \end{aligned}\right\} \quad (18)$$

For systems with PID-regulator the system of equations has obtained similarly based on expressions (13), (14), (16) which has the form:

$$\left.\begin{aligned} eq_1 &:= \operatorname{Re}\left[\Delta(s)\right]_{s=s_1} = 0, \\ eq_2 &:= \operatorname{Im}\left[\Delta(s)\right]_{s=s_1} = 0, \\ eq_3 &:= \operatorname{Re}\left[\frac{d\Delta(s)}{ds}\right]_{s=s_1} = 0, \\ eq_4 &:= \operatorname{Im}\left[\frac{d\Delta(s)}{ds}\right]_{s=s_1} = 0. \end{aligned}\right\}$$

The given system of equations can be solved using numerical methods. As a result, it is possible to determine the parameters of the regulator $K_p$, $T_i$, $T_d$, and the own frequency of the system with a continuous PID-regulator. The resulting solution also ensures the fulfillment of the condition $[(K_p/T_i)_{\max}]_{\max}$, which is the minimization of the variance of the regulated quantity at low-frequency perturbations.

The considered method for calculating continuous systems containing objects without delay has certain limitations, which consists in the fact that the order of the characteristic equation of the system must be not less than the number of settings for adjusting the regulator (SAR) to be determined, including its system frequency. That is for systems with P- and I-regulators, the order of the object must be at least two, and when using PI- or PD- regulators, the mathematical model of the object must have at least a second order. If the object is characterized by some delay, these restrictions are removed because the characteristic equation, in this case, has an infinite number of roots.

The method of the multicriterial optimum is also suitable for the calculation of systems containing objects with derivatives in the numerator of the transfer function. In this case, the order of the characteristic equation does not increase, since the numerator order in the transfer function of the object is less than the order of the denominator.

The method of multicriteria optimum can be extended also to systems with digital controllers. A necessary prerequisite for this is the transformation of the digital system into an equivalent to it continue using the notion of an equivalent continuous controller. Besides, we must take into account the presence in the digital system of additional parameters – a period of discretion $T_0$.

For systems with digital controllers, there are no restrictions on the application of this method, since the characteristic equation, in this case, is always transcendental to a complex variable $s$.

An example of a program for synthesizing digital systems with a PI- regulator developed in the MAPLE 14 is shown:

The calculation of the system with digital PI- regulator and objects with self-alignment and delay at a given value of the period of discreteness

> restart;

Transmitting function of the regulator

> pr1:=$K_p$:ir1:=$K_p$/($T_i$*s):$W_{pi}$:=pr1+ir1;

$$W_{pi} := K_p + \frac{K_p}{T_i s}$$

> ir11:=$K_p$*($T_o$/$T_i$)*(1/(1-z^(-1))):ir11:=simplify(ir11):$W_r$:=pr1+ir11;

$$W_r := K_p + \frac{K_p T_0 z}{T_i(z-1)}$$

Equivalent regulator

> $W_{re}$:=$W_r$*(1/$T_0$)*(1-z^(-1))/s;$W_{re}$:=simplify($W_{re}$);

$$W_{re} := \frac{\left(K_p + \frac{K_p T_0 z}{T_i(z-1)}\right)\left(1 - \frac{1}{z}\right)}{T_0 s}$$

$$W_{re} := \frac{K_p(T_i z - T_i + T_0 z)}{s z T_0 T_i}$$

> $W_{re}$:=subs(z=exp($T_o$*s),$W_{re}$):$W_{re}$:=simplify($W_{re}$);

$$W_{re} := \frac{K_p\left(T_i e^{(T_0 s)} - T_i + T_0 e^{T_0 s}\right) e^{(-T_0 s)}}{s T_0 T_i}$$

> $W_{pi}$:=limit($W_{re}$,$T_o$=0):$W_{pi}$:=simplify($W_{pi}$);

$$W_{pi} := \frac{(T_i s + 1)K_p}{T_i s}$$

Object transmitting function

> $W_o$:=$K_o$*exp(-tau*s)/((T*s+1)^n);

$$W_0 := \frac{K_0 e^{(-\tau s)}}{(Ts+1)^n}$$

Characteristic quasipolinom

>ds:=1+$W_{re}$*$W_o$:ds1:=numer(1+$W_{re}$*$W_o$):ds1:=simplify(ds1):

System of equations

> s:=-m*Omega-I*Omega;

$$s := -m\Omega - \Omega I$$

> eq1:=evalc(Re(ds1))=0:eq1:=simplify(eq1):
> eq2:=evalc(Im(ds1))=0:eq2:=simplify(eq2):
> s:=-m*Omega;

$$s := -m\Omega$$

> eq3:=ds1=0:eq3:=simplify(eq3):

Parameters of the object

> s:='s':Ko:=1;T:=100;tau:=10;n:=1;Wo:=Wo;
> Omega:='Omega':Digits:=20:

The solution of the system of equations

```
> m:=0.5;To:=0.1*tau;
```

$$m := 0.5$$

$$T_0 := 1.0$$

```
> eq1:=evalf(eq1):eq2:=evalf(eq2):eq3:=evalf(eq3):
>resPI:=fsolve({eq1,eq2,eq3},{Kp,Ti,Omega},{Kp=1..10,Ti=1..200,
Omega=0.001..0.1});
```

$$resPI := \{K_p = 6.3752129316126079975, T_i = 33.977404316793602031, \Omega = 0.085867026135281126399\}$$

```
> assign(resPI);
```

Checking the correctness of the solution

```
> eq1:=eq1;eq2:=eq2;eq3:=eq3;
```

$$eq1 := -0.138 \cdot 10^{-16} = 0.$$

$$eq2 := -0.1 \cdot 10^{-18} = 0.$$

$$eq3 := -0.97 \cdot 10^{-17} = 0.$$

## 4 Conclusions

The proposed method of multicriteria parametric optimization allows minimizing the dispersion of established processes in automatic control systems of technological processes.

The experience of calculating automatic systems with typical regulators makes it possible to identify several features associated with the practical use of this method. First of all, we must pay attention to the complexity of the formation of systems of equations that describe the conditions for the achievement of the multicriterial optimum, and their solution. However, these problems are virtually eliminated by the use of modern information technologies, such as MAPLE, as shown in the example above (the calculation of the system with digital PI- regulator and objects with self-alignment and delay at a given value of the period of discreteness).

## References

1. Åström, K.J., Hägglund, T.: Avanced PID Control. ISA-The Instrumentation, Systems and Automation Society (2006). 460 p.
2. O'Dwyer, A.: Handbook of PI and PID Controller Tuning Rules, 3rd edn. Imperial College Press, London (2009). 599 p.

3. Korobiichuk, I., Ladanyuk, A., Shumyhai, D., Boyko, R., Reshetiuk, V., Kamiński, M.: How to increase efficiency of automatic control of complex plants by development and implementation of coordination control system. In: Recent Advances in Systems, Control and Information Technology. Advances in Intelligent Systems and Computing, vol. 543, pp. 189–195 (2017). https://doi.org/10.1007/978–3-319-48923-0_23
4. Zurawski, R.: Integration Technologies for Industrial Automated Systems. Industrial information technology Series. CRC Press, Taylor & Francis Group, Boca Raton (2007). 539 p.
5. Johnson, M.A., Moradi, M.H.: PID Control. New Identification and Design Methods. Springer, London (2005). 543 p.
6. Masum Jujuli, M.: Analytical Design of PID Controller for Enhanced Disturbance Rejection of Processes without Time Delay. Graduate School of Yeungnam University (2010)
7. Korobiichuk, I., Siumachenko, D., Smityuh, Y., Shumyhai, D.: Research on automatic controllers for plants with significant delay. In: Advances in Intelligent Systems and Computing, vol. 519, pp. 449–457 (2017) https://doi.org/10.1007/978-3-319-46490-9_60
8. Åström, K.J., Hägglund, T.: PID Controllers, Theory, Design and Tuning, 2nd edn. Instrument Society of America, Research Triangle (1995)
9. Ho, W.K., Lim, K.W., Xu, W.: Optimal gain and phase margin tuning for PID controllers. Automatica **34**(8), 1009–1014 (1998)
10. Kuzmych, L., Kobylianskyi, O., Duk, M.: Current state of tools and methods of control of deformations and mechanical stresses of complex technical systems. In: Proceedings of SPIE 10808, Photonics Applications in Astronomy, Communications, Industry, and High-Energy Physics Experiments 2018, p. 108085J (2018). https://doi.org/10.1117/12.2501661
11. Kuzmych, L., Kvasnikov, V.: Study of the durability of reinforced concrete structures of engineering buildings. In: Advances in Intelligent Systems and Computing, vol. 543, pp. 659–663 (2017)
12. Korobiichuk, I., Lobok, A., Goncharenko, B., Savitska, N., Sych, M., Vihrova, L.: The problem of the optimal strategy of minimax control by objects with distributed parameters. In: Szewczyk, R., Zieliński, C., Kaliczyńska, M. (eds.) Automation 2019. AUTOMATION 2019. Advances in Intelligent Systems and Computing, vol. 920, pp. 77–85. Springer, Cham (2020). https://doi.org/10.1007/978-3-030-13273-6_8
13. Babich, V., Dovbenko, V., Kuzmych, L., Dovbenko, T.: Estimation of flexures of the reinforced concrete elements according to the National Ukrainian & European standards. In: MATEC Web of Conferences, vol. 116, Article 02005 (2017)
14. Dyakonov, V.P.: Maple – 7. SPB, Petersburg (2002). 672 p. (in Russian)
15. Krasovskiy, A.A.: Handbook of the Theory of Automatic Control. Nauka, Moskva (1987). 712 p. (in Russian)
16. Shavrov, A.V., Soldatov, V.V.: Multi-criteria management under conditions of static uncertainty. Mashinostroenie, Moskva (1990). 160 p. (in Russian)
17. Korobiichuk, I., Ladaniuk A., Ivashchuk V.: Features of control for multi-assortment technological process. In: Szewczyk, R., Krejsa, J., Nowicki, M., Ostaszewska-Liżewska, A. (eds.) Mechatronics 2019: Recent Advances Towards Industry 4.0. MECHATRONICS 2019. Advances in Intelligent Systems and Computing, vol. 1044, pp. 214–221. Springer, Cham (2020). https://doi.org/10.1007/978-3-030-29993-4_27

# Modeling Thermal Modes of Induction Installation for Heating Dispersive Plant Materials

B. I. Kotov[1], Vitaliy Lysenko[2], Dmitruy Komarchuk[2(✉)], Taras Lendiel[2(✉)], Volodymyr Hryshchenko[2], Viktor Trokhaniak[2], and Roman Kalinichenko[2]

[1] State Agrarian and Engineering University of Podilya, Kamenets-Podilsky, Ukraine
[2] National University of Life and Environmental Sciences of Ukraine, Kiev, Ukraine
dmitruyk@gmail.com, taraslendel@gmail.com

**Abstract.** On the basis of the analysis of heat transfer processes in the moving layer of a dispersed material with immersed heat-generating elements placed in a cylindrical inductor, a mathematical model of the non-stationary process of heating of a grain material is formulated. Analytical dependences of grain temperature distribution in the direction of motion are obtained, unevenly distributed thermal power in steady state were taken into account. The static and transient characteristics of the induction heater, the distribution of the temperature of the core loading and the change of its temperature in time are analytically determined and verified according to the experiments. The dynamic characteristics of the installation as a control object are defined as transmission functions via control and perturbation channels.

**Keywords:** Inductor · Core loading · Dispersed (grain) material · Heating dynamics · Power distribution

## 1 Introduction

### 1.1 Topic

Heating of dispersed agricultural materials (grain, seed, oil) is used for disinfestation, intensification of drying processes, removal of oil by pressing, extraction of pulp (cake) and other processing operations.

Existing plants for heating grain materials in the agro-industrial production traditionally use convective heat supply to the material, which causes significant losses of heat energy with the spent heat carrier. The method of heat treatment of grain material based on conductive heating of the material in combination with the use of an induction method of supplying energy to the surfaces of heating has been increasingly offered lately. A significant advantage of the latter method is the possibility of using a branched surface heat transfer, as well as providing them with various kinds of movements (vibrational, rotational, vibrational). To analyze the process of determining the rational


R. Szewczyk et al. (Eds.): AUTOMATION 2020, AISC 1140, pp. 98–111, 2020.
https://doi.org/10.1007/978-3-030-40971-5_10

modes and the prospects of using the heat treatment of different grains and seeds material, it is necessary to have a mathematical description of the process, and preferably analytical dependencies that determine the interaction of individual elements of the equipment installation.

### 1.2 Previous Works

The advantages of induction heating flow media over other methods of electric heating are quite fully presented in [1–3]. The description of the work and the results of experimental studies of the induction heating installation are given in papers [2, 4, 5]. The mathematical model and method of its construction are presented in the paper [7]. The possibilities of parametric identification are highlighted in the papers [8, 9]. In modeling the dynamics of the heating process of the material from the loading of the inductor obtained a uniform distribution of the temperature of the loading surface, although in papers [4, 6] the distribution of power and temperature has a substantially non-uniform distribution. More research is needed to identify the impact of power distribution over load length.

### 1.3 The Purpose of the Work

Creation and research of mathematical models of induction heating of dispersed material.

## 2 Results of the Research

Heating of wet dispersed plant materials from heat-generating ferromagnetic elements (HGE) uniformly placed in the electromagnetic field of a cylindrical inductor [13–15] occurs when the advantages of several methods of heat dissipation are combined [2]. The heat-releasing rod ferromagnetic elements, when heated in a magnetic field, transfer heat (from a branched surface of heat transfer) to the moving or non-moving material by thermal conductivity, radiation and convection through the pores in the layer of dispersed material. Consider the dynamics of complex heat transfer between the heat-generating elements of the rods of the small section evenly spaced in the cross section of the cylindrical body and the moving (not moving) layer of the dispersed material, which is in the middle of the housing of the installation (Fig. 1).

Suppose that with this complex mechanism of heat transfer, the particle heating of the dispersed material occurs without gradient. Temperature gradient in the process of heat transfer in the radial direction of the grain layer and in the rods is not yet taken into account. The coefficients of heat transfer, thermophysical characteristics of the material and structural elements are independent of temperature and equal to the average value for the process. We also accept the one-dimensional distribution of heat in the direction of motion of the material, that is, the axis $x$. The case with the inductor is considered as one element.

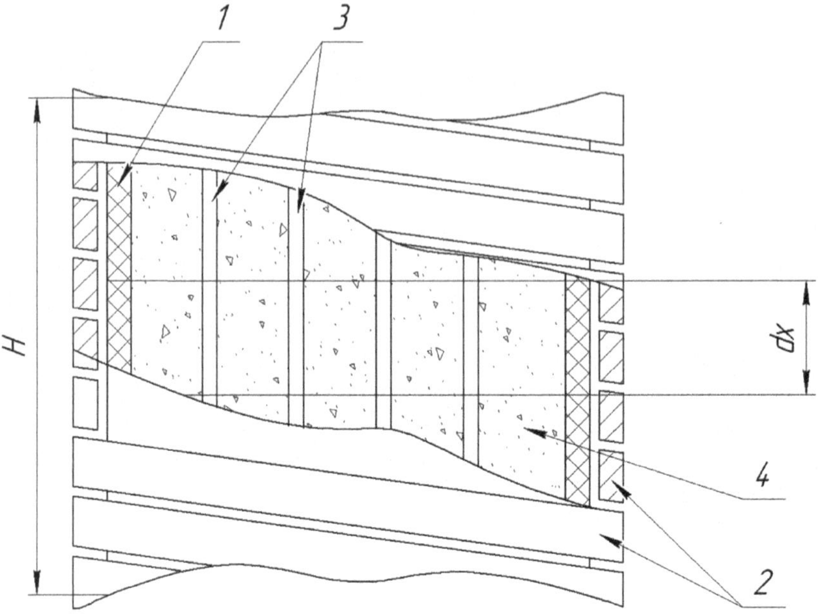


**Fig. 1.** Calculation scheme of induction heater of dispersed material: 1 – corps; 2 – inductor; – heat-emitting elements; 4 – dispersed material.

Let's consider the installation element (channel with moving material and fixed rods) height *dx*. Putting the heat balance equation, which describes the change in temperature of the rod for the selected unit area *dx* per hour *dτ*:

$$\begin{aligned}&\frac{1}{H}[\alpha_1 F_1(\theta_1-\theta_2)+\alpha_2 F_1(\theta_1-\theta_2)+\alpha_3 F_1(\theta_1-\theta_2)+\alpha_4 F_1(\theta_1-t)]dxd\tau\\&=-c_1\rho_1 S_1 ndxd\theta_1+P(x)dxd\tau\end{aligned} \quad (1)$$

where $\alpha_1$ – coefficient of contact heat transfer, W/(m$^2$ °C);

$\alpha_2$ – coefficient of heat exchange by convection, W/(m$^2$ °C);
$\alpha_3$ – coefficient of heat exchange by radiation, W/(m$^2$ °C);
$\alpha_4$ – coefficient of heat exchange by convection, W/(m$^2$ °C);
$F_1$ – surface of fuel elements, m$^2$;
$H$ – core inductor height, m;
$\theta_1$ – the temperature of the fuel elements, °C;
$\theta_2$ – grain temperature, °C;
$t$ – air temperature, °C;
$c_1$ – specific heat capacity of the IFE rods, J/(kg °C);
$\rho_1$ – the density of the IFE rods, kg/m$^3$;
$S_1$ – IFE rods area, m$^2$;
$n$ – the number of rods HGE;
$P(x)$ – total power, W;

$x$ – coordinate of the grain movement direction, m;
$\tau$ – time, sec.

The first term of the left part of the equation determines the amount of heat of the radiation flux absorbed by the material, the second term of the equation determines the amount of heat of the transmitted to material thermal conductivity, the third and fourth terms of the equation take into account the amount of heat transmitted by the convection method to the material and air (in the pores of the layer of dispersed material). All the amount of heat received by the material is spent on increasing the heat content of the material, which is determined by increasing its temperature.

The magnitudes of the heat transfer coefficients that characterize the intensity of heat fluxes can be determined from theoretical ratios that take into account all the components and physical features of the complex heat transfer process in the layer of dispersed material with submerged heat-productive elements, which are given in the next sources [10, 11].

In practical calculations of complex heat transfer processes, as a rule, the coefficient of effective heat exchange is determined, which is determined by the formula:

$$\alpha_{ef} = \frac{\alpha_1 F_1 + \alpha_2 F_2 + \alpha_3 F_3}{F_1 + F_2 + F_3}. \quad (2)$$

It is almost impossible to calculate the value of the heat transfer coefficient according to formula (2), so its value is determined by experiments data, from the solutions of the heat balance equations given in [8].

Using the effective coefficient of heat transfer $\alpha_{ef}$ (attributed to the heat transfer Surface $F_1$) the equation of heat balance for the fuel output element would be written in the form:

$$c_1 S_1 \rho_1 dx d\theta_1 = \eta P(x) dx d\tau - \frac{1}{H}\left[\alpha_{ef} F_1(\theta_1 - \theta_2) - \alpha_k F_1(\theta_1 - t)\right] dx d\tau, \quad (3)$$

where: $\eta$ – efficiency of the inductor.

The complete differential temperature of the HGE had been revealed:

$$d\theta_1 = \frac{\partial \theta_1}{\partial \tau} d\tau + \frac{\partial \theta_1}{\partial x} dx, \quad (4)$$

and using obvious ratios: $dx/d\tau = V$, $V = \frac{G}{\rho S}$, $m = Sdx\rho$, where: $V$ – speed of movement of the material, m/s; $G$ – material mass flow, kg/s, we obtain from Eq. (3):

$$m_1 c_1 \frac{\partial \theta_1}{\partial \tau} = \eta_1 P(x) - \alpha_{ef} F_1(\theta_1 - \theta_2) - \alpha_k F_1(\theta_l - t). \quad (5)$$

Similarly, we formulate the heat balance equation for other elements of the simulation object:

for the flow of grain material:

$$m_2c_2\frac{\partial\theta_2}{\partial\tau}+Gc_2H\frac{\partial\theta_2}{\partial x}=\alpha_{ef}F_1(\theta_1-\theta_2)-\alpha'_{ef}F_4(\theta_2-\theta_4)+\alpha_kF_2(t-\theta_2), \tag{6}$$

for air in between the grain volume:

$$m_3c_3\frac{\partial t}{\partial\tau}=\alpha_kF_1(\theta_l-t)-\alpha_kF_4(t-\theta_4)-\alpha_kF_2(t-\theta_2), \tag{7}$$

for housing with inductor:

$$m_4c_4\frac{\partial\theta_4}{\partial\tau}=P(1-\eta)+\alpha'_{ef}F_4(\theta_2-\theta_4)-\alpha'_kF'_4(\theta_4-t_{oc})-\alpha_kF_4(\theta_4-t). \tag{8}$$

Equations (5–8) indicate: $\theta_4$ – housing temperature with inductor, °C; $P$ – inductor thermal power, W; $F_2, F_4F'_4$ – the surface of the grain, the inner and outer shells, m$^2$; $m_1, m_2, m_3, m_4$ – the mass of fuel elements, grain, air and body with inductor kg; $c_2, c_3, c_4$ – specific heat of grain, air, inductor material, J/(kg °C); $G$ – grain consumption, kg/sec; $t_{oc}$ – ambient temperature, °C; $\alpha_{ef}, \alpha'_{ef}, \alpha_k, \alpha'_k$ – coefficients of heat exchange of the core with the grain, the inner surface of the housing with the grain, the air with the grain and the housing, the surface of the inductor with air, W/(m$^2$ °C).

In the equations were taken into account: the amount of heat released by the inductor $P(1-\eta)$, coefficient of effective heat transfer $\alpha'_{ef}$, related to the inner surface of the housing $F_4$, coefficient $\alpha'_k$ external heat exchange with the environment is determined by the equation:

$$\alpha'_k=\alpha_k+c_{np}\left(T_k^2-T_p^2\right)(T_k-T_p)10^{-8}, \tag{9}$$

where: $c_{np}$ – the radiation factor is given; $T_k$, $T_p$ – the surface temperature of the housing and surrounding surfaces of structure.

Thus, a mathematical model of the dynamic modes of the induction heater of dispersed materials is obtained, in the form of a system of fourth order differential equations. Since the system of Eqs. (5–8) has no analytical solution, let us consider the steady and transient modes of operation of the induction system separately.

Consider steady-state (installed) mode, that is, the mode in which at each point in the volume of the dispersed material and structural elements loaded with the flow all temperature parameters are unchanged over time and the temperature distribution takes place only by coordinate.

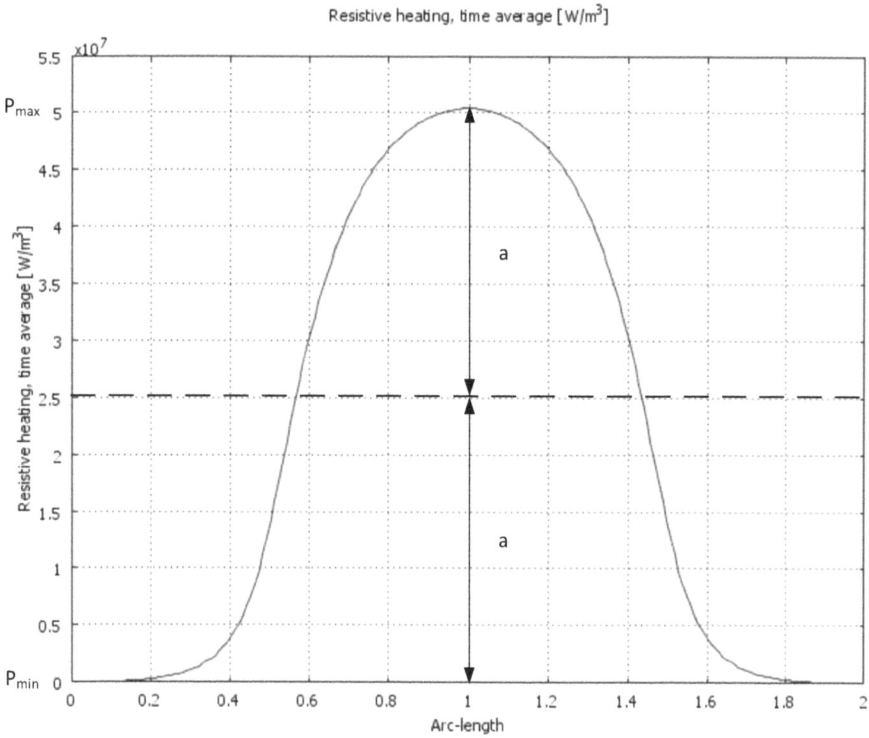


**Fig. 2.** The distribution of the specific thermal power of HGE by coordinate.

As established theoretically and experimentally determined [4], the distribution of thermal power in HGE is uneven and is described by the dependence (Fig. 2):

$$Q = -3.59 \cdot 10^5 + 1.83 \cdot 10^7 x + 9.62 \cdot 10^7 x^2 - 2.95 \cdot 10^8 x^3 + 1.83 \cdot 10^8 x^4, \quad (10)$$

which in a given approximation, can be approximated as a part of the sine curve:

$$\eta P(x) = a(1 + \sin \omega x), \quad (11)$$

where $a = 0.5(P_{\max} - P_{\min})$ – power amplitude value, W;

$P_{\max}$ – maximum thermal output, watts;
$P_{\min}$ – minimum thermal output, watts;
$\omega = \pi/H$, $\mathrm{m}^{-1}$.

By equating time derivatives to zero and excluding (to simplify calculations with 2% error) considering the temperature of the air, considering its contribution to the total heat transfer is taken into account by the convection component in the complex heat transfer by increasing the coefficient of thermal conductivity of the air we obtain a system of equations of steady state:

$$\eta P(x) = \alpha_{ef} F_1(\theta_1 - \theta_2), \tag{12}$$

$$Gc_2 H \frac{\partial \theta_2}{\partial x} = \alpha_{ef} F_1 \theta_1 + \alpha'_{ef} F_4 \theta_4 - \left(\alpha_{ef} F_l + \alpha'_{ef} F_4\right)\theta_2, \tag{13}$$

$$P(1-\eta) = \left(\alpha_{ef} F_4 + \alpha_k F'_4\right)\theta_4 - \alpha'_{ef} F_4 \theta_2 - \alpha_k F'{}_4 t_{oc}. \tag{14}$$

Reducing Eqs. (12)–(14) to one with respect to the dependence (11) after the transformations we will have:

$$T_x \frac{d\theta_2(x)}{dx} + b\theta_2(x) = P_1 + a \sin \omega x, \tag{15}$$

where

$$T_x = \frac{Gc_2 H}{\alpha'_{ef} F_4 + \alpha_{ef} F_1}; \; P_1 = k_2(k_5 t_{oc} + k_3(1-\eta)P) \; + \; k_1 k_2 a; \; b = 1 - k_1 - k_2 k_4;$$

$$k_1 = \frac{\alpha_{ef} F_1}{k_{01}}; \; k_2 = \frac{\alpha'_{ef} F_4}{k_{02}}; \; k_3 = \frac{1}{k_{02}}; \; k_4 = \frac{\alpha'_{ef} F_4}{k_{02}}; \; k_5 = \frac{\alpha_k F'_4}{k_{02}};$$
$$k_{01} = \alpha_{ef} F_1 + \alpha'_{ef} F_4; \; k_{02} = \alpha_k F'_4 + \alpha'_{ef} F_4.$$

We obtain the solution of the inhomogeneous differential equation as the sum of the homogeneous equation:

$$\theta_2(x) = c_1 e^{-\frac{b}{T_x}x} + \frac{P_1}{b} + A \sin \omega x + B \cos \omega x, \tag{16}$$

where $c_1$ – permanent integration; $A$, $B$ – constant coefficients are determined by the method of indeterminate multipliers.

Permanent of integration is determined from the initial condition: $x = 0$; $\theta_2 = \theta_{oz} = t_{oc}$ – material temperature at the entrance to the installation. Omitting the intermediate transformation, we write down the values of coefficients: $A = \frac{ab}{T_x^2 \omega^2 + b^2}$, $B = -\frac{aT\omega}{T^2\omega^2 + b^2}$, $c_1 = \theta_{o2} - \frac{P}{B} + \frac{aT\omega}{b^2 + T_x^2 \omega^2}$.

The temperature of the HGE rods:

$$\theta_1(x) = \frac{hP(x)}{\alpha_{ef} F_1} + \theta_2(x). \tag{17}$$

The temperature of the housing:

$$\theta_4(x) = k_2 P + k_4 t_{oc} + \theta_2(x). \tag{18}$$

Substituting the values $\theta_2(x)$ from Eq. (16) into Eqs. (17), (18) one determine the temperature distribution of the rods and the housing by height in the direction of the grain flow. The grain temperature at the outlet of the installation is determined from Eq. (16) by the substitution $x = H$.

In Fig. 3 the graph of temperature distribution of HGE by altitude is given.

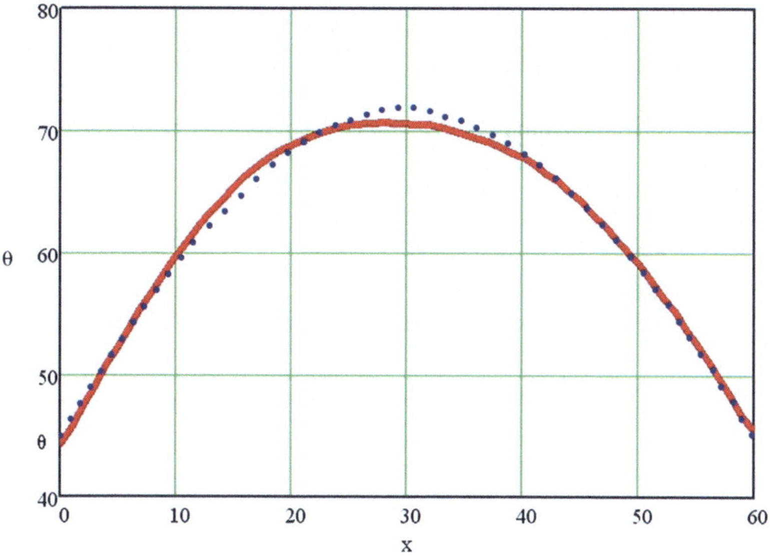


**Fig. 3.** Distribution of HGE temperature by coordinate: • – experimental data; ▬ – estimated data.

It is quite difficult to determine the temperature distribution of the grain material as it moves in a channel filled with ferromagnetic loading rods.

Determine the temperature field (one-dimensional) loading in static and dynamic modes in the absence of grain material in the working channel of the installation and compare the obtained data with the data of the experiment [13].

We will describe the non-stationary thermal regime of the installation (heating period) by the system of differential equations for three thermal capacities:
downloads:

$$m_1 c_1 \frac{d\theta_1}{d\tau} = \eta P(x) - \alpha'_{ef} F_1(\theta_1 - t_2) - \alpha'_{ef} F_4(\theta_l - \theta_4), \tag{19}$$

air:

$$m_3c_3\frac{dt}{d\tau} = \alpha'_{ef}F_1(\theta_l - t) - \alpha'_{ef}F_4(t - \theta_4), \tag{20}$$

housing with inductor:

$$m_4c_4\frac{d\theta_4}{d\tau} = P(1-\eta) - \alpha_{ef}(\theta_4 - t_{oc})F'_4 + \alpha'_{ef}F_4(\theta_l - \theta_4) + \alpha_{ef}F_4(t - \theta_4). \tag{21}$$

The value of the coefficient of effective heat exchange takes into account heat exchange by radiation and convection and is calculated [9] by the formula (9). The heat exchange occurs between the load and the housing. Since the heat capacity of the air $m_3c_3$ is much smaller than other capacities $(m_4c_4 \gg m_3c_3 \ll m_1c_1)$ its magnitude can be neglected.

From Eq. (20) we determine:

$$t = K_1\theta_1 + K_2\theta_4, \tag{22}$$

where $K_1 = \dfrac{\alpha'_{ef}F_1}{\alpha'_{ef}F_1 + \alpha_2F'_4}$; $K_2 = \dfrac{\alpha_{ef}F_4}{\alpha'_{ef}F_1 + \alpha_2F_4}$; $\alpha_2 = \alpha_k$.

Substituting the values from Eq. (22) into (19) and (21) after transformations we obtain:

$$T_1\frac{d\theta_1}{d\tau} + a_1\theta_1 - b_1 = \theta_4, \tag{23}$$

$$T_2\frac{d\theta_4}{d\tau} + a_2\theta_4 - b_2 = \theta_1, \tag{24}$$

where: $a_1 = \dfrac{K_5}{K_3}$; $b_1 = \dfrac{P(x)}{K_3}$; $T_1 = \dfrac{m_1c_1}{K_3}$; $K_3 = \alpha'_{ef}(F_4 + F_1K_2)$; $a_2 = \dfrac{K_6}{K_4}$;

$$b_2 = \frac{P(1-\eta)}{K_4} + \frac{\alpha'_{ef}F_4}{K_4}t_{oc};\; T_2 = \frac{m_4c_4}{K_4};\; K_4 = \alpha_{ef}(F_1 + F_4K_1);$$
$$K_5 = \alpha'_{ef}(F_1 + F_4) - \alpha'_{ef}F_1K_1;\; K_6 = \left(\alpha_2 + \alpha'_{ef}\right)F_4 - \alpha_{ef}F_4(1 + K_2).$$

Substituting the values $\theta_4$ from Eq. (23) and the derivative $\dfrac{d\theta_4}{d\tau}$ into Eq. (24) we get:

$$A\frac{d^2\theta_1}{d\tau^2} + B\frac{d\theta}{d\tau} + C\theta = D(x), \tag{25}$$

where: $A = T_1T_2$; $B = (a_1T_2 + a_2T_1)$; $C = a_2a_1 - 1$; $D = b_2 + b_1a_2$;

$$D(x) = \frac{P(1-\eta) + \alpha'_{ef}F_4t_0}{K_4} + \frac{K_6P(x)}{K_4K_3}.$$

We obtain the solution of the inhomogeneous Eq. (25) as the sum of a homogeneous equation and a partial solution:

$$\theta_1(\tau, x) = c_1e^{r_1\tau} + c_2e^{r_2\tau} + \frac{D(x)}{C}. \tag{26}$$

Determining the permanent integration $c_1$ and $c_2$ on initial condition $\tau = 0$; $\theta_1(\tau, x) = \theta_{01}$ ($\theta_{01}$ – the initial temperature HGE) after the transformations we get the dependence:

$$\theta_1(\tau, x) = \frac{\theta_{01}C - D(x)}{C(r_1 - r_2)}(r_1e^{r_2\tau} - r_2e^{r_1\tau}) + \frac{D(x)}{C}, \tag{27}$$

where $r_1, r_2 = \dfrac{-B \pm \sqrt{B^2 + 4AC}}{2A}$ – the roots of the characteristic equation.

In Fig. 4 there are presented the graphical dependences of the temperature change of the HGE over the time.

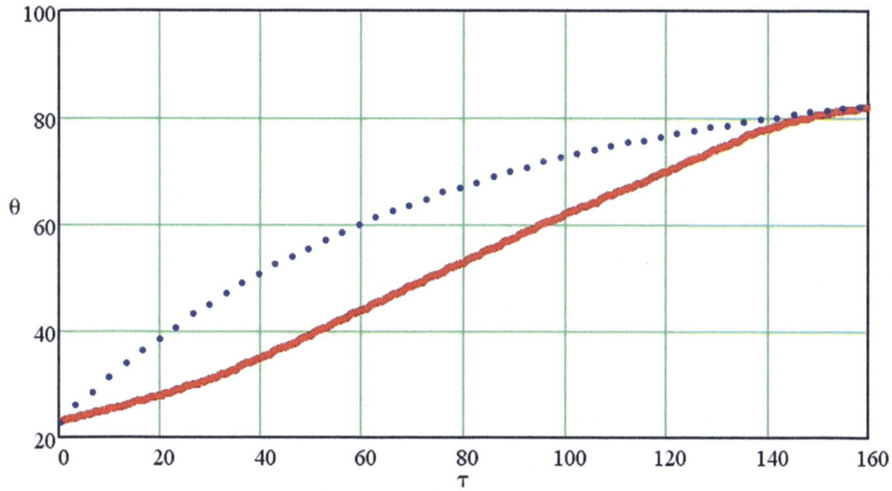


**Fig. 4.** Change in temperature of the HGE (middle of the rod) in time: • – experimental data; ■ – estimated data.

Stationary mode temperature distribution over the length of IFE is obtained from Eqs. (23) and (24) at $d\theta_1/d\tau = d\theta_4/d\tau = 0$:

$$\theta_1(x) = \frac{a_2}{a_1a_2 - 1}\left[\frac{P_0}{K_4} + \frac{\alpha'_{ef}F_4t_{oc}}{K_4} + P\left(\frac{1-\eta}{K_4} + \frac{1}{K_3}\sin\frac{\pi}{H}x\right)\right]. \quad (28)$$

To create an effective system of automatic control of the mode of induction heating of the dispersed material, it is necessary to have a mathematical model of dynamic properties of the installation, which would take into account the inertia of all the elements (units). Traditionally (in terms of control tasks), we get a dynamic model of the control object in the form of transfer functions through the control and perturbation channels. For this purpose, the component that takes into account the change in the flow temperature of the grain material in Eq. (6) is replaced by the coordinate derivative by the coordinate from Eq. (16), that is:

$$\frac{d\theta_2(x)}{dx} = -\frac{b}{T_x}C_1 + A\omega\cos\omega x - B\omega\sin\omega x = f(x). \quad (29)$$

Dependence (29) does not change in time and treated as a "gain coefficient" inertial component "grain stream". In this case, the system of differential equations in partial derivatives (5–8) becomes a system of equations in ordinary derivatives.

Neglecting the thermal capacity of the air ($m_3c_3 = 0$) and applying the Laplace transform to the system of Eqs. (5–8) at zero initial conditions, after the corresponding transformations we obtain a system of linear equations in the operator form:

$$\begin{cases} (T'_1p + 1)\theta_1(p) = a_1 + b_1\theta_2(p) + c_1\theta_4(p) \\ (T'_2p + 1)\theta_2(p) = a_2 + b_2\theta_4(p) + c_2\theta_4(p), \\ (T'_4 + 1)\theta_4(p) = a_4 + b_4\theta_2(p) \end{cases} \quad (30)$$

and the equation of communication:

$$t(p) = K_7\theta_1(p) + K_8\theta_2(p), \quad (31)$$

where $p = (d\tau)^{-1}$ – Laplace operator; $T'_1 = \frac{m_1c_1}{K_{01}}$; $a_1 = \frac{\eta P(x)}{K_{01}}$; $b_1 = \frac{\alpha_{ef}F_1}{K_{01}}$;

$$c_1 = \frac{\alpha_kF_1F'_4}{K_{01}};\ K_{01} = F_1\left(\alpha_{ef} + \alpha_kF'_1\right);\ F'_1 = \frac{F_1}{F_l + F_4};\ F'_4 = \frac{F_4}{F_1 + F_4};$$

$$F'_2 = \frac{m_2c_2}{K_{02}};\ a_2 = \frac{G_2c_2H}{K_{02}}f(x);\ b_2 = \frac{\alpha_{ef}F_1}{K_{02}};\ c_2 = \frac{\alpha_{ef}F_4}{K_{02}};$$

$$K_{02} = \alpha_{ef}F_1 + \alpha'_{ef}F_4;\ T'_4 = \frac{m_4c_4}{K_{04}};\ a_4 = \frac{P(1-\eta) + \alpha_{ef}F'_4t_{oc}}{K_{04}};$$

$$b_4 = \frac{\alpha_{ef}F_4}{K_{04}};\ K_{04} = \alpha_{ef}\left(F_4 + F'_4\right);\ K_7 = \frac{\alpha_kF_1}{\alpha_kF_1 + \alpha_kF_2};\ K_8 = \frac{\alpha_kF_4}{\alpha_kF_1 + \alpha_1F_2}.$$

Solving the system of Eq. (30) relatively $\theta_2(P)$ we obtain:

$$A_4\left(A_1p^3+A_2p^2+A_3p+1\right)\theta(p)=B_3\left(B_1p^2+B_2p+1\right), \quad (32)$$

where $A_1=T_1'T_2'T_4'\frac{1}{A_4}$; $A_2=\left(T_1'T_4'+T_2'T_4'+T_2'T_1'\right)\frac{1}{A_4}$;

$$A_3=\left(T_1'+T_2'+T_4'\right)\frac{1}{A_4};\ A_4=c_2b_4+b_1b_2+b_2^2;$$

$$B_1=a_2T_1T_4;\ B_2=(a_2+a_1b_2)T_4+(a_2+a_2a_4)T_1;\ B_3=a_2+a_4(c_2+c_1b_2)+c_1b_2$$

The coefficient $B_3$ includes a managed variable $P \to P(p)$ and a disturbing variable $t_{oc}(p)$. Revealing the values of the coefficients for the variables, we take into consideration:

$$B_3(p)=k_1'P(p)+k_2't_{oc}(p), \quad (33)$$

where $k_1'=\frac{1-\eta}{K_{04}}(c_2+c_1b_2)+\frac{\eta b_2}{K_{01}}\sin\omega x$; $k_2'=\frac{\alpha_{ef}'F_4'}{K_{04}}$.

Taking into account (33) the transmission functions on the corresponding channels can be written in the form:

$$W_1(p)=\frac{\theta_2(p)}{P(p)}=\frac{B_1p^2+B_2p+1}{A_4(A_1p^3+A_2p^2+A_3p+1)}k_1'P(p), \quad (34)$$

$$W_2(p)=\frac{\theta_2(p)}{t_{oc}(p)}=\frac{B_1p^2+B_2p+1}{A_4(A_1p^3+A_2p^2+A_3p+1)}k_2't_{oc}(p). \quad (35)$$

The block diagram of the control object will look like the one shown in Fig. 5.

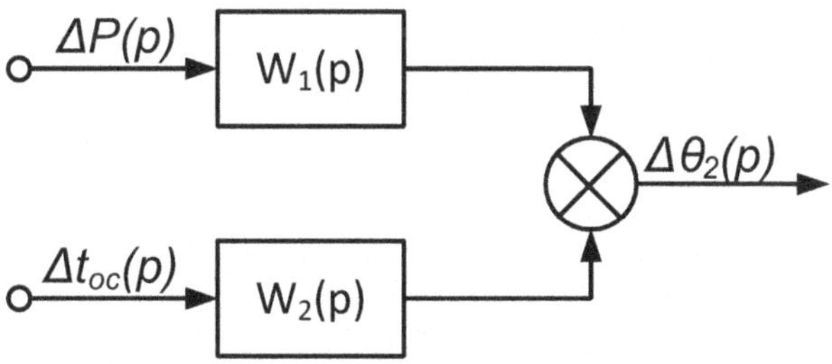


**Fig. 5.** Block diagram of the dynamics of the control object.

## 3 Conclusions

1. Formulated mathematical models of temperature modes of induction installation of continuous action for heating grain material that take into account the unequal distribution of power along the length of the core fuel elements, heat loss in the environment, radiation, convective and conductive heat transfer in a layer.
2. The analysis of the obtained models made it possible to determine the influence of heat dissipation unevenness on the static and dynamic characteristics of the simulation object and to obtain analytical dependencies that determine the temperature distribution of the HGE and grain material in the stationary and transient modes.

## References

1. Horbatkov, S.A., Kuvaldyn, A.B., Myneev, V.E., y dr.: Khymycheskye aparaty s ynduktsyonnym obohrevom. Khymyia, Moscow (1985). 176p.
2. Komarchuk, D.S.: Rezhymy roboty elektrotekhnolohichnoho kompleksu z systemoiu avtomatychnoho upravlinnia dlia teplovoi obrobky zerna ripaku. Dysertatsiia kand. tekhn. nauk. NUBiP Ukrainy, Kiev (2014). 205p.
3. Cherniakov, Y.E.: Yssledovanye protsessa sushky semian v plotnom sloe pry konvektyvno-konduktyvnom teplopodvode y obosnovanye parametrov selektsyonnoi sushylky. Avtoreferat dys. kand. tekhn. nauk. Novosybyrsk (1981). 18p.
4. Bereziuk, A.O.: Elektromahnitni i teploviprotsesy v induktsiinykh ustanovkakh nahrivannia teplonosiiv. Avtoreferat kand. tekhn. nauk. NUBiP Ukrainy, Kiev (2013). 21p.
5. Kondratenko, I.P., Lysenko, V.P., Bereziuk, A.O., Komarchuk, D.S.: Induktsiina ustanovka dlia termoobrobky zerna ripaku. Visnykahrarnoi nauky, no. 12, pp. 50–58 (2012)
6. Kondratenko, I.P., Bereziuk, A.O.: Vyznachennia heometrychnykh parametriv rozrakhunkovoi modeli tsylindrychnoho induktora dlia nahrivu feromahnitnykh trub. Visnyk NUBiP Ukrainy **174**(1), 125–129 (2012)
7. Lysenko, V.P., Komarchuk, D.S.: Matematychnemo deliuvannia protsesu termoobrobky zerna ripaku v shchilnomu shari z induktsiinymnahrivom. Bioresursy i pryrodokorystuvannia, Kiev, no. 5–6, pp. 119–129 (2013)
8. Lysenko, V.P., Kalinichenko, R.A., Komarchuk, D.S.: Identyfikatsiia protsesu nahrivu zerna ripaku (chap. 1). Naukovyivisnyk NUBiP Ukrainy, no. 174, pp. 98–100 (2012)
9. Kotov, B.I., Lysenko, V.P., Komarchuk, D.S., Kalinichenko, R.A.: Identyfikatsiia dynamiky elektrychnykh ustanovok termoobrobky furazhnoho zerna pry vyrobnytstvi kombikormiv. Naukovyi visnyk Tavriiskoho derzhavnoho ahrotekhnolohichnoho universytetu **4**(2), 3–8 (2010)
10. Krysher, O.: Nauchnыe osnovы tekhnyky sushky. Yz-vo Ynostrannoi lyteraturы, Moscow (1961). 28p.
11. Chudnovskyi, A.F.: Teplofyzycheskye kharakterystyky dyspersnыkh materyalov. Hos. Yzd-vo fyz-mat. lyt, Moscow (1962). 456p.
12. Antonyshyn, N.V., Nykytyn, V.S.: O teploobmene yzuchenyem v zasыpke dyspersnoho materyala. teplo y massoperenos v protsesakh sushky y termoobrabotke. Mynsk «Nauka y tekhnyka», pp. 161–169 (1970)
13. Kondratenko, I.P., Lysenko, V.P., Komarchuk, D.S.: Doslidzhennia rozpodilu temperatury v zavantazhenni tsylindrychnoho induktora. Naukovyivisnyk Natsionalnoho universytetu

bioresursiv i pryrodokorystuvannia Ukrainy. Seriia: Tekhnika ta enerhetyka APK, vol. 184, no. 1, pp. 74–82 (2013). http://nbuv.gov.ua/UJRN/nvnau_tech_2013_184(1)__13
14. Kawaguchi, H., Enokizono, M., Todaka, T.: Thermal and magnetic field analysis of induction heating (2), 193–198
15. Codecasa, L., Alotto, P., Moro, F.: Fast solution of induction heating problems by structure-preserving nonlinear model order reduction. IEEE Trans. Magn. **52**(3) (2016). https://doi.org/10.1109/tmag.2015.2491640

# A Proactive Approach to Resistant UAV Mission Planning

Grzegorz Radzki[1], Peter Nielsen[2], Grzegorz Bocewicz[1](✉), and Zbigniew Banaszak[1]

[1] Faculty of Electronics and Computer Science, Koszalin University of Technology, Koszalin, Poland
radzki.grzegorz@gmail.com, bocewicz@ie.tu.koszalin.pl, zbigniew.banaszak@tu.koszalin.pl

[2] Department of Materials and Production, Aalborg University, Aalborg, Denmark
peter@mp.aau.dk

**Abstract.** Proactive mission planning for a fleet of Unmanned Aerial Vehicles (UAVs) can be seen as a sequence of routing problems mathematically formalized as 0-1 knapsack problems. Taking into account the fact that weather conditions change during a mission, the time horizon of the planned mission is subdivided into time windows corresponding to periods of stable weather. Also, keeping in mind fuel constraints, each knapsack problem is formulated as follows: Given is the fleet size and a set of spatially dispersed target points specified by the volume of expected deliveries and the coordinates of their location, which allow to determine the amount of fuel consumed during flight along a particular route segment/from one drop-off point to another. Determine a subset of locations so that the total fuel required to cover the total distance traveled by the UAV fleet is less than or equal to the given limit e.g. determined by battery capacity, and the total volume of deliveries is as large as possible. In this context, policies aimed at minimizing the total travel time and/or the total distance traveled are considered. Some potential directions of future research on resistant UAV mission planning are discussed.

**Keywords:** UAV mission planning · Weather-resistant planning

## 1 Introduction

The use of unmanned aerial vehicles (UAVs) for transportation tasks gives rise to new challenges related to the organization of transport. Typical limitations of air transport systems that use UAVs include the need for periodical battery replacement (which implies the need to determine in advance the distance range that can be covered given the battery capacity), a limited payload capacity, the possibility of collision between UAVs performing their missions, etc. [1, 2, 5]. It is worth noting that a UAV's performance is highly influenced by the payload it is carrying, the weather conditions in which it is operating (wind may limit the number of target locations served, and low


R. Szewczyk et al. (Eds.): AUTOMATION 2020, AISC 1140, pp. 112–124, 2020.
https://doi.org/10.1007/978-3-030-40971-5_11

temperatures may adversely affect battery performance), and the non-linearity of fuel consumption.

It should be noted that the UAV routing approaches found in the literature fail to take into account the variability of weather conditions and the associated non-linearity of energy consumption [8, 13, 16]. This is mainly due to difficulties associated with the need for ongoing tuning of motion trajectories during flight progress. The process of adjustment is very complex and does not protect the mission plan against potential failure due to battery depletion or other causes. An alternative solution is to take into account the uncertainty/variability of weather conditions and the related level of energy consumption already at the stage of planning flight missions.

In this context, the approach developed in this study, which is implemented in a constraint programming environment (IBM ILOG), assumes that energy consumption is a non-linear function that depends on weather conditions, carrying payload, UAV "geometry" [4] and the flight trajectory [6]. The goal is to find proactive flight mission plans (routings and schedules) which are collision-free and allow to maximize the quantity of goods delivered to customers by a fleet of UAVs with an energy capacity limit, flying in given weather conditions. The solutions sought are sequences of routing problems formulated as 0-1 knapsack problems. In other words, the time horizon of a planned mission is subdivided into time windows corresponding to periods of unchanging weather. Given is the fleet size and a set of spatially dispersed target points specified by the volume of expected deliveries and the coordinates of their location, which allow to determine the amount of fuel consumed during flight from one drop-off point to another. These assumptions allow to determine, for each time window, a subset of locations that satisfy the fuel constraints assumed for the UAV fleet while maximizing the total volume of deliveries.

Apart from maximizing customer satisfaction (e.g., the total volume of deliveries), other criteria can also be considered. The models most frequently found in the literature focus on how to minimize the total travel time and/or the total distance traveled [4, 14, 16]. Criteria aimed at minimizing the total fuel consumed by UAVs (UAV operation costs, such as battery consumption) or minimizing the UAV fleet are used relatively less often [7, 14]. The present paper is organized as follows: Sect. 2 provides an example illustrating the routing problem under consideration. Sections 3 and 4 present the proposed declarative model and formulation of the problem, respectively. Section 5 reports experiments which verify the correctness of the model developed and the possibilities of its practical application. The key conclusions are formulated and the main directions of future research are suggested in Sect. 6.

## 2 A Motivation Example

Let's consider a company that provides air transport services using a fleet of UAVs. In the case in point, the fleet consists of three UAVs with identical technical parameters (Fig. 1). The UAVs deliver goods to six customers located in an area covering

10 km$^2$ – the network of connections is shown in Fig. 1. Node $N_1$ represents the location of the company (the base which the UAVs take off from/land at) and nodes $N_2 - N_7$ represent the locations of the individual customers. Known is the demand of the individual customers for the goods transported by the UAVs, which is the same for each customer and equals 30 kg. It is assumed that the UAVs must deliver to each customer the exact quantity of goods they demand. It is additionally assumed that the goods are transported in various weather conditions, which affect the rate of battery discharge (wind speed $v_w = 10$ m/s and the direction of wind $\theta = 110°$ are both taken into account). Considering the fact that weather conditions may affect the possibility of implementing a given flight mission plan, the problem under consideration boils down to seeking the answer to the following question: *Is the given fleet of UAVs sufficiently large to cater to customers' needs (deliver the required quantity of goods) in the given transport network and within the assumed time horizon under the specific weather conditions*?

In other words what we are looking for is a proactive flight mission plan (UAV routes and schedules) which will allow a fleet of UAVs with an energy capacity limit, flying in given weather conditions, to deliver the required quantity of goods to customers.

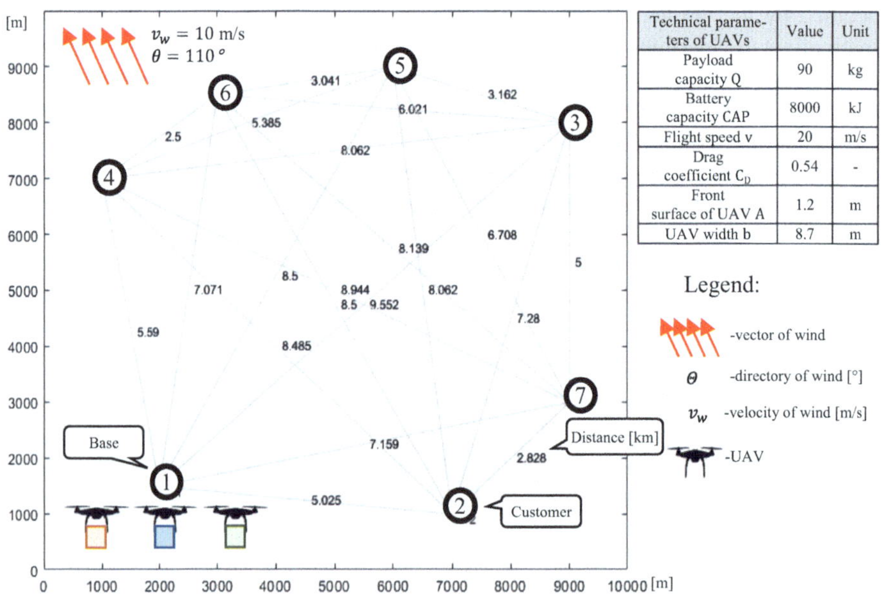


| Technical parameters of UAVs | Value | Unit |
|---|---|---|
| Payload capacity Q | 90 | kg |
| Battery capacity CAP | 8000 | kJ |
| Flight speed v | 20 | m/s |
| Drag coefficient $C_D$ | 0.54 | - |
| Front surface of UAV A | 1.2 | m |
| UAV width b | 8.7 | m |

**Fig. 1.** Transportation network

Two main strategies for delivering goods under changing weather conditions have been proposed in the literature [9, 11–13, 16]. Their principles are illustrated in Fig. 2. The first strategy (Fig. 2a) assumes that the UAV travels at a constant flight speed $v_{i,j} = 20\,\mathrm{m/s}$. In addition, it is assumed that the UAV moves along route $\pi_1 = (N_1, N_2, N_3, N_4, N_1)$ and carries 90 kg of goods. To be able to maintain the assumed flight speed for each route segment under the given weather conditions ($v_w = 10\,\mathrm{m/s}$), the UAV motor must generate thrust (speed vector $\overrightarrow{va_{i,j}}$) to compensate for the effects of wind. This results in variable energy consumption, which depends on UAV speed $\overrightarrow{va_{i,j}}$ and the weight of the freight $f_{i,j}$ transported by the UAV. Power $e_{i,j}$, which defines the amount of energy consumed along segment $(i,j)$ is described by the formula below [16]:

$$e_{i,j} = \frac{1}{2} C_D A D \left(va_{i,j}\right)^3 + \frac{\left(ep + f_{i,j}\right)^2}{D b^2 va_{i,j}}, \tag{1}$$

where: $C_D$, $A$, $D$, $b$, and $ep$ stand, respectively, for the following constant parameters: drag coefficient, front surface of UAV, air density, UAV width, and UAV weight. The parameters $f_{i,j}$ and $va_{i,j}$ are the weight of the pay load transported along segment $(i,j)$ and motor thrust, respectively.

Speed vector $\overrightarrow{va_{i,j}}$ should compensate for the effects of wind in such a way that the UAV can move between nodes at speed $20\,\mathrm{m/s}$. The value of parameter $va_{i,j} = \left|\overrightarrow{va_{i,j}}\right|$ is determined from the following relationship:

$$va_{i,j} = \sqrt{\left(v_{i,j} cos\vartheta_{i,j} - v_w cos\theta\right)^2 + \left(v_{i,j} sin\vartheta_{i,j} - v_w sin\theta\right)^2}, \tag{2}$$

where: $\vartheta_{i,j}$ – angle of inclination of the flight speed vector $\overrightarrow{v_{i,j}}$, $\theta$ – angle of inclination of the wind speed vector $\overrightarrow{v_w}$.

The approach presented graphically in Fig. 2a guarantees a constant flight speed and hence a constant flight time along a specific route ($T = 1268\,\mathrm{s}$); however, to maintain this speed it is necessary to continuously adjust the flight to the current weather conditions (energy consumption varies depending on weather conditions, in particular the strength and direction of wind). For this complex strategy, the energy consumption associated with delivering goods along route $\pi_1$ is $E = 89\%$.

The approach illustrated in Fig. 2b, in turn, assumes that engine thrust is constant throughout the mission ($va_{i,j} = 20\,\mathrm{m/s}$). This results in different flight speeds ($v_{i,j}$) for different segments of the route and a different total flight time ($T = 1605\,\mathrm{s}$). When motor thrust speed is constant, power $e_{i,j}$ (1) for each route segment is independent of weather conditions ($va_{i,j}$ is constant). The energy consumption is then dependent on flight time $t_{i,j}$ and the weight of the freight $f_{i,j}$. With this strategy, to fly the entire route $\pi_1$, the UAV has to use up $E = 72\%$ of all (pre-stored) energy.

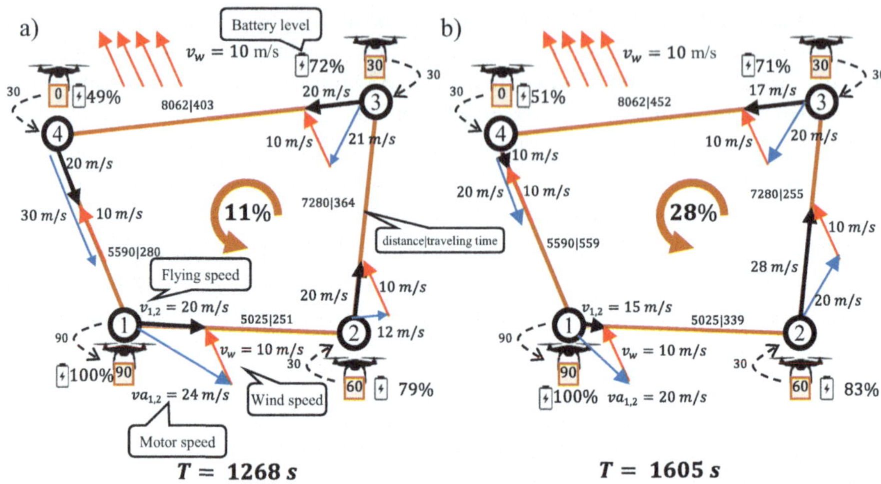


**Fig. 2.** Strategies for delivering goods following a constant flight speed (a), a constant engine thrust (b).

As it is easily seen, the second strategy ensures a lower energy consumption at the expense of a longer flight time, and it is this strategy that is widely used in flight mission planning [1, 14, 15]. A characteristic feature of the first strategy is that flight time remains constant independent of weather conditions. This feature is particularly important in situations where the goods have to be transported within specified time windows and/or when they are to be delivered to customers just-in-time. There are few publications [13, 14] regarding UAV mission planning that consider this type of delivery strategy. It is also worth emphasizing that most of the models encountered in the literature assume that the weight of the UAV does not change during flight [10, 13]. Figures 3a and b present such a situation – the total weight of the UAV remains unchanged during flight along route $\pi_1$; energy consumption is $E = 96\%$. Figures 3c and d illustrate situations in which the weight of an UAV changes as the cargo is successively unloaded at sequential delivery points (the UAV delivers 30 kg of goods each to nodes $N_2$, $N_3$, $N_4$). It should be emphasized that in this case, the energy consumption depends on the direction of flight and is $E = 89\%$ when the UAV flies counterclockwise and $E = 81\%$ when the UAV flies clockwise.

This example shows that models in which additional features, such as variability of UAV weight, have been taken into account, can be used to generate routes (and route directions) with a lower energy consumption than the models proposed so far [15, 16]. The introduction of new features, however, involves the need to take into account additional decision variables, which significantly increases the computational complexity of the problem.

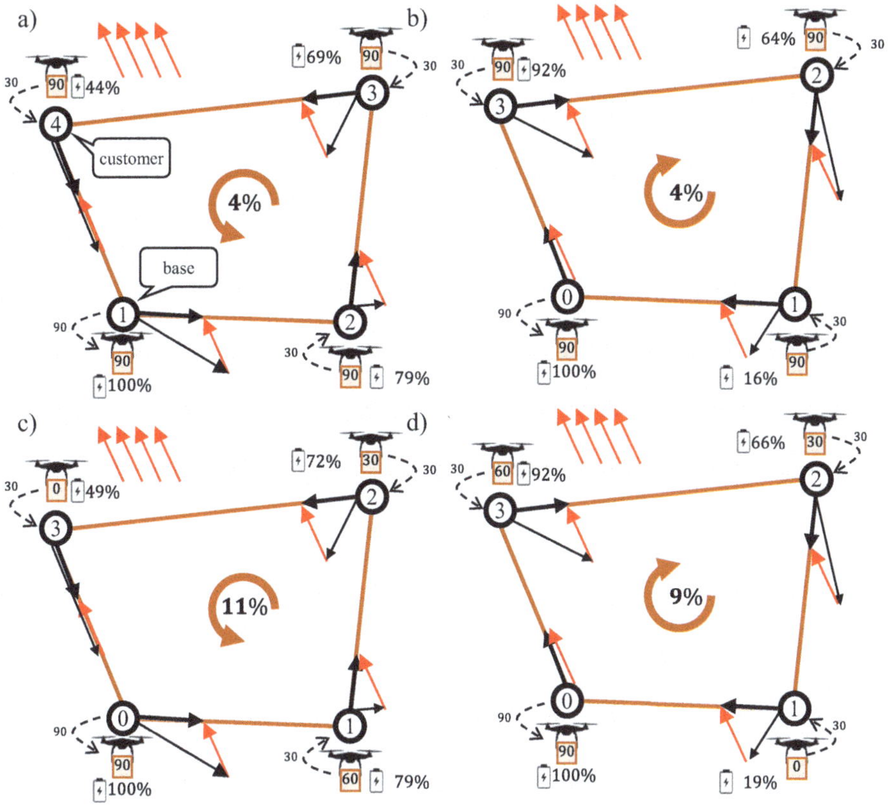


**Fig. 3.** Energy consumption under a constant UAV weight (a) and (b), and under a variable UAV weight (c) and (d)

In the next section, we present a declarative UAV mission planning model which assumes that flight speed remains unchanged throughout the mission (strategy from Fig. 2a) and that UAV weight changes during the flight (approach from Figs. 3c, d).

## 3 Declarative Model

The problem considered assumes that the structure of the goods distribution network (the number and location of customers and customer demand) is known. The goods are transported by UAVs. Also known is the time horizon in which all flights should be completed. Moreover, the following assumptions are taken into account:

- The weather conditions are known and constant (wind speed $v_w$ and direction $\theta$).
- All UAVs are in the base before the start of the delivery mission.
- The same type of freight is delivered to all customers.
- During the flight, the total weight of the UAV decreases as portions of the load are left at drop-off locations).

- The UAVs travel at a constant speed ($v_{1,2} = v_{1,3} = \cdots = v_{7,6} = 20\,\mathrm{m/s}$).

The goal is to find collision-free flight plans and routes that guarantee the highest level of customer satisfaction. The model is defined as follows:

| Parameters | |
|---|---|
| Network: | Technical parameters of the UAV fleet: |
| $I$ – number of nodes<br>$t_{i,j}$ – flight time between nodes $N_i$ and $N_j$<br>$m_i$ – demand for goods at the $i$-th node<br>$w_i$ – priority of the i-th node $i = 1..I$, $w_1 = 0$<br>$TN$ – node occupation time<br>$TS$ – time interval at which UAVs can take off from the base<br>$block_{\{i,j\};\{a,b\}}$ – binary variable corresponding to intersecting edges:<br>$\mathrm{block}_{\{i,j\};\{a,b\}} = \begin{cases} 1 & \text{when } \{i,j\} \text{ and } \{a,b\} \text{intersect} \\ 0 & \text{otherwise} \end{cases}$ | $K$ – size of the UAV fleet<br>$Q$ – payload capacity of a UAV<br>$CAP$ – UAV battery capacity<br>$C_D$ – drag coefficient<br>$A$ – front surface of a UAV<br>$b$ – UAV width<br>$ep$ – weight of an empty UAV<br>$va_{i,j}$ – motor thrust speed |
| Environmental parameters: | |
| $H$ – planning horizon $H = [0, t_{max}]$<br>$D$ – air density | $v_w$ – wind speed<br>$\Theta$ – wind direction |
| Decision variables | Sets |
| $x_{i,j}^k$ – a binary variable describing whether the k-th UAV travels from $N_i$ to $N_j$<br>$x_{i,j}^k = \begin{cases} 1 & \text{when k - th UAV travels from } N_i \text{ to } N_j \\ 0 & \text{otherwise} \end{cases}$<br>$f_{i,j}^k$ – weight of freight transported by the $k$-th UAV travelling from $N_i$ to $N_j$<br>$e_{i,j}^k$ – energy per unit of time, consumed by $k$-th UAV during a flight from $N_i$ to $N_j$<br>$s^k$ – take-off time of the $k$-th UAV<br>$y_i^k$ – time at which the $k$-th UAV arrives at $N_i$<br>$c_i^k$ – weight of freight delivered to node $N_i$ by the k-th UAV<br>$cp_i$ – total weight of freight delivered to $N_i$<br>$bat^k$ – battery level of the $k$-th UAV | $Y^k$ – set of times $y_i^k$; schedule of the k-th UAV<br>$Y$ – family $Y^k$; schedule of the UAV fleet<br>$C^k$ – set $c_i^k$; weight of freight delivered by the k-th UAV<br>$C$ – family $C^k$<br>$\Pi$ – set of UAV fleet routes |

**Constraints**

Routes. Relationships between the variables describing UAV take-off times/mission start times and task order:

$$s^k \geq 0,\ k = 1 \ldots K \tag{3}$$

$$(k \neq q) \Rightarrow \left(\left|s^k - s^q\right| \geq TS\right),\ k,\ q = 1 \ldots K \tag{4}$$

$$\sum_{j=1}^{I} x_{1,j}^{k} = 1,\ k = 1 \ldots K \tag{5}$$

$$\left(x_{1,j}^{k} = 1\right) \Rightarrow \left(y_{j}^{k} = s^{k} + t_{1,j}\right),\ j = 1 \ldots I,\ k = 1 \ldots K \tag{6}$$

$$\left(k \neq q \wedge y_{i}^{k} \neq 0 \wedge y_{i}^{q} \neq 0\right) \Rightarrow \left(\left|y_{i}^{k} - y_{i}^{k}\right| \geq TN\right),\ i = 1 \ldots I,\ k,\ q = 1 \ldots K \tag{7}$$

$$\left(x_{i,j}^{k} = 1\right) \Rightarrow \left(y_{j}^{k} = y_{i}^{k} + t_{i,j} + TN\right),\quad j = 1 \ldots I,\ i = 2 \ldots I,\ k = 1 \ldots K \tag{8}$$

$$y_{i}^{k} \geq 0,\ i = 1 \ldots I,\ k = 1 \ldots K \tag{9}$$

$$\sum_{j=1}^{I} x_{i,j}^{k} = \sum_{j=1}^{I} x_{j,i}^{k},\ i = 1 \ldots I,\ k = 1 \ldots K \tag{10}$$

$$y_{i}^{k} \leq H \times \sum_{j=1}^{I} x_{i,j}^{k} \forall i \in I,\ \forall k \in K \tag{11}$$

$$x_{i,i}^{k} = 0,\ i = 1 \ldots I,\ k = 1 \ldots K \tag{12}$$

Collision avoidance. Intersecting edges $(b_{\{i,j\};\{a,b\}} = 1)$ cannot be occupied by more than one UAV at the same time $\left(x_{i,j}^{k} = 1,\ x_{i,j}^{q} = 1\right)$.

$$\begin{gathered}\left(block_{\{i,j\}\{a,b\}} \wedge x_{i,j}^{k} = 1 \wedge x_{a,b}^{q} = 1\right) \Rightarrow \left(y_{b}^{q} \leq y_{j}^{k} - t_{i,j}\right) \vee \left(y_{j}^{k} \leq y_{b}^{q} - t_{a,b}\right) \\ i,\ j = 1 \ldots I,\ k,\ q = 1 \ldots K,\ k \neq q\end{gathered} \tag{13}$$

Delivery of freight. Relationships between the variables describing the amount of freight delivered to nodes by UAVs and the demand for goods at a given node:

$$c_{i}^{k} \geq 0,\ i = 1 \ldots I,\ k = 1 \ldots K \tag{14}$$

$$c_{i}^{k} \leq Q * \sum_{j=1}^{I} x_{i,j}^{k},\ i = 1 \ldots I,\ k = 1 \ldots K \tag{15}$$

$$\sum_{i=1}^{I} c_{i}^{k} \leq Q,\ k = 1 \ldots K \tag{16}$$

$$\left(x_{i,j}^{k} = 1\right) \Rightarrow c_{j}^{k} \geq 1,\ k = 1 \ldots K,\ i = 1 \ldots I,\ j = 2 \ldots I \tag{17}$$

$$\sum_{k=1}^{K} c_{i}^{k} = cp_{i},\ i = 1 \ldots I \tag{18}$$

$$cp_{i} \leq m_{i},\ i = 1 \ldots I \tag{19}$$

$$\sum_{i=1}^{I} c_{i}^{k} = cs^{k},\ k = 1 \ldots K \tag{20}$$

$$\left(x_{1,j}^{k}=1\right)\Rightarrow\left(fc_{j}^{k}=cs^{k}\right),\,j=1\ldots I,\,k=1\ldots K \tag{21}$$

$$\left(x_{i,j}^{k}=1\right)\Rightarrow\left(fc_{j}^{k}=fc_{i}^{k}-c_{i}^{k}\right),\,i,j=1\ldots I,\,k=1\ldots K \tag{22}$$

$$\left(x_{1,j}^{k}=1\right)\Rightarrow\left(f_{1,j}^{k}=cs^{k}\right),\,j=1\ldots I,\,k=1\ldots K \tag{23}$$

$$\left(x_{i,j}^{k}=1\right)\Rightarrow\left(f_{i,j}^{k}=fc_{j}^{k}\right),\;\;i,j=1\ldots I,\,k=1\ldots K \tag{24}$$

Battery consumption. The amount of energy needed to complete a task may not exceed the maximum battery capacity of an UAV.

$$bat^{k}\leq CAP,\,k=1\ldots K \tag{25}$$

$$\sum\nolimits_{i=1}^{I}\sum\nolimits_{j=1}^{I}x_{i,j}^{k}\times t_{i,j}\times e_{i,j}^{k},\,k=1\ldots K \tag{26}$$

Where: $e_{i,j}^{k}$ is calculated from fixed values of: $C_D$, $A$, $D$, $b$, $ep$, $v_w$, and $\theta$.

Customer satisfaction. Customer satisfaction is expressed as the sum of the product of the variables $w_i$ and $cp_i$. Customer satisfaction should be understood as the ratio of the amount of goods delivered to the demand for goods at a given node expressed as a percentage. Customer satisfaction should be higher than or equal to $CSL$.

$$\frac{\left(\sum_{i=1}^{I}w_{i}\times cp_{i}\right)}{\left(\sum_{i=1}^{I}w_{i}\times m_{i}\right)}\times 100\%\geq CSL \tag{27}$$

## 4 Problem Formulation

In the context of the proposed model, the problem under consideration can be formulated as follows: Given is a fleet of $K$ UAVs moving in a transport network consisting of $I$ nodes (where the priorities and demands for goods are known). Do there exist routes $\Pi$ in that network that guarantee customer satisfaction $CSL$ while following the constraints associated with routing and collision avoidance (3)–(13), delivery of the required quantity of goods and the reduction in UAV weight (14)–(24), and energy consumption (25)–(27)?

This problem can be considered as a Constraint Satisfaction Problem (CSP) [3]:

$$\mathrm{CSP}=(\mathcal{V},\mathcal{D},\mathcal{C}) \tag{28}$$

where: $\mathcal{V}=\{\Pi,\,Y,\,C\}$ – a set of decision variables: $\Pi$ – variables describing UAV routes, $Y$ – variables describing the schedule of UAVs fleet tasks, $C$ – variables describing the quantity of goods delivered by UAVs.

$\mathcal{D}$ – a finite set of descriptions of the decision variables,

$\mathcal{C}$ – a set of constraints describing the relationships between routes, the flight schedule and the transported loads (3)–(27).

To solve *CSP* (28), one has to determine the values of the decision variables for which all the constraints are satisfied. By implementing *CSP* (28) in a constraint programming environment, such as IBM ILOG, one can build a computational engine to answer the formulated above question.

## 5 Computational Experiments

Let's consider the network from Fig. 1, in which the weather conditions: $v_w = 10\ \frac{\text{m}}{\text{s}}$; $\Theta = 110°$ are assumed to be constant within the time window of 2000 s. Given is a fleet of three UAVs, whose parameters are described in Fig. 1. The battery capacity of each UAV is 80000 kJ. It is assumed that during the flight, the total weight of an UAV decreases as portions of the cargo are dropped-off at a delivery location and the UAV flight speed is constant at 20 m/s.

The answer to the following question is sought: *Do there exist routes $\Pi$ that can ensure 100% customer satisfaction in the given weather conditions?*

To answer this question, one must solve problem (28). This problem was implemented and solved in the declarative programming environment IBM ILOG (Intel Core i7-M4800MQ 2.7 GHz, 32 GB RAM). The solution was obtained in 723 s. Figure 4 shows the computed routes and flight schedule. A set-up like this guarantees that the demanded quantity of goods are delivered to customers in the given weather conditions. Customer satisfaction at all nodes is 100%. Battery consumption for the UAVs travelling along these routes under the given weather conditions is 41%, 45%, 76%.

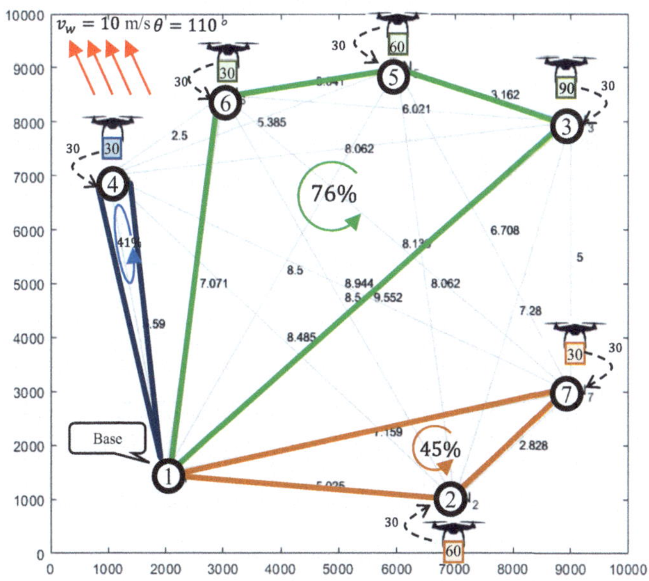


**Fig. 4.** Obtained solution

To compare, the model proposed in [13], which does not take into account the constraints associated with decreasing UAV weight (21–24), returns solutions in a shorter time of 36 s. However, the routes obtained in this way are characterized by a higher energy consumption: 46%, 50%, 81%.

In addition to the experiments reported above, we compared the effectiveness of the proposed model. The results of the analysis are shown in Table 1. The present results confirm the observation that a model that takes into account the variable weight of UAVs allows to obtain better results in terms of energy consumption; however, the time in which a solution is calculated is on average 20 times longer than for the model that assumes that UAV weight is constant. The complexity of the problem (the number of constraints and decision variables) is 2.5 times larger in this situation.

The above relationship is important from the point of view of decision support in flight mission planning. Interactive (online: $t < 600$ s) support can be provided for (1) a network consisting of no more than 8 nodes and a fleet of 2 UAVs when UAV weight is assumed to be constant throughout the mission. In practice, this means that “fast” decisions require energy consumption allowances to be made a priori. The presented results were obtained assuming that the UAVs included in the fleet always have a constant battery capacity ($CAP$ = 160 000 kJ), and the percentage energy consumption $E$ refers to the total energy of the entire fleet. The consequence of such assumptions is lower average energy consumption of one UAV for large fleets, e.g. composed of 4 UAVs, in relation to the energy consumption of small fleets, e.g. composed of 2 UAVs.

**Table 1.** Results of selected experiments

| $I$[1)] | $K$ | assumptions | $v_w = 10$ [m/s] | | | | $v_w = 15$ [m/s] | | | | $NC$ | $NDV$ |
|---|---|---|---|---|---|---|---|---|---|---|---|---|
| | | | $\theta = 30≥$ | | $\theta = 120≥$ | | $\theta = 30≥$ | | $\theta = 120≥$ | | | |
| | | | $E$ | $TC$ | $E$ | $TC$ | $E$ | $TC$ | $E$ | $TC$ | | |
| 4 | 2 | variable weightof UAV | 27,24 | >600 | 27,49 | >600 | 38,10 | >600 | 39,15 | >600 | 1004 | 3468 |
| | | constant weight of UAV | 28,96 | 4,31 | 29,62 | 4,02 | 39,70 | 3,48 | 41,24 | 3,86 | 356 | 828 |
| | 3 | variable weightof UAV | 24,14 | >600 | 21,03 | >600 | 37,38 | >600 | 29,98 | >600 | 1302 | 4438 |
| | | constant weight of UAV | 25,31 | 4,34 | 22,97 | 4 | 39,59 | 4,07 | 31,76 | 4,22 | 654 | 1774 |
| | 4 | variable weightof UAV | 22,25 | >600 | 21,68 | >600 | 32,64 | >600 | 31,00 | >600 | 1708 | 5860 |
| | | constant weight of UAV | 23,95 | 4,66 | 23,39 | 5,03 | 34,07 | 5,06 | 32,68 | 4,55 | 1036 | 3076 |
| 6 | 2 | variable weightof UAV | 35,24 | >600 | 31,95 | >600 | 51,30 | >600 | 44,39 | >600 | 3190 | 12254 |
| | | constant weight of UAV | 37,56 | 5,94 | 34,45 | 4,88 | 53,40 | 6,49 | 46,74 | 5,27 | 910 | 3014 |
| | 3 | variable weightof UAV | 32,36 | >600 | 29,77 | >600 | 47,17 | >600 | 38,46 | >600 | 4152 | 16656 |
| | | constant weight of UAV | 34,03 | 73,25 | 29,77 | 27,5 | 48,70 | 71,91 | 40,24 | 30,19 | 1902 | 7476 |
| | 4 | variable weightof UAV | 27,07 | >600 | 25,17 | >600 | 39,38 | >600 | 35,11 | >600 | 5538 | 23270 |
| | | constant weight of UAV | 28,79 | 249,49 | 26,95 | 115,35 | 40,93 | 283,02 | 37,10 | 181,15 | 3259 | 13910 |
| 8 | 2 | variable weightof UAV | 37,23 | >600 | 34,89 | >600 | 54,91 | >600 | 52,00 | >600 | 7704 | 31312 |
| | | constant weight of UAV | 40,00 | 60,44 | 37,84 | 27,25 | 57,19 | 112,94 | 52,67 | 13,32 | 2104 | 8688 |
| | 3 | variable weightof UAV | 31,96 | >600 | 32,99 | >600 | 54,23 | >600 | 65,43 | >600 | 10470 | 45818 |
| | | constant weight of UAV | 34,79 | >600 | 36,18 | >600 | 51,64 | >600 | 71,43 | 188,39 | 4926 | 23306 |
| | 4 | variable weightof UAV | 26,06 | >600 | 26,46 | >600 | 41,43 | >600 | 37,66 | >600 | 14760 | 68520 |
| | | constant weight of UAV | 28,12 | >600 | 29,14 | >600 | 43,43 | >600 | 40,46 | 212,04 | 8924 | 44776 |
| 10 | 2 | variable weightof UAV | 42,49 | >600 | 39,78 | >600 | 71,49 | >600 | 55,41 | >600 | 15810 | 66402 |
| | | constant weight of UAV | 46,44 | 177,35 | 42,67 | 57,03 | 76,07 | >600 | 58,20 | 230,46 | 4290 | 19962 |
| | 3 | variable weightof UAV | 37,20 | >600 | 33,12 | >600 | 61,84 | >600 | 43,15 | >600 | 22109 | 101500 |
| | | constant weight of UAV | 39,38 | >600 | 35,54 | >600 | 64,92 | >600 | 45,45 | 293,44 | 10782 | 55600 |
| | 4 | variable weightof UAV | 37,14 | >600 | 33,89 | >600 | 58,02 | >600 | 45,73 | >600 | 31702 | 155146 |
| | | constant weight of UAV | 39,31 | >600 | 36,64 | >600 | 60,10 | >600 | 48,02 | >600 | 20182 | 108346 |

[1)] $I$ – number of nodes; $K$ – size of the UAV fleet; $E$ –total consumed energy [%]; $TC$ – time of computation [s]; $NC$- number of constraints; $NDV$ – number of decision variables

The differences between the amount of energy consumed under the assumption of constant/variable weight UAV range from 1% to 5% and increase with the size of the transport network. The sensitivity of energy consumption $E$ due to wind direction ($\Theta = 30°$, $\Theta = 120°$) is on average 5% points and in extreme cases reaches 19% points. Much greater sensitivity of energy consumption $E$ is observed in situations related to changes in wind speed ($v_w = 10$, $v_w = 15$ [m/s]). Changes in energy consumption are on average 15% points and in extreme situations they are even 30% points, e.g. the network of 10 nodes and 2 UAVs (see marked in yellow cells distinguished in the Table 1).

## 6 Conclusions

Because fleet sizes are limited and UAV flight range is constrained by battery capacity, large-scale vehicle routing problems, which are an essential part of adequate UAV mission planning, are best formalized mathematically as specific multicriteria knapsack problems. Since knapsack problems are NP-hard, their solutions in real-life cases are only approximate. This means that approximate calculation techniques derived from artificial intelligence methods have to be used. Because mission planning problems have a declarative representation, the techniques which are based on a constraints programming paradigm are best suited to solving those problems.

In a model that takes into account changing weather conditions, routes must be determined considering changes in wind direction and strength. The process of designing relevant proactive plans for UAV fleet missions boils down to dividing the planned time horizon into time windows corresponding to unchanging weather conditions, as predicted in a weather forecast.

The computer experiments performed in the present study confirmed the efficiency of the proactive and knapsack-problem-like modelling concepts implemented in UAV mission planning. Two policies aimed at minimizing total travel time at the cost of saving battery power were considered. Special attention was paid to research focused on the sensitivity of energy consumption due to wind speed and direction and changes.

In our future research on resistant UAV mission planning we plan to explore the relationships linking the total distance traveled with the total travel time and the cost of saving battery power of an UAV fleet. Particular attention will be paid to the pick-up delivery problem with time windows and to planning the size of fleets composed of non-identical UAVs.

## References

1. Amorosi, L., Chiaraviglio, L., D'Andreagiovanni, F., Blefari-Melazzi, N.: Energy-efficient mission planning of UAVs for 5G coverage in rural zones. In: Conference: IEEE International Conference on Environmental Engineering, 12–14 March 2018 (2018). https://doi.org/10.1109/ee1.2018.8385250
2. Berntzen, L., Florea, A., Molder, C., Bouhmala, N.: A strategy for drone traffic planning dynamic flight-paths for drones in smart cities. In: The Eighth International Conference on Smart Cities, Systems, Devices and Technologies, pp. 12–18 (2019)

3. Boukoberine, M.N., Zhou, Z., Benbouzid, M.: A critical review on unmanned aerial vehicles power supply and energy management: solutions, strategies, and prospects. Appl. Energy **255**, 1 (2019). https://doi.org/10.1016/j.apenergy.2019.113823
4. Chaari, T., Chaabane, S., Aissani, N., Trentesaux, D.: Scheduling under uncertainty: survey and research directions. In: IEEE 3rd International Conference on Advanced Logistics and Transport, pp. 229–234 (2014). https://doi.org/10.1109/icadlt.2014.6866316
5. Evers, L., Dollevoet, T., Barros, A.I., Monsuur, H.: Robust UAV mission planning. Ann. Oper. Res. **222**(1), 1–17 (2011). https://doi.org/10.1007/s10479-012-1261-8
6. Guerriero, F., Surace, R., Loscrí, V., Natalizio, E.: A multi-objective approach for unmanned aerial vehicle routing problem with soft time windows constraints. Appl. Math. Model. **38**(3), 839–852 (2014). https://doi.org/10.1016/j.apm.2013.07.002
7. Khoufi, I., Laouiti, A., Adjih, C.: A survey of recent extended variants of the traveling salesman and vehicle routing problems for unmanned aerial vehicles. Drones **3**(3), 66 (2019). https://doi.org/10.3390/drones3030066. www.mdpi.com/journal/drones
8. Kim, S.J., Lim, G.J., Cho, J.: Drone flight scheduling under uncertainty on battery duration and air temperature. Comput. Ind. Eng. (2018). https://doi.org/10.1016/j.cie.2018.02.005
9. Luo, H., Liang, Z., Zhu, M., Hu, X., Wang, G.: Integrated optimization of unmanned aerial vehicle task allocation and path planning under steady wind. PLoS ONE **13**(3), e0194690 (2018). https://doi.org/10.1371/journal.pone.0194690
10. Radzki, G., Thibbotuwawa, A., Bocewicz, G.: UAVs flight routes optimization in changing weather conditions – constraint programming approach. Appl. Comput. Sci. **15**(3), 5–20 (2019). https://doi.org/10.23743/acs-2019-17
11. Rucco, A., Aguiar, A.P., Pereira, F.L., Sousa, J.B.D.: A predictive path-following approach for fixed-wing unmanned aerial vehicles in presence of wind disturbances. Advances in Intelligent Systems and Computing, pp. 623–634 (2016)
12. Techy, L., Woolsey, C.A.: Minimum-time path planning for unmanned aerial vehicles in steady uniform winds. J. Guid. Control Dyn. **32**(6), 1736–1746 (2009)
13. Thibbotuwawa, A., Bocewicz, G., Nielsen, P., Banaszak, Z.: Planning deliveries with UAV routing under weather forecast and energy consumption constraints. IFAC-PapersOnLine **52**(13), 820–825 (2019). https://doi.org/10.1016/j.ifacol.2019.11.231
14. Thibbotuwawa, A., Nielsen, P.: Unmanned aerial vehicle routing problems: a literature review. J. Logist. Res. (submitted, under review)
15. Thibbotuwawa, A., Nielsen, P., Banaszak, Z., Bocewicz, G.: Factors affecting energy consumption of unmanned aerial vehicles: an analysis of how energy consumption changes in relation to UAV routing. In: Advances in Intelligent Systems and Computing, pp. 228–238. Springer (2018)
16. Thibbotuwawa, A., Nielsen, P., Zbigniew, B., Bocewicz, G.: Energy consumption in unmanned aerial vehicles: a review of energy consumption models and their relation to the UAV routing. In: Świątek, J., Borzemski, L., Wilimowska, Z. (eds.) Information Systems Architecture and Technology: Proceedings of 39th International Conference on Information Systems Architecture and Technology – ISAT 2018. Advances in Intelligent Systems and Computing. Springer, Cham (2019)

# Monte-Carlo Based Lateral Thruster Parameters Optimization for 122 mm Rocket

Mariusz Jacewicz, Robert Glebocki, and Rafal Ozog(✉)

Division of Mechanics, Faculty of Power and Aeronautical Engineering, Warsaw University of Technology, Nowowiejska 24, 00-665 Warsaw, Poland
{mjacewicz, rglebocki, rozog}@meil.pw.edu.pl

**Abstract.** One of the current tendencies is to equip the unguided munition designed for ballistic shooting with low cost, solid propellant lateral thrusters based, actuators to achieve the controlled flight functionality and reduce the collateral damage. The basic technical challenge connected with this type of pulsed control system is that each from the thrusters could be used only once which results in projectile low control authority. The thruster parameters have crucial impact on the achieved hit accuracy. The main goal of this article was to investigate and understand the influence of control force magnitude on the guidance process using six-degree-of-freedom numerical simulation. 122 mm artillery rocket controlled with the aim of 30 solid propellant thrusters mounted before center of mass was chosen as a test object. Single channel control was considered. The impact point prediction algorithm based on point mass model was developed and implemented into a Matlab software. Using Monte-Carlo simulations the optimum lateral thruster force amplitude was obtained for shots at low elevation angle. The numerical experiments showed that with the proposed method the circular error probable of the projectile might be reduced 10 times when compared to unguided case.

**Keywords:** Monte-Carlo · Lateral thruster · Impact point prediction guidance

## 1 Introduction

One of the modern requirements is to improve the delivery accuracy of the existing unguided munitions. The object of interest of this work is concentrated on rocket artillery. The main advantages of this kind of weapon are: long range, high firepower and high mobility of the launcher [1]. The main disadvantage is high dispersion of the impact points which makes its area weapon. There exists a tendency to equip the unguided projectiles with a various kind of actuators like body-fixed canards [2–4], roll decoupled fins [5], lateral thrusters [6, 7], foldable drag plates and movable internal parts to achieve the controlled flight functionality. Using this approach, the collateral damage and the number of projectiles required to eliminate the target could be reduced.

On the other hand, surface to surface rockets have low control authority which limits the amount of possible control strategies. The guidance methods designed for guided munition fall into three main categories [5]: reference trajectory tracking, impact point prediction and trajectory shaping. The main requirements for control


R. Szewczyk et al. (Eds.): AUTOMATION 2020, AISC 1140, pp. 125–134, 2020.
https://doi.org/10.1007/978-3-030-40971-5_12

algorithms intended for smart weapons are the guidance accuracy and low computational burden due to limited resources available onboard of the rocket. The trajectory tracking method [8] relies on flight along prespecified ballistic curve. The path definition is preloaded onto memory of the onboard computer before the projectile launch. The advantage of this method is its low cost. The predictive guidance [9, 10] uses a model of the projectile implemented into a control system to obtain the impact point location coordinates. In this method the control effort rely on driving the impact point to the target. The trajectory shaping algorithm is based on the following non-ballistics trajectories [11]. This method allows to for e.g. extend the range and control the impact angle of the projectile but the main drawback of this approach is that generally requires more control authority than predictive guidance.

The main goal of the article was to further extend the study on pulse thrusters to minimize the dispersion of the rocket and obtain the most suitable thrust force amplitude of the single lateral thruster when the predictive guidance is applied.

## 2 Rocket Mathematical Model

### 2.1 The Test Platform

The analyzed object (see Fig. 1) is 122 mm in caliber, 2.6 m long rocket with four wrap-around fins. The projectile spins slowly (<30 Hz) during the flight due to small fin cant angle.

**Fig. 1.** Surface to surface lateral thruster steered 122 mm rocket.

The parameters before/after main motor burnout was assumed to be: center of mass location $x_{cg0} = 1.446/x_{cgk} = 1.645$ m from the rocket base, mass $m_0 = 64.75/m_k = 38.5$ kg, axial moments of inertia $I_{x0} = 0.137/I_{xk} = 0.092$ kgm$^2$ and lateral moments of inertia $I_{y0} = I_{z0} = 38.80/I_{yk} = I_{zk} = 29.16$ kgm$^2$. In the basic unguided configuration, the rocket parameters were designed for ballistic flight.

The functionality of controlled flight was obtained with equipping the rocket with a guidance unit composed from power supply, inertial navigation system, onboard computer and a set of 30 small solid propellant lateral thrusters mounted before the center of mass and located 1.9 m from the rocket base. The nozzles of thrusters are spaced equally around the fuselage. In this solution the thrusters generate not only force perpendicular to the rocket longitudinal axis but also the moment around its center of mass. The main difficulty connected with this type of actuator is the fact that each thruster could be fired only once. The limited energy stored in the grain of the thrusters and limited number of firings provides very low control authority of the rocket. It means that the achievable lateral accelerations are much lower than in case of high maneuver air-to-air missiles and the resulting the impact point could be translated only by several hundred meters when comparing to ballistic, unguided flight. To overcome this difficulty the appropriate control algorithm must be designed.

## 2.2 Model of the Rocket

The rocket was modeled as a six degree-of-freedom (6DoF) rigid body with variable mass and the center of mass location. Flat Earth approximation was applied. The equations of motion of the rocket were as follows [12]:

$$\begin{bmatrix} \dot{U} \\ \dot{V} \\ \dot{W} \end{bmatrix} = \frac{1}{\mathrm{m}} \begin{bmatrix} X_b \\ Y_b \\ Z_b \end{bmatrix} - \begin{bmatrix} 0 & -R & Q \\ R & 0 & -P \\ -Q & P & 0 \end{bmatrix} \begin{bmatrix} U \\ V \\ W \end{bmatrix} \tag{1}$$

$$\begin{bmatrix} \dot{P} \\ \dot{Q} \\ \dot{R} \end{bmatrix} = \begin{bmatrix} I_x & 0 & 0 \\ 0 & I_y & 0 \\ 0 & 0 & I_z \end{bmatrix}^{-1} \begin{bmatrix} L_b \\ M_b \\ N_b \end{bmatrix} - \begin{bmatrix} I_x & 0 & 0 \\ 0 & I_y & 0 \\ 0 & 0 & I_z \end{bmatrix}^{-1} \begin{bmatrix} 0 & -R & Q \\ R & 0 & -P \\ -Q & P & 0 \end{bmatrix} \begin{bmatrix} I_x & 0 & 0 \\ 0 & I_y & 0 \\ 0 & 0 & I_z \end{bmatrix} \begin{bmatrix} P \\ Q \\ R \end{bmatrix} \tag{2}$$

$$\begin{bmatrix} \dot{x}_n \\ \dot{y}_n \\ z_n \end{bmatrix} = \begin{bmatrix} \cos\Theta\cos\Psi & \sin\Phi\sin\Theta\cos\Psi - \cos\Phi\sin\Psi & \cos\Phi\sin\Theta\cos\Psi + \sin\Phi\sin\Psi \\ \cos\Theta\sin\Psi & \sin\Phi\sin\Theta\sin\Psi + \cos\Phi\cos\Psi & \cos\Phi\sin\Theta\sin\Psi - \sin\Phi\cos\Psi \\ -\sin\Theta & \sin\Phi\cos\Theta & \cos\Phi\cos\Theta \end{bmatrix} \begin{bmatrix} U \\ V \\ W \end{bmatrix} \tag{3}$$

$$\begin{bmatrix} \dot{\Phi} \\ \dot{\Theta} \\ \dot{\Psi} \end{bmatrix} = \begin{bmatrix} 1 & \sin\Phi\tan\Theta & \cos\Phi\tan\Theta \\ 0 & \cos\Phi & -\sin\Phi \\ 0 & \sin\Phi/\cos\Theta & \cos\Phi/\cos\Theta \end{bmatrix} \begin{bmatrix} P \\ Q \\ R \end{bmatrix} \tag{4}$$

where $U$, $V$, $W$ are rocket linear velocities in body-fixed coordinate frame, $P$, .., $R$ are rocket roll, pitch and yaw rates, $\Phi$, $\Theta$, $\Psi$ are Euler angles (roll, pitch and yaw), $x_n$, $y_n$, $z_n$ are position coordinates of the rocket center of mass in the north-east-down frame. The $m$ is projectile mass. $I_x$ is the longitudinal and $I_y$, $I_z$ lateral moments of inertia. The dot symbol above variables at left side of abovementioned equations is a first derivative with respect to time. $X_b$, $Y_b$, $Z_b$ are sum of gravity, propulsive, aerodynamic and control forces and moments in body fixed frame [12]. In similar way, $L_b$, $N_b$, $Z_b$ are moments acting on the rocket body.

The main motor thrust force was obtained from the ground tests. The operation time of the motor was 3 s and the total impulse was 60000 Ns. Thrust force misalignment angles in pitch $\Theta_T$ and yaw planes $\Psi_T$ were included into the model.

The aerodynamic characteristics of the rocket were obtained with the aim of theoretical and engineering, semi-empirical methods [13, 14]. The nondimensional forces and moments coefficients were implemented as a function of Mach number, angles of attack and sideslip into the software using lookup table methodology. Base drag variations due to main motor state (powered/unpowered phase of flight) were included. Air thermodynamic properties were used according to U.S. Standard Atmosphere [15].

## 3 Impact Point Prediction Guidance

At first the firing logic of the thrusters was developed. The thrusters were assumed to be fired according to scheme presented in Fig. 2a. Using the single channel control the actuation system generates forces only over a part of single roll cycle (Fig. 2b). In result, the lateral thruster must be fired at appropriate roll angle.

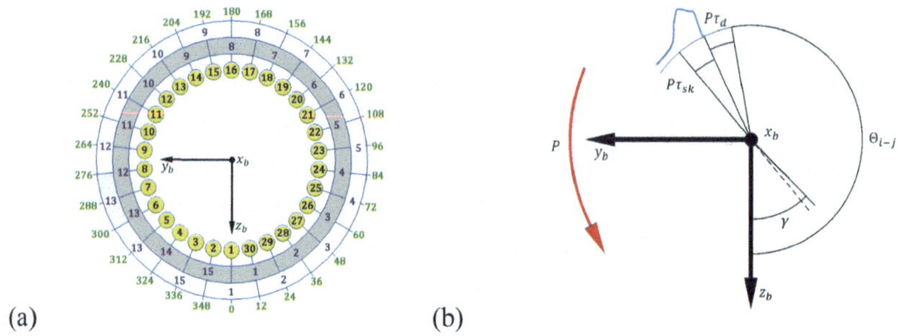

**Fig. 2.** (a) Lateral thrusters firing sequence (b) control force direction as a function of roll angle

The thrusters firings conditions were formulated as follows [12, 16, 17]:

- the lateral thruster was not already consumed
- the time $t$, from the last thruster firing which has occurred at time $t_{last}$ is longer than $\tau \in (0; \infty)$

$$t - t_{last} > \tau \tag{5}$$

- lateral thruster must be fired when is at opposite side than desired flight direction

$$|\gamma - \Phi_i - \pi - P(\tau_d + \tau_{sk})| < \varepsilon \tag{6}$$

where $\gamma$ is angle defined in Fig. 2b, is the lateral thruster location angle $\Phi_i$. $\tau_d$ is ignition delay (≈0.001 s) and $\tau_{sk}$ is half time of the lateral pulse operation time. $\varepsilon$ is angular tolerance of firing error and it was chosen to be 2°.
- the pitch angle $\Theta$ of the rocket is smaller than $\Theta_g$ or the time from launch $t$ is greater than $t_g$

$$\Theta \leq \Theta_g \bigwedge t \geq t_g \tag{7}$$

- impulses are generated only when

$$\sqrt{(x_t - x_p)^2 + (y_t - y_p)^2} \geq r_{2tt} \tag{8}$$

where $x_t$, $y_t$ are target coordinates, $x_p$, $y_t$ are predicted impact point location and $r_{2tt}$ is a small threshold (e.g. 1 m) responsible for guidance accuracy. It prevents from the undesired consumption of thrusters when the hit error is relatively small (below several meters) from practical point of view.

Various models could be used to determine the impact point location [9]. Linear theory based predictive guidance was investigated in [18]. It uses analytical closed form solution to evaluate the future projectile states. To extend the capabilities of this approach, modified projectile linear theory intended for fast calculation of the impact point when projectile is launched at high elevation angles was presented in [19]. The basic constraint connected with the predictive control is the limited computational power of the onboard computer. The simple point mass model was used to calculate the impact point of the rocket. The equations of motion in the navigation $O_n x_n y_n z_n$ frame were as follows [20]:

$$\frac{dx_n}{dt} = U_n \tag{9}$$

$$\frac{dy_n}{dt} = V_n \tag{10}$$

$$\frac{dz_n}{dt} = W_n \tag{11}$$

$$\frac{dU_n}{dt} = -\frac{\pi \rho d^2 C_{X0}}{8m} \sqrt{U_n^2 + V_n^2 + W_n^2} U_n \tag{12}$$

$$\frac{dV_n}{dt} = -\frac{\pi \rho d^2 C_{X0}}{8m} \sqrt{U_n^2 + V_n^2 + W_n^2} V_n \tag{13}$$

$$\frac{dW_n}{dt} = -\frac{\pi \rho d^2 C_{X0}}{8m} \sqrt{U_n^2 + V_n^2 + W_n^2} W_n + g \tag{14}$$

where $U_n$, $V_n$, $W_n$ are velocities with respect to navigation frame. The air density $\rho$ was assumed to be the same as in 6 DOF model and $g$ is gravity acceleration. Drag coefficient $C_{X0}$ as a function of Mach number for unpowered phase of flight was used. The equations of motion are numerically integrated during the rocket flight.

When the target location $x_t$, $y_t$ and estimated impact point coordinates $x_p$, $y_p$ are known the range and crossrange errors were computed [5, 21, 22]:

$$\Delta x = x_p - x_t \qquad \Delta y = y_p - y_t \tag{15}$$

The angle of correction $\gamma$ was calculated as [22–24]:

$$\gamma_p = mod\left(-atan2\frac{\Delta y}{\Delta x}, 2\pi\right) \tag{16}$$

No guidance process takes place in the ascending portion of trajectory because the roll rate of the rocket is too high. The controlled flight starts after vertex of the projectile trajectory and last to the rocket touchdown. The developed mathematical model of the rocket with the guidance algorithm was implemented in the MATLAB 2018a software. The equations of motion were integrated numerically using ode4 algorithm with fixed step size 1e-4 s. In the simulation the impact points were predicted with the frequency 10 Hz.

## 4 Results and Discussion

Monte-Carlo methodology and the developed mathematical model of the rocket were used to obtain the dispersion patterns. The rocket dispersion is a result of combination of a various errors such as initial aiming inaccuracies, motor thrust misalignments, launcher vibrations and wind disturbances, which influence the location of impact point. The thrust force of the single thruster was modelled as a rectangular pulse. The operation time of the single thruster was fixed and assumed to be 0.02 s. Other predictive algorithm parameters were: $\tau = 0.5$ s, $\Theta_g = 0°$, $t_g = 7$ s and $r_{2tt} = 1$ m. To investigate the influence of the lateral thruster parameters on rocket delivery accuracy a series of simulations for various force magnitudes was evaluated. The simulation inputs (Table 1) were assumed to be an Gaussian random independent variables pseudo-generated with the aim of Marsenne-Twister algorithm. The standard deviations were assumed using data from [6, 10, 17].

**Table 1.** Monte-Carlo parameters used in simulation.

| Parameter | Mean value | Standard deviation | Unit |
|---|---|---|---|
| $m_0$ | 64.75 | 0.1 | kg |
| $m_k$ | 38.5 | 0.1 | kg |
| $I_{x0}$ | 0.137 | 0.005 | kgm$^2$ |
| $I_{xk}$ | 0.092 | 0.005 | kgm$^2$ |
| $U_0$ | 32 | 1 | m/s |
| $V_0$ | 1 | 0.2 | m/s |
| $W_0$ | 3 | 0.2 | m/s |
| $P_0$ | 1350 | 4 | °/s |
| $Q_0$ | −6 | 0.2 | °/s |
| $R_0$ | 0 | 0.2 | °/s |
| $\Phi_0$ | 0 | 2 | ° |
| $\Theta_0$ | 20 | 0.1 | ° |
| $\Psi_0$ | 0 | 0.2 | ° |
| $x_n$ | 0 | 0.03 | m |
| $y_n$ | 0 | 0.05 | m |
| $z_n$ | 0 | 0.05 | m |
| $\Theta_T$ | 0 | 0.01 | ° |
| $\Psi_T$ | 0 | 0.01 | ° |

In the each case the 400 simulations was evaluated using desktop PC with Intel i7 and 16 GB RAM. Circular error probable (CEP) was used as a measure of the rocket impact

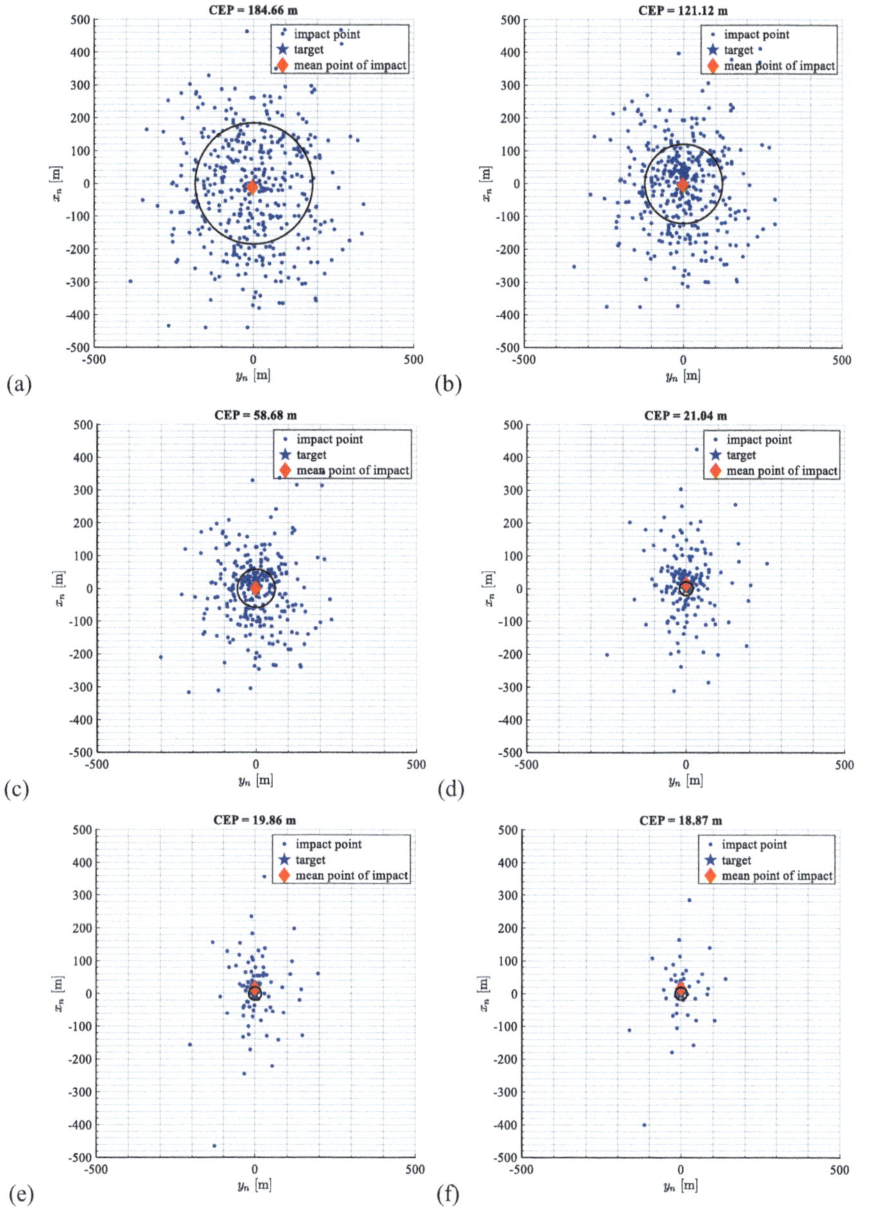


**Fig. 3.** The impact point patterns (a) unguided flight and guided flights for thrust force magnitudes: (b) 200 N (c) 400 N (d) 600 N (e) 800 N (f) 1000 N.

points dispersion. CEP is a radius of circle centered on the target which include 50% of the impact points. At first, the unguided case was considered. Next, the simulations for guided flights were evaluated. The obtained hitting patterns were presented in Fig. 3.

For unguided flight the lateral dispersion is bigger than longitudinal because of low launch elevation angle and the CEP is 184.66 m. When the impact point guidance was applied the CEP values were reduced rapidly. The smallest dispersion was obtained for the thrust amplitude 1000 N. No bias error in mean impact point location was introduced when the control algorithm was used.

The obtained hitting errors were presented in the form of histograms in Fig. 4.

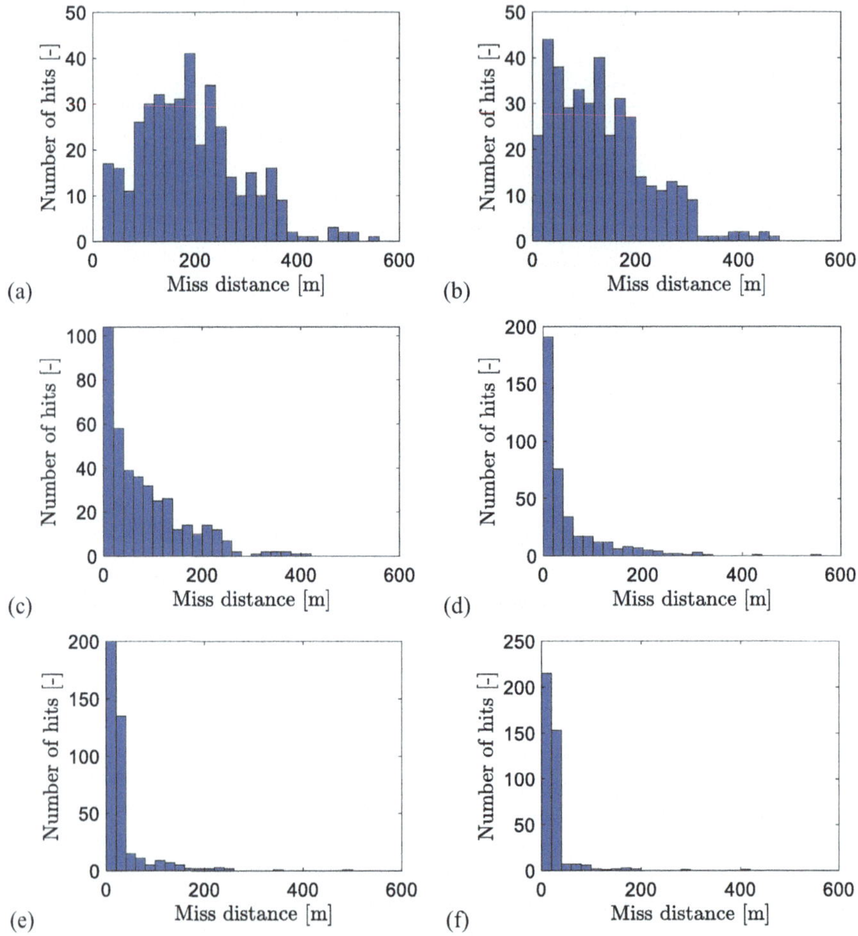


**Fig. 4.** The rocket miss distance distributions (a) unguided flight and guided flights for thrust force magnitudes: (b) 200 N (c) 400 N (d) 600 N (e) 800 N (f) 1000 N.

For the unguided case the miss distances forms a shape similar to Gauss curve. Several rockets achieved error more than 500 m. For subsequent guided cases a large number of hit errors below several meters was obtained.

In the Fig. 5 the CEP (vertical axis in logarithmic scale) as a function of lateral thruster force magnitude was presented.

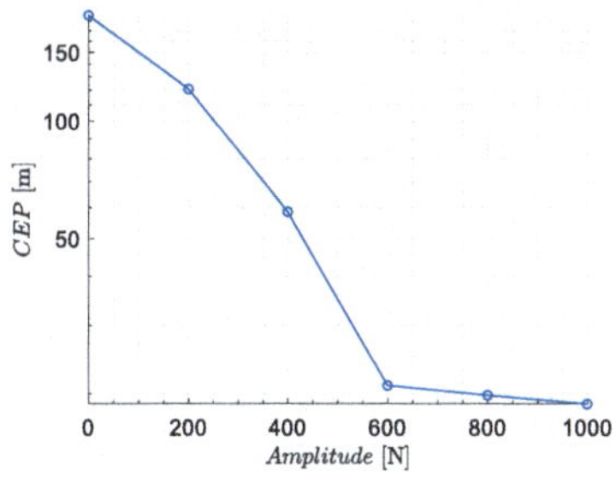


**Fig. 5.** Circular error probable of the rocket for various thrust force magnitudes

The obtained relationship shows that for values larger than 600 N the profit from increasing the amplitude is negligible. The obtained results proved that with the aim of carefully selected lateral thruster parameters and a suitable guidance law, the accuracy of the projectile might be improved significantly.

## 5 Conclusion

In this article the influence of solid propellant lateral motor side force magnitude on the dispersion of 122 mm artillery projectile was presented. 6DoF mathematical model in Matlab was used to investigate behavior of the rocket. The characteristic feature of lateral thrusters controlled projectile is its low control authority. Predictive algorithm based on point mass model was used to steer the rocket precisely to the target. Using the calculated impact error, the control is applied to reduce the predicted miss distance. Even relatively small 200 N side force allows to reduce the dispersion significantly. Numerical Monte-Carlo simulations proved, that with the presented approach the circular error probable could be reduced approximately ten times when compared to unguided rocket.

**Acknowledgements.** This work was supported by The National Centre for Research and Development (NCBiR) under project DOB-BIO8/10/01/2016 "Projectiles control system technology development".

## References

1. Pavković, B., Pavić, M., Ćuk, D.: Enhancing the precision of artillery rockets using pulsejet control systems with active damping. Sci. Tech. Rev. **62**(2), 10–19 (2012)
2. Gamble, A., Jenkins, P.: Low cost guidance for the multiple launch rocket system (MLRS) artillery rocket. IEEE Aerosp. Electron. Syst. Mag. **16**(1), 33–39 (2001)
3. Mandić, S.: Dispersion reduction of artillery rockets guided by flight path steering method. Aeronaut. J. **120**(1225), 435–456 (2016)

4. Guo, Q.-W., Song, W.-D., Gao, M., Fang, D.: Advanced guidance law design for trajectory-corrected rockets with canards under single channel control. Eng. Lett. **24**(4), 469–477 (2016)
5. Gagnon, E., Vachon, A.: Efficiency analysis of canards-based course correction fuze for a 155-mm spin-stabilized projectile. J. Aerosp. Eng. **29**(6), 1–10 (2016)
6. Pavkovic, B., Pavic, M., Cuk, D.: Frequency-modulated pulse-jet control of an artillery rocket. J. Spacecr. Rockets **49**(2), 286–294 (2012)
7. Drescher, T., Nielson, J.: Rocket trajectory correction using strap-on GPS guided thrusters. In: IEEE 1998 Position Location and Navigation Symposium (Cat. No. 98CH36153), Palm Springs (1998)
8. Jitpraphai, T., Burchett, B., Costello, M.: A comparison of different guidance schemes for a direct fire rocket with a pulse jet control mechanism. In: AIAA Atmospheric Flight Mechanics Conference and Exhibit, Guidance, Navigation, and Control and Co-located Conferences (2001)
9. Fresconi, F., Cooper, G., Costello, M.: Practical assessment of real-time impact point estimators for smart weapons. J. Aerosp. Eng. **24**(1), 1–11 (2011)
10. Gross, M., Costello, M., Fresconi, F.: Impact point model predictive control of a spin-stabilized projectile with instability protection. In: AIAA Atmospheric Flight Mechanics (AFM) Conference, Boston (2013)
11. Gagnon, E., Lauzon, M.: Course correction fuze concept analysis for in-service 155 mm spin-stabilized gunnery projectiles. In: AIAA Guidance, Navigation and Control Conference and Exhibit, Honolulu (2008)
12. Jitpraphai, T., Costello, M.: Dispersion reduction of a direct-fire rocket using lateral pulse jets. J. Spacecr. Rockets **38**(6), 929–936 (2001)
13. Krasnov, N.F.: Rocket Aerodynamics (NASA Technical Translation). National Aeronautics and Space Administration (1971)
14. Department of Defense: MIL-HDBK-762 Design of Aerodynamically Stabilized Free Rockets. United States Department of Defense (1990)
15. National Aeronautics and Space Administration: U.S. Standard Atmosphere 1976, Washington, D.C. (1976)
16. Gupta, K.: Trajectory correction flight control system using pulsejet on an artillery rocket. Def. Sci. J. **58**(1), 15–33 (2008)
17. Gao, M., Zhang, Y., Yang, S.: Firing control optimization of impulse thrusters for trajectory correction projectiles. Int. J. Aerosp. Eng. **2015**, 1–11 (2015)
18. Slegers, N.: Predictive control of a munition using low-speed linear theory. J. Guid. Control Dyn. **31**(3), 768–775 (2008)
19. Hainz, L., Costello, M.: Modified projectile linear theory for rapid trajectory prediction. J. Guid. Control Dyn. **28**(5), 1006–1014 (2005)
20. Wang, Z., Chang, S.: Impact point prediction and analysis of lateral correction analysis of two-dimensional trajectory correction projectiles. Def. Technol. **9**(1), 48–52 (2013)
21. Zhang, Y., Gao, M., Yang, S., Fang, D.: Optimization of trajectory correction scheme for guided mortar projectiles. Int. J. Aerosp. Eng. **2015**, 1–14 (2015)
22. Park, W., Yun, J., Ryoo, C.-K., Kim, Y.: Guidance law for a modern munition. In: International Conference on Control, Automation and Systems 2010, Gyeonggi-do (2010)
23. Fresconi, F.: Guidance and control of a projectile with reduced sensor and actuator requirements. J. Guid. Control Dyn. **34**(6), 1757–1766 (2011)
24. Burchett, B., Costello, M.: Model predictive lateral pulse jet control of an atmospheric rocket. J. Guid. Control Dyn. **25**(5), 860–867 (2002)

# Tuning the Fractional Order PID Controller in the Forced Air Heating System Using Biologically Inspired algorithms

Klaudia Dziedzic(✉)

AGH University of Science and Technology, Krakow, Poland
kdz@agh.edu.pl

**Abstract.** The main objective of this research is to present the tuning of the fractional order PID controller for the forced air heating system using the particle swarm optimization algorithm (PSO), cockroach swarm optimization algorithm (CSO), grey wolf optimizer algorithm (GWO). In preliminaries, fractional calculus is discussed. Then, all three biological algorithms are briefly presented. Obtained simulation results allow comparison of individual algorithms in terms of overshoot, settling time and performance criteria (IAE, ITAE).

**Keywords:** Fractional order systems · FOPID · PSO algorithm · CSO algorithm · GWO algorithm

## 1 Introduction

Proportional Integral Derivative controllers (PID) are widely used in the industrial process, because of its simplicity of implementation. However, it is worth taking care of improving their quality and robustness. That is why, in 1999, the Fractional Order PID (FOPID) controller was proposed by Podlubny in [12]. The use of the fractional calculus in automation has recently been very popular. These methods are applied in the identification processes or in the control of various kinds of physical systems [6,7,11,13]. The controller proposed by Podlubny has two additional parameters the order of fractional integration $\alpha$ and the order of fractional derivative $\beta$. Therefore, it has five parameters that are made the FOPID more flexible.

To achieve acceptable performance it must be found an optimal set of values for $Kp$, $Ki$, $Kd$, $\alpha$ and $\beta$. There are many optimization methods to find the optimal solution. Recently, the development of an evolutionary computation technique, mostly inspired by the behavior of animals, is increased [4]. They are based on herd behavior and have some characteristics such as simply implementation of their concept and a well-balanced mechanism to enhance global exploration abilities. They are based on herd behavior and have some characteristics such as simply implementation of their concept and a well-balanced


R. Szewczyk et al. (Eds.): AUTOMATION 2020, AISC 1140, pp. 135–146, 2020.
https://doi.org/10.1007/978-3-030-40971-5_13

mechanism to enhance global exploration abilities. In this article, there are discussed three herd algorithms: the particle swarm optimization algorithm (PSO), the cockroach swarm optimization algorithm (CSO) and the grey wolf optimizer algorithm (GWO).

The rest of the paper is organized as follows. First, the used definitions in this article of the fractional calculus and the FOPID controller will be presented. Then, the considered control plant and its integer order transfer function model are given. Next, an overview of individual optimization methods is prepared. Finally, a comparison between the results obtained by the heuristic methods via simulation is presented.

## 2 Fractional Order Systems

### 2.1 The Differential and Integral Operator

The basic definition for non-integer systems is the differential-integer operator ${}_aD_t^{\alpha}$, defined as follows:

$$ {}_aD_t^{\alpha} = \begin{cases} \frac{d^{\alpha}}{dt^{\alpha}} & \text{when } \alpha > 0 \\ 1 & \text{when } \alpha = 0 \\ \int_a^t (d\tau)^{\alpha} & \text{when } \alpha < 0 \end{cases} \tag{1} $$

There is a lot of definitions of fractional derivative-integral operator, e.g. [1,2]. In this article, I will present two, commonly used. The definition of Riemann-Liouville for $\alpha > 0$ is the function defined below

$$ {}_aD_t^{\alpha}(t) = \frac{d^{\alpha} f(t)}{dt^{\alpha}} = \frac{1}{\Gamma(p-\alpha)} \frac{d^p}{dt^p} \int_a^t \frac{f(\tau)}{(t-\tau)^{\alpha+1-p}} d\tau, \tag{2} $$

Where $\Gamma()$ is the gamma function of Euler Eq. (3), $p$ is a positive integer satisfying limitation $p - 1 < \alpha < p$.

$$ \Gamma(x) = \int_0^{\infty} t^{x-1} e^{-t} dt \tag{3} $$

Another definition is the definition of Caputo, for $p - 1 < \alpha < p$, which is given as

$$ {}_aD_t^{\alpha}(t) = \frac{d^{\alpha} f(t)}{dt^{\alpha}} = \frac{1}{\Gamma(p-\alpha)} \int_0^t \frac{f^p(\tau)}{(t-\tau)^{\alpha+1-p}} d\tau, \tag{4} $$

### 2.2 Methods of Describing Fractional Order Control Systems

The fractional system SISO (one input and one output) is described by a non-integer differential equation of

$$ \sum_{i=0}^{n} a_i D_t^{\alpha_i} y(t) = \sum_{k=0}^{m} b_k D_t^{\beta_i} u(t) \tag{5} $$

To get the transfer function of the fractional system, it is applied the Laplace transform to both sides of Eq. (5) with zero initial conditions [8]:

$$G(s) = \frac{Y(s)}{U(s)} = \frac{b_m s^{\beta_m} + b_{m-1} s^{\beta_{m-1}} + ... + b_0 s^{\beta_0}}{a_n s^{\alpha_n} + a_{n-1} s^{\alpha_{n-1}} + ... + a_0 s^{\alpha_0}} \tag{6}$$

### 2.3 Methods of Approximation

When it comes to implementation of Eq. (6) in Matlab, it is needed to have the finite, integer-order transfer function. The proposed approximation method by [10] allows to achieve an elementary FO transfer function $s^\alpha$ as follows:

$$s^\alpha \approx k_f \prod_{n=1}^{N} \frac{1 + \frac{s}{\mu_n}}{1 + \frac{s}{\nu_n}} = \frac{L_{ORA}(s)}{D_{ORA}(s)}, \tag{7}$$

where:

$$\begin{aligned} \mu_1 &= \omega_l \sqrt{\eta} \\ \nu_n &= \mu_n \gamma,\ n = 1, .., N \\ \mu_{n+1} &= \nu_n \eta,\ n = 1, .., N-1 \end{aligned} \tag{8}$$

$$\begin{aligned} \gamma &= \left(\frac{\omega_h}{\omega_l}\right)^{\frac{\alpha}{N}} \\ \eta &= \left(\frac{\omega_h}{\omega_l}\right)^{\frac{1-\alpha}{N}} \end{aligned} \tag{9}$$

In Eqs. (7) and (9) $N$ is the order of the approximation, $k_f$ is the steady state gain. $\omega_l$ and $\omega_h$ describe the range of angular frequency.

### 2.4 Fractional Order PID Controller

The controller is part of the closed-loop control system shown in Fig. 1. The transfer function of the FOPID controller is described by the following equation:

$$G_c(s) = k_p + k_i s^{-\alpha} + k_d s^\beta, \tag{10}$$

where $k_p$, $k_i$ and $k_d$ are parameters describing the proportional, integral and derivative parts, $\alpha, \beta$ are fractional orders of integral and derivative actions.

## 3 Optimization Methods

### 3.1 PSO - Particle Swarm Optimization

First a heuristic optimization method used in this work is Particle Swarm Optimization (PSO). This method is based on swarm intelligence. It is an iterative algorithm. A particle, which is a single solution, has the position and velocity.

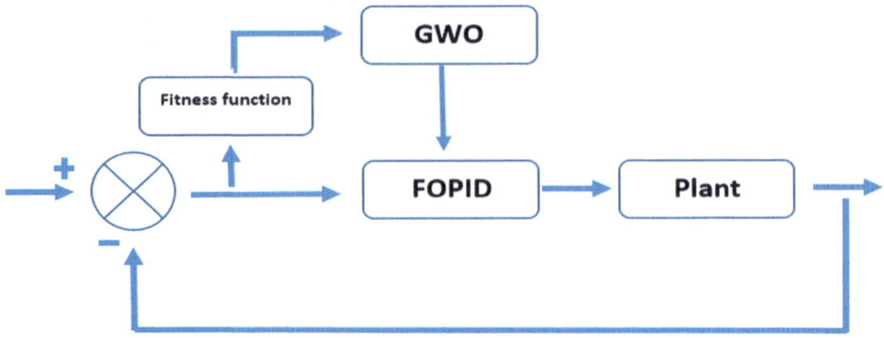


**Fig. 1.** The closed-loop control system.

During each iteration, a particle flies in space to find food, with respect to the parameters of its neighbors and the global best solution. The position of the food is the optimal value. The smaller the value is, the better performance is achieved. The objective function, which is considered is called the fitness function of the specific problem [14].

In the beginning, the initial positions are selected for each particle in a swarm at a random number from a predetermined search range. Finding the best solution is done in the main loop. In each iteration, for each particle is calculated the cost function. Later, the algorithm search for the global best solution, which is comparing calculated cost function for each particle, in case if the better result was achieved. At the end of the iteration, the position and the velocity are updated according to Eqs. (11) and (12), and the calculations are repeated.

$$P_j(m+1) = P_j(m) + V_j(m) \tag{11}$$

$$V_j(m+1) = c_1 r_1 V_j(m) + c_2 r_2 (P_{Best} - P_j(m)) + c_3 r_3 (G_{Best} - P_j(m)), \tag{12}$$

where: $P_j(m)$ i $V_j(m)$ means the position and velocity of the jth particle in the mth iteration, $P_{Best}$ is the particle best solution, $G_{Best}$ is the best global solution achieved so far, $r1$ and $r2$ are random numbers between (0,1), $c1$ and $c2$ are learning rates, $j$ = 1, ..., J - number of particles, $m$ = 1, ..., M - number of iterations.

### 3.2 Cockroach Swarm Optimization

Cockroach Swarm Optimization (CSO) is the next biologically inspired optimization method. It is developed from the three characteristic behavior of cockroaches [5]. The algorithm is iterative. In the preliminary activities, we search for the initial values of unknown parameters of the specific problem. Each cockroach, which is a single solution, stores its solution. The three typical for cockroach phases are repeated in each iteration [9].

1. **Swarm**
   In population, each solution hunt for a cockroach with a better one. If he found, then cockroach changes his location towards an insect with a better solution. If not, the cockroach moves towards the current global solution. The following relations are obtained in the form of

$$x_i = \begin{cases} x_i + step * rand * (P_{Best} - x_i) \text{ where } x_i \neq P_{Best} \\ x_i + step * rand * (G_{Best} - x_i) \text{ where } x_i = P_{Best} \end{cases}, \tag{13}$$

   where $x_i$ and $x_j$ are positions of cockroaches in the ith iteration, *step* is a set at the beginning coefficient, $P_{Best}$ is the particle best solution, $G_{Best}$ is the best global solution achieved so far.
   In this phase, there is an added parameter of the visibility of the neighborhood Eq. (14), defined by the *visual* parameter. This parameter tells how many neighbors, each cockroach can see.

$$x_i = Opt * (x_j, |x_i - x_j| \leq visual) \tag{14}$$

2. **Dispersing**
   In this phase, each insect performs a random move in the solution space as follows

$$x_i = x_i + rand(1, D), \tag{15}$$

   where $D$ is the dimension of problem.
   Equation (16) shows one additional coefficient of *hungerthreshold*. Once it is reached, the cockroach migrates towards food.

$$x_i = x_i + (x_i - ct) + x_{food}, \tag{16}$$

   where $c$ is a set at the beginning coefficient, and $x_{food}$ is a parameter of the hunger threshold.
3. **Ruthless behavior**
   In this phase, a stronger cockroach eats the weaker one. It is completed by taking the current global optimum by the random cockroach.

### 3.3 Grey Wolf Optimizer

Grey Wolf Optimizer (GWO) is the third heuristic method presented in this paper. It is based on the behavior of the grey wolf living in the herd [3]. There are two interesting features. One is a strictly defined hierarchy and second is group hunting. There are three types of main wolves. The leader is the alpha wolf who is a decision-maker. The next one is the beta wolf. The beta helps the alpha with various activities related to the herd. The third level is for the delta wolves, they have different positions in the pack like scouts, guards, and hunters. The lowest rank is occupied by the omega wolves.

The algorithm is iterative. In each iteration, the hunting process consists of three phases.

1. **Tracking and racing the victim**
   Grey wolves encircle the prey during the hunt. Each wolf updates his position in the direction of the victim Eqs. (19, 20). In the algorithm, this is presented as follows:
   $$A = 2a * r_1 - a \tag{17}$$
   $$C = 2 * r_2 \tag{18}$$
   $$D = |C * X_p(t) - X(t)| \tag{19}$$
   $$X(t+1) = X_p(t) - A * D, \tag{20}$$
   where $t$ is the number of current iteration, $A$ and $C$ mean coefficients, $Xp$ is the victim position vector, and $X$ is the wolf position, $a$ is parameter, which is linearly reduced from 2 to 0 in relation to iteration, $r1$ and $r2$ are a random number between (0.1).
2. **Surrounding the victim**
   It is assumed that the best information about the victim is available to the leader, which shares knowledge with the beta and the delta. Therefore, every wolf should update its position in given directions Eqs. (19, 20). In the end, the location of the agent is calculated according to
   $$X(t+1) = \frac{X_\alpha + X_\beta + X_\delta}{3}, \tag{21}$$
   where $X(t+1)$ is the position of agent calculated for next iteration, $X_\alpha$, $X_\beta$, $X_\delta$ are the position of the alpha, the beta and the delta wolves.
3. **Attack of prey**
   When the prey stops moving, the herd is attacking it. Mathematically, the $a$ decreases as the iteration progresses.

## 4 Experimental System

In this article, I am considering the isothermal room used for example in a pharmaceutical factory. The control system is demand to assure the temperature about 37 °C in the room with cubature about 75m$^3$. The system is shown by the SCADA screen in Fig. 2. The control signal is the power of the heater 2E1 expressed in the percent of its maximal range. The controlled temperature is measured inside the return duct by the sensor Pt 1000 (element 6TE1 in the scheme). The return duct sucks the cold air from the room to the heater. Later the heated and filtered by the HEPA filter is going to the isothermal room via supply duct. The disturbance in this system is the outside temperature.

The second order transfer function model of the temperature in the room is as follows:
$$G(s) = \frac{0.0002493s + 0.0000002117}{s^2 + 0.002379s + 0.0000005053} \tag{22}$$

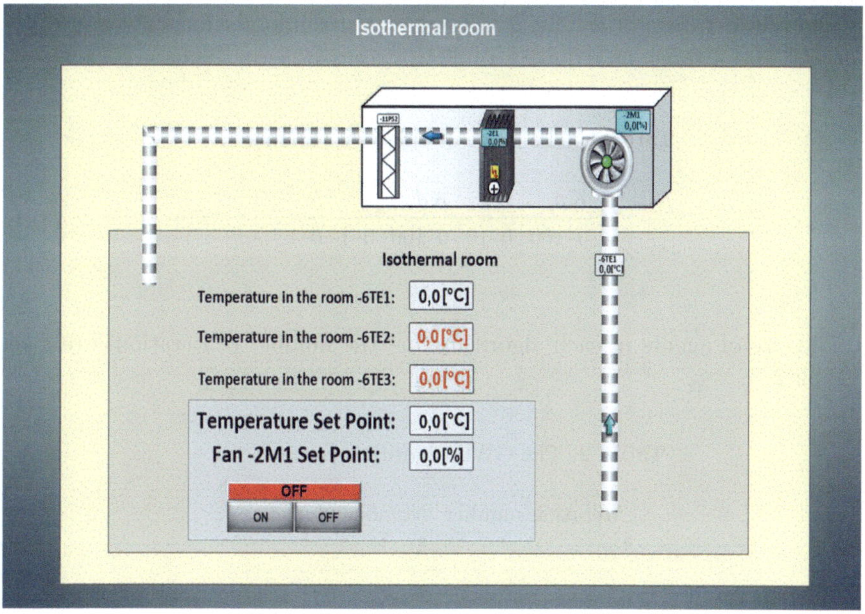


**Fig. 2.** The forced air heating system

## 5 Cost Function

In the case of fine-tuning the fractional PID controller, we deal with control performance. The closed-loop control system can be described using parameters: overshoot $y_{max}$ expressed in of reference value and settling time $Ts$. In addition, integral criteria will be used in this work, which are shown below.

- Integral Time Absolute Error (ITAE)

$$ITAE = \int_{0}^{\infty} t|e(t)|dt. \tag{23}$$

- Integral Square Error (ISE)

$$ISE = \int_{0}^{\infty} e^2(t)dt. \tag{24}$$

Optimization algorithms look to minimize the cost function of the process. In this design, the method focuses on minimizing the ITAE criterion.

## 6 Numerical Tests

Simulations took place on the change of the temperature of 10 °C. This section will present some examples of research results.

The initial values of the agent are generated randomly for PSO and GWO from the initial range as required Table 1.

**Table 1.** The range of unknown parameters.

| $Kp$ | $Ki$ | $Kd$ | $\alpha$ | $\beta$ |
|---|---|---|---|---|
| 0–100 | 0–10 | 0–100 | 0–1 | 0–1 |

The size of agents in each algorithm and the number of iterations are given in Table 2.

**Table 2.** The GWO algorithm parameters.

| Iteration number | Number of agent |
|---|---|
| 20 | 5 |

The parameters of ORA approximation applied in simulations are shown in Table 3.

**Table 3.** The ORA approximation parameters.

| $N$ | $\omega_l$ | $\omega_h$ |
|---|---|---|
| 8 | $10^{-4}$ | $10^{4}$ |

Each presented simulation result is shown with the PID control. The PID settings have been received by the automatic Matlab function and have following transfer function

$$G_c(s) = 15.9621 + 0.02858\frac{1}{s} + -2282.176s \tag{25}$$

### 6.1 Particle Swarm Optimization

Initial values for swarm parameters are taken from Table 1. Coefficient $c_1$ and $c_1$ from Eq. 12 are set both to 1.4 (Figs. 3 and 4).

$$G_c(s) = 530.5978 + 2.9312s^{-0.2995} + 10.3283s^{0.2568} \tag{26}$$

$$G_c(s) = 361.8502 + 5.5218s^{-0.4815} + 27.5396s^{0.7417} \tag{27}$$

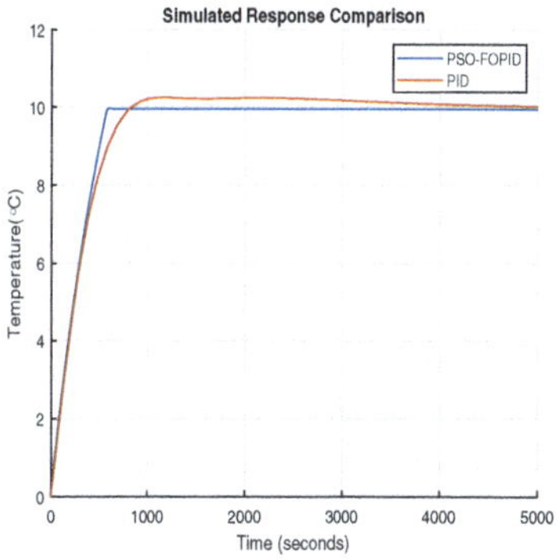


Fig. 3. Simulated response with FOPID-PSO (blue) and PID (red). Transfer function Eq.(26).

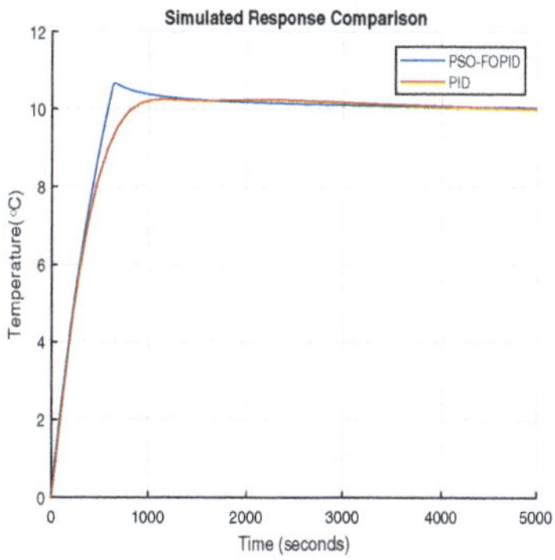


Fig. 4. Simulated response with FOPID-PSO (blue) and PID (red). Transfer function Eq.(27).

### 6.2 Cockroach Swarm Optimization

The values of initial vector have the form as follows:

$$G_c(s) = 50 + 5s^{-0.5} + 50s^{0.5}, \tag{28}$$

where the parameters have the value of half from range taken from Table 1. In the case of calculations in CSO, the algorithm parameters must be set according to Eqs. 13, 15, 16 and 14 (Figs. 5 and 6, Table 4).

**Table 4.** The CSO parameters.

| *visual* | *step* | $x_{food}$ | *ct* |
|---|---|---|---|
| 3 | 2 | 4 | 1.4 |

$$G_c(s) = 1.0448 + 0.2851s^{-0.6210} + 6.2103s^{0.0498} \tag{29}$$

$$G_c(s) = 0.3925 + 0.3039s^{-0.6153} + 0.6336s^{0.6934} \tag{30}$$

### 6.3 Grey Wolf Optimizer

Initial values for swarm parameters are chosen from Table 1 (Figs 7 and 8).

$$G_c(s) = 671.9424 + 9.2116s^{-0.2917} - 11.4012s^{0.6277} \tag{31}$$

$$G_c(s) = 853.0783 + 1.4688s^{-0.4845} + 632.0790s^{0.0072} \tag{32}$$

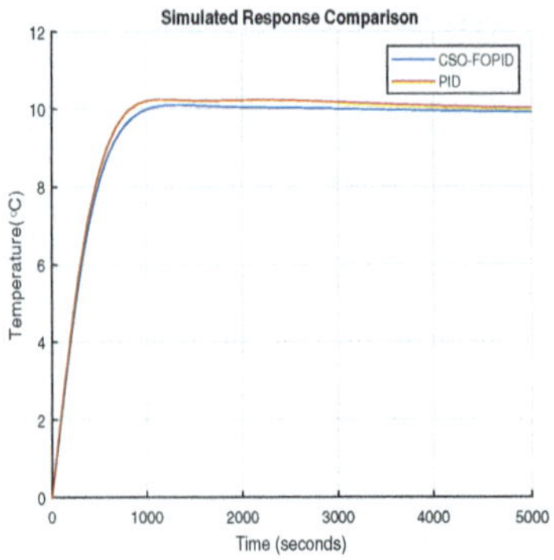


**Fig. 5.** Simulated response with FOPID-CSO (blue) PID (red). Transfer function Eq.(29).

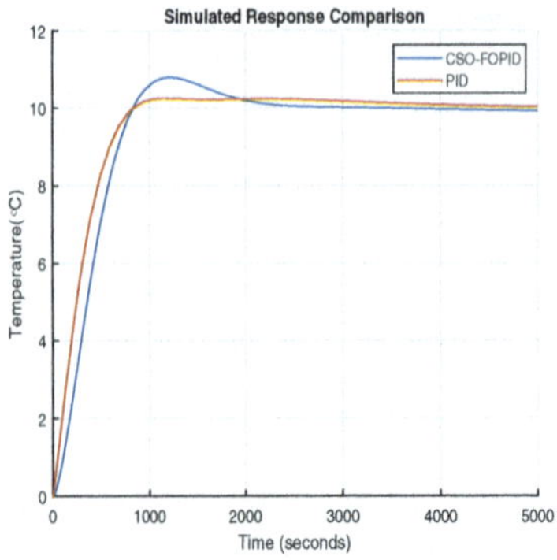


**Fig. 6.** Simulated response with FOPID-CSO (blue) and PID (red). Transfer function Eq.(30).

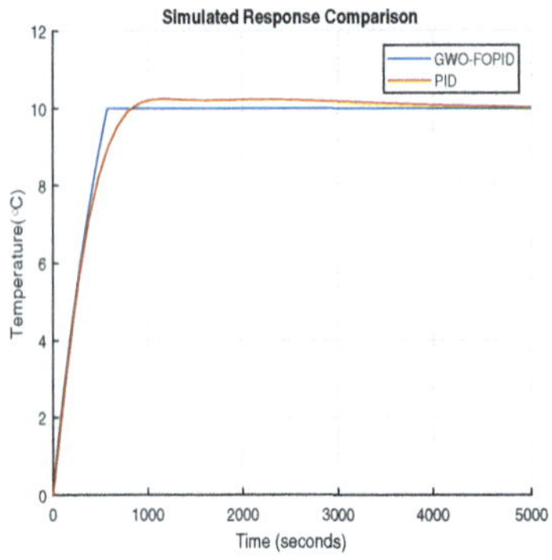


**Fig. 7.** Simulated response with FOPID- GWO (blue) and PID (red). Transfer function Eq. (31).

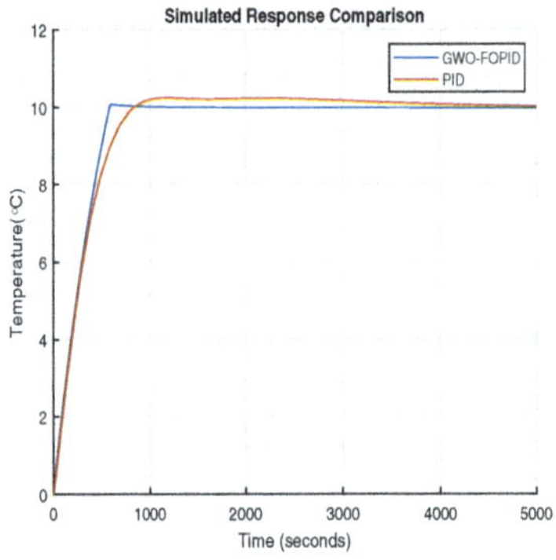


**Fig. 8.** Simulated response with FOPID-GWO (blue) and PID (red). Transfer function Eq.(32).

### 6.4 Simulation Analysis

Figure 9 shows a comparison of the best results for all three algorithms. It can be seen that CSO control lasts much longer than two other algorithms.

Table 5 shows a comparison of used optimization algorithms. The simulations were carried out on the same number of iterations and the same population size. Thanks to this, comparing time gives a clear view of the duration of individual algorithms. A row comparing the number of initial parameters indicates that GWO needs the least amount of work to prepare the algorithm. Given that the two parameters were set in advance, only the range remained, which is often given as requirements. The biggest problem is with CSO, which needs 12 parameters. This algorithm requires finding an initial vector also, which determines the optimization process because a wrong one can lengthen the search process or not lead to a satisfactory result.

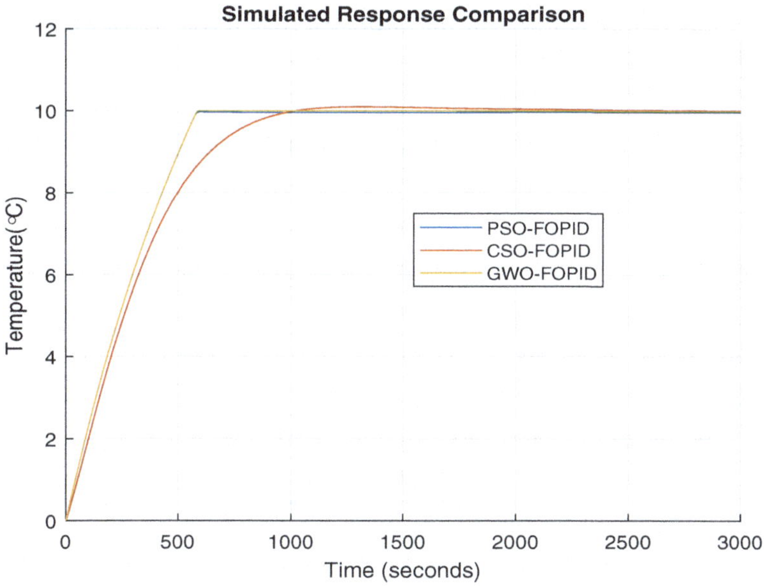


**Fig. 9.** Simulated responses comparison.

**Table 5.** Summary of the results of the considered algorithms

| Parameter | Optimization method | | | | | |
|---|---|---|---|---|---|---|
| | *PSO* (26) | *PSO* (27) | *CSO* (29) | *CSO* (30) | *GWO* (31) | *GWO* (32) |
| ITAE | 19493 | 15752 | 6.8404*e*+07 | 1.2909*e*+08 | 4777.3 | 1091.3 |
| ISE | 5467.4 | 4971.9 | 3.2119*e*+05 | 4.3969*e*+05 | 2739 | 1308.1 |
| Overshoot | 0 | 3.3626 | 1.3626 | 8.9772 | 0 | 0.8935 |
| Settling time | 28.65 | 38.59 | 854.705 | 2199 | 21.4317 | 10.21 |
| Time | 163.94 s | 83.55 s | 208.97 s | 125.67 s | 41.22 s | 45.81 s |
| Initial vector | | – | *YES* | *YES* | – | – |
| Number of initial parameters | 6 | 6 | 12 | 12 | 3 | 3 |

When it comes to control, I receive the largest repeatability of correct settings using GWO. Using this algorithm, performance parameters of control are the best for various tests. For CSO and PSO there are overshoots and the values of integral criteria are higher.

## 7 Conclusions

- The optimization methods used in this work give satisfactory results. They allow finding the optimal tuning parameters for the controller with a non-integral order.
- The table with a comparison of used optimization algorithms presents that the GWO algorithm is the most suitable one. Using the other methods, there

is a need for additional work at the beginning to find coefficients characteristic for a particular method.

- The last row of the comparison table shows the most important problem, which is setting the appropriate initial vector for a population. In CSO we need to find it. In other, we need range, which probably will be given by the requirements.

## References


1. Das, S.: Functional Fractional Calculus for System Identication and Controls. Springer, Heidelberg (2008)
2. Kaczorek, T.: Selected Problems of Fractional System Theory. Springer, Berlin (2011)
3. Jangir, P., Bhesdadiya, R.,Ladumor, D., Trivedi, I.: A multi-objective grey wolf optimization algorithm for economic/environmental dispatch. In: PNFE-2016At: St. Peters Engineering college, Hyderabad, India, October 2016. https://doi.org/10.13140/RG.2.2.21536.79364
4. Kwiecien, J., Filipowicz, F.: Algorytmy stadne w problemach optymalizacji. Automatyka, Tom 11, Zeszyt 2 (2011)
5. Kwiecien, J., Filipowicz, F.: Comparison of firefly and cockroach algorithms in selected discrete and combinatorial problems. Bull. Pol. Acad. Sci. Tech. **62**(4), 794–804 (2014)
6. Majka, Ł.: Applying a fractional coil model for power system ferroresonance analysis. Bull. Pol. Acad. Sci. Tech. Sci. **66**(4), 467–474 (2018). https://doi.org/10.24425/124263
7. Matignon, D.: Stability results for fractional differential equations with applications to control processing . In: IMACS-SMC Proceedings, Lille, France, July 1996
8. Meng, L., Wang, D., Han, P.: Identication of fractional order system using particle swarm optimization. In: 2012 International Conference on Machine Learning and Cybernetics, July 2012. https://doi.org/10.1109/ICMLC.2012.6359551
9. Obagbuwa, I., Adewumi, A.: An improved cockroach swarm optimization, May 2014. https://doi.org/10.1155/2014/375358
10. Oustaloup, A., Levron, F., Mathieu, B., Nanot, F.M.: Frequency-band complex noninteger differentiator: characterization and synthesis. IEEE Trans. Circ. Syst. I Fundam. Theor. Appl. **47**(1), 25–39 (2000)
11. Petras, I.: Fractional derivatives, fractional integrals and fraction dierential equations. Technical University of Kosice (2012)
12. Podlubny, I.: Fractional Differential Equations. Academic Press, Cambridge (1999)
13. Teplakov, A.: Fractional-order Modeling and Control of Dynamic Systems. Tallin Unviersity of Technology, Tallin (2015)
14. Shi, Y., Eberhart, R.: A modied particle swarm optimizer. In: IEEE World Congress on Computational Intelligence, June 1998. https://doi.org/10.1109/ICEC.1998.699146

# The Use of CUSUM-Charts for Identification the Technological Process Disorder at the Initial Stage

Yevhen Volodarsky[1], Igor Pototskiy[2(✉)], and Zygmunt Lech Warsza[3]

[1] Department of Automation of Experimental Studies, National Technical University of Ukraine «KPI», Kiev, Ukraine
vet-l@ukr.net

[2] SE «Ukrmetrteststandart», Kiev, Ukraine
pototskiy@ukrcsm.kiev.ua

[3] Industrial Research Institute of Automation and Measurement (PIAP), Warsaw, Poland
zlw@op.pl

**Abstract.** The accumulated sum control charts use the data obtained at the current and at all previous control stages. The using of these charts allows to determine the moment when the process goes out of statistically controlled state with less delay than the Shewhart charts. However, the presence of a time delay between the occurred event and its detection on the control chart can lead to large production losses (sometimes irreparable losses). The article considers a new approach that allows to identify violations in the process at the initial stage. For this it was suggested to consider not the average values of sample results at control points, but the standard deviations at these points. Using Pearson statistics, the probabilities of finding the standard deviation in elementary intervals are determined, into which the area of permissible deviations is divided in a statically controlled process. The initial premise is the assumption that the output of the sample dispersion outside the range of tolerances at the control point is evidence that there are violations in the process. The probability of a control point falling beyond acceptable limits is the upper limit in the presence of violations in the process. This probability is a measure to detect violations at the initial stage. As a criterion, the number of consecutive control points falling in elementary intervals was selected. The probability of such a complex event should be less than the value of the measure to identify violations in the process. A concept is introduced and a critical number of points is determined. The efficiency of identifying process violations is determined. Recommendations on the use of the results are given.

**Keywords:** Quality control · Technological process · Cumulative sum control charts CUSUM · Standard deviation · Pearson probability distribution

## 1 Introduction

To ensure the normal course of the technological process and identify possible violations, in practice, the most widely used is Shewhart control charts [1]. Despite the simplicity of construction, Shewhart control charts have one significant drawback – they


R. Szewczyk et al. (Eds.): AUTOMATION 2020, AISC 1140, pp. 147–156, 2020.
https://doi.org/10.1007/978-3-030-40971-5_14

have low sensitivity to process violations and practically do not allow to establish the time of process disorder [2]. In this case, sensitivity is understood as the time delay between the moment of violations occurrence in the technological process and the detection of this event using a control chart. Accumulated sum control charts or CUSUM-charts [3–5], display not only the measurement result at the current control point. They use the information obtained at all previous control points, and thereby reflect the change in the process as a whole [6]. This makes it possible to detect process violations with a lower time delay. Moreover, the more significant the disorder of the process, the faster the increase (decrease) of the curve that reflects the change of accumulated sum in time. The article proposes a new approach for identifying process disorder at the initial stage of its appearance. At the beginning of the article, the above example shows the identification of violations (bias) using the V-mask method, the choice of mask parameters and its use in the technological process analysis. Accumulated sum control charts use sample averages defined at checkpoints. However, as you know, the rms value is more sensitive to changes than the average value. Using Pearson statistics, the probabilities of finding the standard deviation value in elementary intervals are determined, into which the area of permissible deviations is divided in a statically controlled technological process. In the final part, according to the results of the analysis, algorithms are considered that allow to identify process disorder at the initial stage.

## 2 Control of Process Average Deviation

Already mentioned international standard [3] for the control of processes by means of CUSUM-charts proposed to use a graphical tool – V-mask. One of the main characteristics of the V-mask is the angle of its rays. The practical aspects of using a V-mask are described in detail in [7]. The graphical V-mask method involves the presence of an operator to control the state of the process, its automation and decision-making without operator involvement is quite a problem. In addition, this method can produce signals with a delay of less than the Shewhart charts, which does not always allow you to make management decisions on time.

As an example, Fig. 1 shows the results of a simulated process of control that has a normal distribution law, a normalized value of $\mu_0 = 100$ units, and a standard deviation of $\sigma_0 = 20$ units. At time $t = 11$, a shift $\Delta = +10$ units ($\delta = \frac{\Delta}{\sigma_0} = +0.5$) is introduced into the process, which remains constant. The offset graph, starting from step 11, is shown on the CUSUM-chart with a dotted line.

The calculated parameter $d$ and the angle $\Theta$ of inclination of the V-mask rays are determined by the formulas [7]:

$$d = -\frac{2}{n\delta^2} \cdot \ln\frac{\alpha}{2} \tag{1}$$

$$\theta = arctg\frac{\delta\sigma_0}{2f}, \tag{2}$$

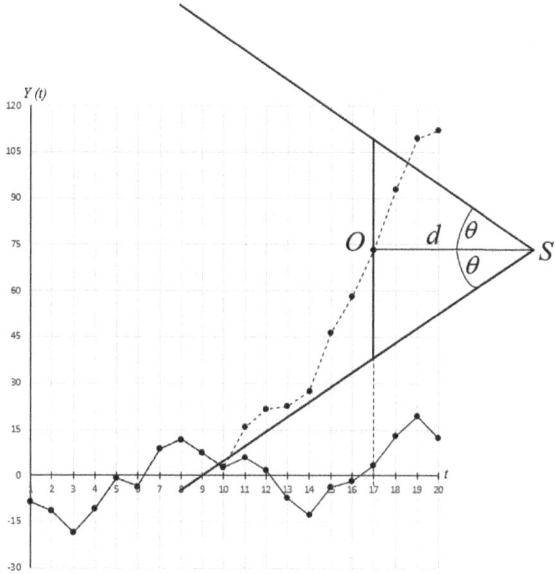


**Fig. 1.** An example of the CUSUM-charts construction with the setting level offset $\delta = +0.5$.

where $n$ – the number of sample observations; $\alpha$ – level of statistical significance; $f$ – scale factor required for graphical control.

As is seen from Fig. 1, a CUSUM-chart using a V-mask showed a process disorder only at the seventh step after the event occurred. Such a response time cannot be considered satisfactory and the received signal is timely.

To reduce the response time of the method to the mean shift, the criterion is to determine the change of the slope angle of the regression line constructed between the last two points plotted on the CUSUM-chart. It is it increase in relation to critical values. The limiting value of the slope angle, as shown in [7], is calculated by the formula:

$$\mathrm{tg}\theta_{\mathrm{lim}} = \frac{\delta\sigma_0}{2}, \tag{3}$$

where $\delta$ – normalized process offset value (average values of measurement results); $\sigma_0$ – normalized standard deviation.

At the same time, exceeding the limit value $\mathrm{tg}\theta_{\mathrm{lim}}$ for the first time, can be considered as a warning signal, while for the second consecutive time, exceeding is a reliable signal that the average value exceeds the set limits and the need to intervene in the process.

## 3 Control of Standard Deviation

When establishing the statistical controllability of technological processes, they often operate with the average values, their displacement and, accordingly, the hit of these values within the pre-set boundaries. However, an equally important indicator is the change in the scattering of observation results. A change in the variance or standard deviation indicates a disorder in the production process or a change in the sensitivity of the measuring instruments used in the control.

The established criteria for the detection of technological process disorder are based on the analysis of average sample values [8]. Because the number of observations $n$ increases with the law of distribution of the averages approaching normal, then the probabilities of hitting averages in the six ranges from $-3\sigma$ to $+3\sigma$ with step $\sigma$ in two-sided control were investigated.

To estimate the change over time of the variance of measurement results by a measuring technique, it is necessary to determine what statistical criterion can be used to characterize the distribution of magnitude. It can be assumed that the sample values of the measured value have a normal distribution law. As is known, for random variables of a normal distribution, when constructing a confidence interval, statistics are used to estimate the variance, the distribution of which approximates to the Pearson distribution with the number of degrees of freedom $v = (n - 1)$ [9], i.e.:

$$\chi^2 = \frac{S_n^2 \cdot (n-1)}{\sigma_0^2}, \tag{4}$$

where: $S_n^2$ – sample variance estimation; $n$ – the number of observations in the sample; $\sigma_0^2$ – dispersion of the aggregate.

The estimation of the hit of statistics in a certain confidence interval is carried out after setting the significance level $\alpha$ and finding the tabulated value of the quantiles of distribution:

$$\chi^2_{1-\frac{\alpha}{2},n-1} \leq \chi^2 \leq \chi^2_{\frac{\alpha}{2},n-1}, \tag{5}$$

where $\chi^2_{1-\frac{\alpha}{2},n-1}$, $\chi^2_{\frac{\alpha}{2},n-1}$ – quantiles of the Pearson distribution, depending on the significance level $\alpha$ and the number of degrees of freedom $v = n - 1$.

Accordingly, the probability of a statistic falling into a certain confidence interval is:

$$P\left(\chi^2_{1-\frac{\alpha}{2},n-1} \leq \chi^2 \leq \chi^2_{\frac{\alpha}{2},n-1}\right) = 1 - \alpha \tag{6}$$

Substituting in (3) the corresponding value (1), we obtain:

$$P\left(\chi^2_{1-\frac{\alpha}{2},n-1} \leq \frac{S_n^2(n-1)}{\sigma_0^2} \leq \chi^2_{\frac{\alpha}{2},n-1}\right) = 1 - \alpha \tag{7}$$

The control of technological processes during production, which apples the criterion $\chi^2$, is based on the known normalized variance value $\sigma_0^2$ and control the current value of sample variance $S_n$. After simple mathematical transformations, inequality (5) is relative $S_n$ will be:

$$\sqrt{\frac{\chi^2_{1-\frac{\alpha}{2},n-1}\cdot\sigma_0^2}{n-1}} \le S_n \le \sqrt{\frac{\chi^2_{\frac{\alpha}{2},n-1}\cdot\sigma_0^2}{n-1}} \tag{8}$$

Based on (6) and (8), the probability of the value $S_n$ getting into the confidence interval is as follows:

$$P\left(\sqrt{\frac{\chi^2_{1-\frac{\alpha}{2},n-1}\cdot\sigma_0^2}{n-1}} \le S_n \le \sqrt{\frac{\chi^2_{\frac{\alpha}{2},n-1}\cdot\sigma_0^2}{n-1}}\right) = 1-\alpha \tag{9}$$

Finding a sample value of the standard deviation within this interval is evidence of the statistical controllability of the process.

The Pearson distribution with the number of degrees of freedom $v = n - 1$ is widely described in the literature and has the following basic parameters:

- mode: $v - 2$ (at $v \geq 2$);
- mathematical expectation: $v$;
- dispersion: $2v$.

For the purpose of research, the control process was simulated with the number of observations $n = 5$ and the normalized standard deviation $\sigma_0 = 20$ units. The Pearson's distribution for the number of degrees of freedom $v = n - 1 = 4$ is shown in Fig. 2.

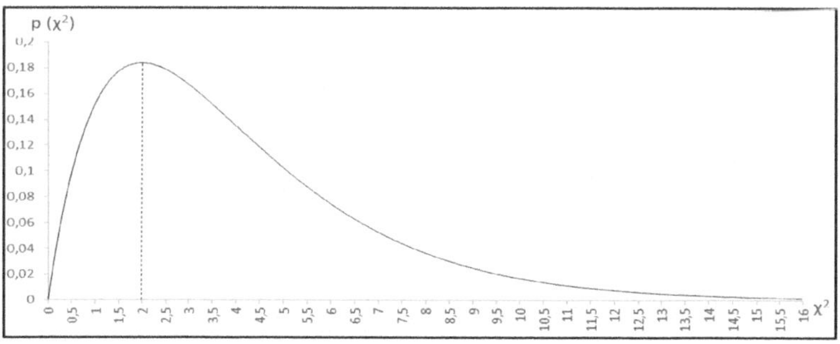


**Fig. 2.** Pearson distribution for the number of degrees of freedom $v = 4$.

The dotted line is showing the distribution mode, which is at the point $\chi^2 = 2$.

To further analyze and construct confidence intervals, it is necessary to set the significance level $\alpha$. As in [7], we have a level of significance $\alpha = 0.01$. This means that there is a 1% risk that the result will be out of the interval $\chi^2_{1-\frac{\alpha}{2},n-1}$ to $\chi^2_{\frac{\alpha}{2},n-1}$ with a stable

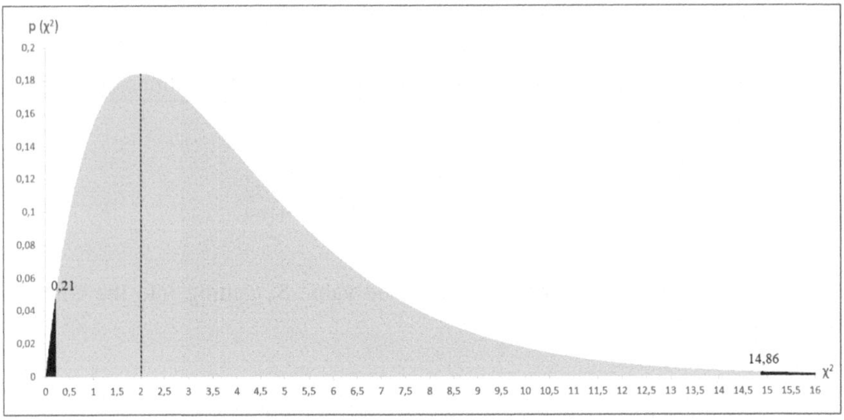


**Fig. 3.** Pearson distribution with bounded values $\chi^2_{1-\frac{0.01}{2},5-1} = 0.21$ and $\chi^2_{\frac{0.01}{2},5-1} = 14.86$.

process. Finding the table values of the quantiles of the distribution $\chi^2_{1-\frac{0.01}{2},5-1} \approx 0.21$ and $\chi^2_{\frac{0.01}{2},5-1} \approx 14.86$, we construct the Pearson distribution with the applied boundary values (Fig. 3).

We calculate the conditional value of the center line for the case $CL$ for $\chi^2_{\text{mod}} = 2$, as well as the bottom $L_{CL}$ and upper $U_{CL}$ limit values of $S_n$ for $\chi^2_{1-\frac{\alpha}{2},n-1}$ and $\chi^2_{\frac{\alpha}{2},n-1}$, respectively:

$$CL = \sqrt{\frac{\chi^2_{\text{mod}} \cdot \sigma_0^2}{n-1}} = \sqrt{\frac{2 \cdot 20^2}{5-1}} = 14.14\,\text{units};$$

$$L_{CL} = \sqrt{\frac{\chi^2_{1-\frac{\alpha}{2},n-1} \cdot \sigma_0^2}{n-1}} = \sqrt{\frac{0.21 \cdot 20^2}{5-1}} = 4.55\,\text{units}; \tag{10}$$

$$U_{CL} = \sqrt{\frac{\chi^2_{\frac{\alpha}{2},n-1} \cdot \sigma_0^2}{n-1}} = \sqrt{\frac{14.86 \cdot 20^2}{5-1}} = 38.55\,\text{units}.$$

To facilitate the analysis of the obtained values of the sample standard deviations $S_n$ plot the expanded Pearson distribution on the control chart with the value of the corresponding values $CL$, $L_{CL}$ and $U_{CL}$ as shown in Fig. 4a. Given that, in the absence of non-random reasons that affect the technological process, the probability of the value output $S_n$ beyond the limit values $L_{CL}$ and $U_{CL}$ equal to only 1%. Even one output for these values is a signal that the standard deviation has changed significantly, and the process requires intervention. To prevent value $S_n$, output beyond the established boundaries and the early warning of the negative consequences caused by it, is advisable to develop a number of precautionary criteria that could advance the exit of the process from a managed state.

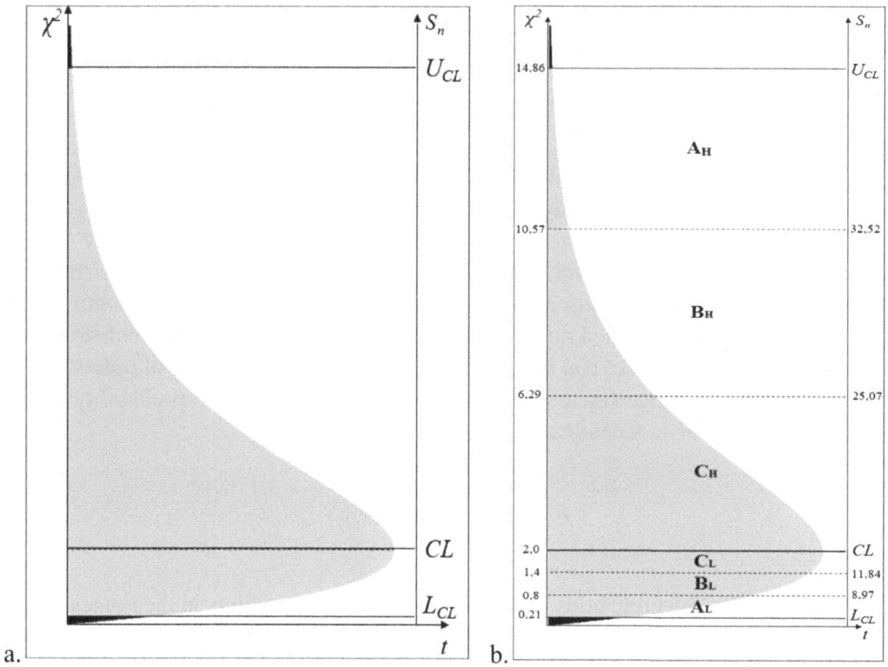


**Fig. 4.** a. Control chart with plotted Pearson distribution and b. Its control ranges.

Let us divide the interval $L_{CL} - CL$ in three levels across the width and denote them as $A_L$, $B_L$ and $C_L$. The width (step) of each range in value $\chi^2$ will be:

$$\frac{\chi^2_{\mathrm{mod}} - \chi^2_{1-\frac{\alpha}{2},n-1}}{3} = \frac{2 - 0,21}{3} = 0.6$$

We obtain the following boundaries of the ranges by $\chi^2$:

– $A_L$: from 0.21 to 0.8; – $B_L$: from 0.8 to 1.4; – $C_L$: from 1.4 to 2.0.

Let us just divide the interval $U_{CL} - CL$ into three equal ranges $A_H$, $B_H$ and $C_H$ and calculate the above limits and draw the boundaries of range zones on the control chart previously constructed. The result is shown in Fig. 4b.

As mentioned above, the probability of a point on the control chart getting out of bounds $L_{CL}$ and $U_{CL}$ is 1% and if this is the case, it is believed that the process is affected by non-random values. In the same way, the sequence of points on the control chart, which are located in certain ranges and the probability of which is less than 1%, may be signaled that the process is affected by non-random variables and necessary intervention. Let us call this sequence as a critical sequence of points.

Critical point sequences can be:

- in a range ($A_L$, $B_L$, $C_L$, $C_H$, $B_H$ or $A_H$);
- in several adjacent ranges (for example ($A_L + B_L$), ($C_L + C_H$) etc.);
- below or above *CL*.

For further research, we determine the probability of a point below the *CL* on the control chart. To do this, we use a one-sided rule:

$$P\left(\chi^2 \leq \chi^2_{\text{mod}}\right) = 1 - \alpha \tag{11}$$

The probability of a point falling below the *CL* level in our case is 26.5%. Subtracting from this value a one-way probability of falling below the control $L_{CL}$, which, based on the accepted value of $\alpha = 0.01$, equal to 0.5%, we get the probability of one-way falling below the *CL* level, which is considered that the process has not lost statistical controllability equal to 26%. Determine how many points on the control chart should fall within the specified range in order for a critical sequence of points to occur.

Critical points sequences can be found by analyzing the probability of points falling into one or another range (or a several adjacent ranges) using the probability multiplication theorem for independent events [10, 11]:

$$P\left(\prod_{i=1}^{n} A_i\right) = \prod_{i=1}^{n} P(A_i) \tag{12}$$

Based on (12), the probability of hitting two consecutive points in the ranges below *CL*, in which the decision on the need to intervene in the process has not yet been decided, will be $P = 0.26 \cdot 0.26 = 0.068$ or 6.8%. Continuation of calculations according to (12) will determine the moment (point) when the probability becomes less than 1%. It will signal the occurrence of a critical sequence of points and the need for intervention to the process. In Fig. 5 the results of the calculations and graphs the probability of consecutive points in the range ($A_L$, $B_L$, $C_L$) are shown.

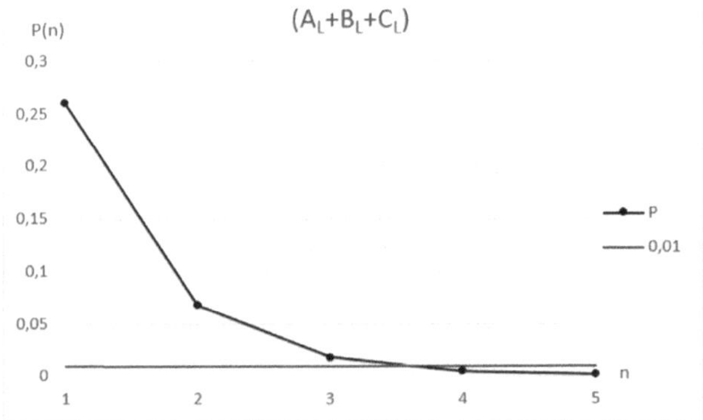


**Fig. 5.** The probability of consecutive points falling into the range ($A_L$, $B_L$, $C_L$).

From Fig. 5 implies that after the fourth consecutive hit of a control chart point in the range ($A_L$, $B_L$, $C_L$), the probability of an event becomes less than 1% (i.e. 0.45%),

**Table 1** Results of probability and number of points in critical sequence calculation

| Range | $A_L$, $B_L$, $C_L$ | $A_H$, $B_H$, $C_H$ | $A_L$ | $B_L$ | $C_L$ | $C_H$ | $B_H$ | $A_H$ | $B_L$, $C_L$ | $B_H$, $C_H$ | $A_L$, $B_L$, | $A_H$, $B_H$, | $C_L$, $C_H$ |
|---|---|---|---|---|---|---|---|---|---|---|---|---|---|
| Probability of falling P | 0.26 | 0.73 | 0.057 | 0.094 | 0.109 | 0.556 | 0.147 | 0.027 | 0.203 | 0.703 | 0.151 | 0.174 | 0.665 |
| Number of points in critical sequence N | 4 | 15 | 2 | 2 | 3 | 8 | 3 | 2 | 3 | 14 | 3 | 3 | 12 |

and a critical sequence of points is formed. This, in turn, may indicate loss of the managed state process.

Similarly, we calculate the probability of occurrence of a critical sequence of points for the level above *CL*, but below the $U_{CL}$ threshold level. Probability of getting to the point range ($A_H$, $B_H$, $C_H$) equal to 73%.

In the following, we calculate the probability of forming critical sequence of points for all ranges, as well as some of their combinations. The results of the calculations are shown in Table 1.

The analysis of the results showed that for the $A_L$, $B_L$, $B_H$, and $A_H$ ranges, only two consecutive points are sufficient for the occurrence of a critical succession of points, whereas the $C_H$ range requires eight points. Similarly, for combinations of ranges ($A_L$, $B_L$), ($A_H$, $B_H$) and ($B_L$, $C_L$), a critical sequence of points occurs after the third consecutive event, and for ($B_H$, $C_H$) after the fourteenth.

Based on the data in Table 1, it is possible to construct an algorithm for detecting the process disorder at the initial stage of its occurrence. For example, at the beginning, two consecutive points are shown to reflect the corresponding values of the sample mean deviations in the $A_L$, $B_L$ or $A_H$ range. If not, three consecutive points in the range, according to Table 1, etc., are displayed. This allows for an early detection of a process disorder.

## 4 Conclusions

A new approach to controlling the statistical controllability of a technological process by analyzing the standard deviation while controlling the state of the process is presented. The approach is based on probabilistic analysis of the hit of the received sample mean deviation in certain Pearson distribution ranges.

The concept of critical sequence of points is introduced, the occurrence of which indicates the influence of non-random variables on the course of technological process or process of measuring product parameters. At the same time, unlike the classic CUSUM-charts application, it allows to detect not only the additive displacement of the process, but also its multiplicative displacement.

## References

1. ISO 7870-2:2013 Control charts – Part 2: Shewhart control charts
2. Nishina, K., Kuzuya, K., Ishi, N.: Reconsideration of control charts in Japan. Front. Stat. Quality Control **8**, 136–150 (2005)
3. ISO 7870-4:2011 Control charts – Part 4: Cumulative sum charts
4. Murdoch, J.: Control Charts: Finance and Statistics (1986)
5. Montgomery, D.C.: Introduction to Statistical Quality Control, 7th edn. Wiley, Hoboken (2012)
6. Volodarsky, E., Kosheva, L., Warsza, Z., Idźkowski, A.: Precautionary statistical criteria in the monitoring quality of technological process. In book Recent Advances in Systems, Control and Information. In: Proceeding of International Conference SCIT 2016, Warsaw, Poland. Technology, pp. 740–750. Springer, Cham (2017). ISBN 978-3-319-48923-0
7. Volodarsky, E., Pototskiy, I.: Theoretical substantiation and application of CUSUM-charts. In: Conference Proceedings 2019 IEEE 8th International Conference on Advanced Optoelectronics and Lasers CAOL*2019, Sozopol, Bulgaria, pp. 636–639 (2019)
8. Uncertainty of Measurement and reliability of decision making on compliance. E. Volodarsky, Z. Warsza, L. Kosheva, M. Klevtsova. Advance in Intelligent systems and computing. Volume 920. Automation 2019. Progress in automation, robotics and measurement techniques, pp. 672–683 (2019)
9. Cramer, H.: Mathematical Methods of Statistics. Princeton University Press, Princeton (1999)
10. Florescu, I.: Probability and Stochastic processes, 1st edn. Wiley, Hoboken (2015)
11. Gallager, R.G.: Stochastic Processes: Theory for Applications. Cambridge University Press, Cambridge (2014)

# System for Detection of Vehicles in Multiple Video Streams in Road Infrastructure Monitoring

Marcin Słomiany(✉), Piotr Gemza, Filip Jędrzejczyk, Mateusz Maciaś, and Jakub Główka

ŁUKASIEWICZ Research Network – Industrial Research Institute for Automation and Measurements PIAP, Warsaw, Poland
{mslomiany,pgemza,fjedrzejczyk,mmacias,jglowka}@piap.pl

**Abstract.** This article describes hardware and software architecture of system for road infrastructure monitoring. The specific emphasis is on camera based vehicle detection based on information coming from multiple sensors. Extensive overview of available visual vehicle detectors is provided and choice made for the Roadsens+ system development is explained. Presented experimental results includes quality rates for two different detection implementations and samples of positive detections as well as false positives. Achieved results allow to assess system usefulness in practical conditions.

**Keywords:** Video processing · Convolutional Neural Networks · Monitoring · Vehicle detector

## 1 Introduction

Speed enforcement systems such as photo radars are means to improve speed limit enforcement and deter speeding. From technological perspective they are using radar systems to determine vehicle speed and optro-electronic sensors to document information about the vehicle. It is clear that success of implementation of such systems depends on actual deployment and use.

New technologies allows to automatically detect multiple types of dangerous behaviours of drivers on the road. Speeding is only one among others such as: entering the crossings at red light, using of bus lanes, not following the turn signs on crossings or overtaking in forbidden sections. Products available on the market use 3D radars and visual systems. Manufacturers are using different techniques to determine crossing of road rules - hence the solutions vary in the way they are integrated with the infrastructure and what kind of data is available to the system operator.

On the Polish market there are several systems available or planned with different characteristics. Examples includes Integra 3D, PolCam Smart Eye, TrafiStar SR590 3D. All mentioned systems allows tracking of 32 objects simultaneously and work on 4 lanes in both directions.


R. Szewczyk et al. (Eds.): AUTOMATION 2020, AISC 1140, pp. 157–169, 2020.
https://doi.org/10.1007/978-3-030-40971-5_15

This paper presents the results of development programme done in cooperation between Sieć Badawcza Łukasiewicz - Przemysłowy Instytut Automatyki i Pomiarów PIAP and largest Polish manufacturer or speed control devices: Zurad Sp. z o. o. Project was called: RoadSens+. The objective of the work was to combine different capabilities into one system and to allow flexible use of the hardware and software solutions in different operating scenarios. All different aspects of detectors were considered including speed control - local and on road sections, red light control, dangerous maneuvers etc. One of the core requirements was to develop modular, fast, efficient system with minimal needs for integration with existing infrastructure. Following sections summarizes technical challenges, design choices and obtained results with specific attention to the topic of vehicle detection.

## 2 System Architecture

### 2.1 Hardware

On of the goals was to design system that can perform functions of speed camera, section control and red lights camera, using same components (with different configuration).

System division into hardware components is shown on Fig. 1. Most important components are:

- central unit,
- radar (3D),
- camera module with IR illuminator,
- flash lamp.

Central unit is responsible for all processing tasks (including vehicle detection). It's also main communication hub, receiving data from all system components and is capable of sending events to authorities.

### 2.2 Software

In this subsection the system software structure and composition are presented. The control (logic) part of the Roadsense+ system is responsible for:

- receiving and storing data from external components: cameras, radar,
- processing of received data:
  - performing detection of vehicle and its position,
  - identifying vehicle features (like: license plate, ADR plate, type, make, model, color),
  - monitoring the status of traffic lights.
- performing fusion of data from all sensors,
- detecting traffic violation.
- communicating with authorities (reporting violations).

Overall system software architecture is shown in Fig. 2.

System has been divided into nodes accordingly. Node is a process which perform single functionality and is connected with other nodes. Each node can communicate with other nodes by publishing and receiving data.

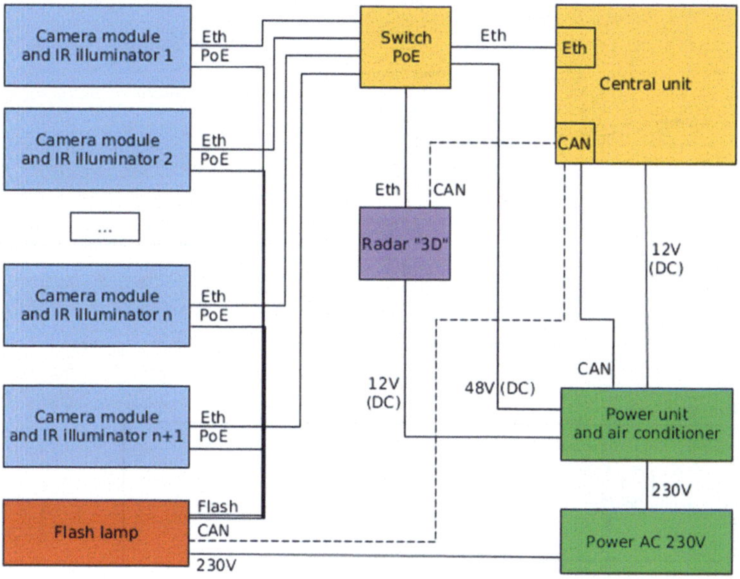


**Fig. 1.** Architecture of Roadsense+ subsystem

## 3 Vehicle Detection

The main goal of vision based vehicle detector used in Intelligent Transportation Systems (ITS) is to process incoming camera frames to detect vehicles passing area and begin it's tracking. By that we can measure vehicle motion parameters: vehicle velocity, acceleration, and trajectory [9]. These informations can be used to determine traffic offense committed by the driver of this vehicle.

Detection accuracy of vision system operating in outdoor environment is affected by many factors such as [4,9]:

- *Vehicle type* various vehicle differs by shape, size and color. Also shadow cast by vehicle can be detected as part of the vehicle's or as other vehicle
- *Traffic conditions* vehicles in close proximity e.g during a traffic jam can be merged into single detection object
- *Lighting condition* illumination differs during the day and especially at night. Additional lighting equipment can be used e.g street lamp, infrared camera to allow detection at nighttime. Also shadows cast by nearby object affect detection accuracy.
- *Atmospheric conditions* camera visibility decreases during rain, fog etc.
- *Measurement based* camera vibration, moving object motion blur effect, vehicle pose changes during maneuvers.
- *Environment based* motion, color and shape variation of background objects.

Vehicle detection methods can be divided into **Feature Detection** and **Motion Detection** approaches [18]. Motion Based approaches detect object by distin-

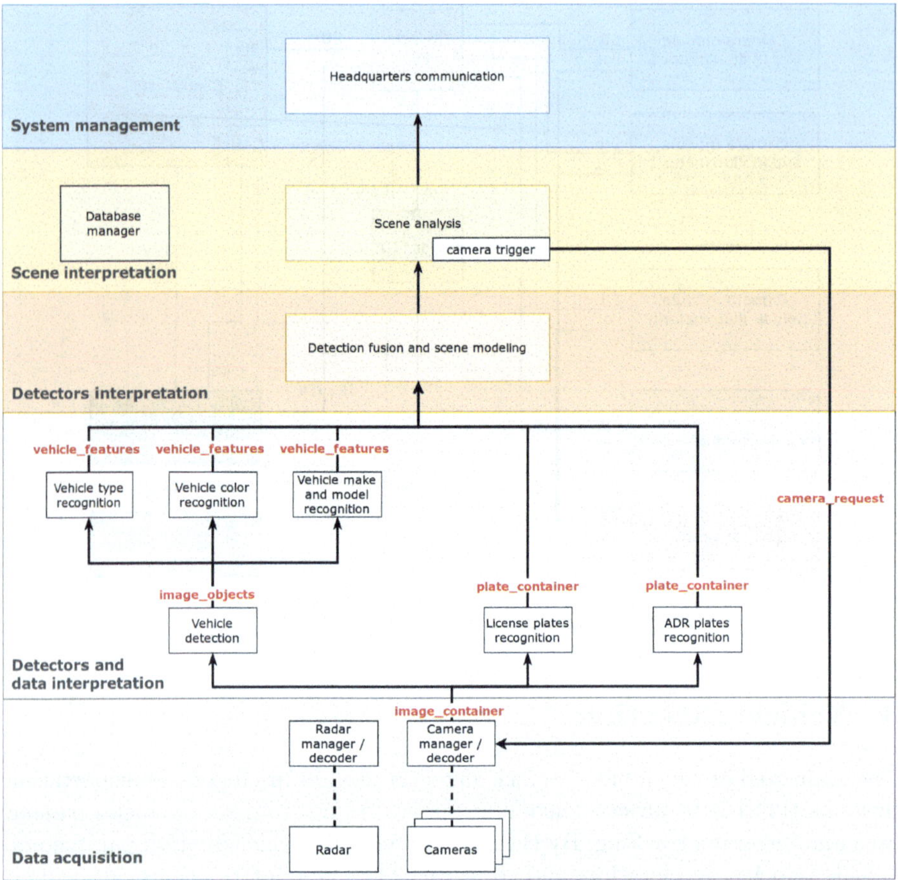


**Fig. 2.** Software architecture

guishing it from static environment, while Feature Based methods search specific combinations patterns (color, texture and shape) on image to detect a vehicle.

## 3.1 Motion Based Detection

In this methods vehicle detection in performed by extracting from static background moving foreground or by computing pixel difference between sequence of images. Methods used in this approach are Background Subtraction and Frame Differencing.

**Background subtraction** consists in determining the difference between the pixels from the current photo and the pixels from the background model. In this method a threshold value is used to determine moving objects (foreground). The threshold value can have a fixed or variable value. In this approach we can list following methods of determining the probability distribution of the value of background pixels [16]:

- based on known background photos (parametric model),
- based on changes in background pixels value (non-parametric model).

Methods based on a parametric model build a background model based on the provided photos and then comparing subsequent frames to the model and classifying the regions of the photos as a background or foreground. **Frame Averaging** method is based on determining average pixel values based on a set of images from a given place. The resulting background model is then subtracted from the incoming frames. **Median Filter** is determining the background by calculating the median for each pixel in the set of photos [16]. **Single Gaussian** - pixels are classified as background or foreground depending on their position in the Gaussian distribution of specific pixel value in a set of past frames. It assumes that the value of each pixel in the photo changes slightly over time and stays within the normal distribution. A rapid change in pixel value (caused by motion) causes them to fall out of the normal distribution. The background model is determined recursively based on changes in pixel values [3]. **Gaussian Mixture Model** assumes that the image contains many different normal distributions, therefore it is better to classify pixels as a background or foreground depending on the resultant position of many distributions [2].

Non-parametric methods do not create a model because they treat pixel values as resulting from some specific distribution, and the probabilistic representation of the environment is created based on the previous pixel value [11]. An example of a method using this type of background model is the **Kernel Density Estimation** in which KDE describes multimodal probability density function. The probability for background pixels is determined from previous n-samples [16].

**Frame Differencing** is based on determining the difference between the values of the corresponding pixels in subsequent photos. A threshold value is applied to the determined values deciding whether the change can be treated as object movement. In these methods detection is performed by using two [15] or more [8] images. The advantage of this type of method is low computational complexity, while the disadvantages are high susceptibility to interference, lighting conditions and low accuracy when detecting very fast/slow objects or many objects at once.

### 3.2 Feature Based Detection

These types of methods use information about features such as the shape, color or texture of the object, on the basis of which the object is detected. Unlike motion-based methods, these methods can also detect non-moving vehicles [9].

**Machine Vision Approach.** In this approach, information on such features as color, symmetry, contours, shadow, vehicle lights, tires or windows are used to detect the vehicle in the photo. For analyzing a photo in terms of its features the following methods can used [16]:

- Haar-like features - pixel intensity in the detection window is summed for each region of the image, and then the difference between adjacent regions is calculated,
- Histogram of Oriented Gradients (HOG) - in the measuring window gradients are determined based on the size and direction of the pixel intensity change, from which next the HOG is created.

Then, based on the obtained features of the object, classification is performed. Methods such as SVM [19] and AdaBoost [17] are used for this purpose.

**Deep learning Approach** is a machine learning method that allows the extraction of features describing an object automatically. Convolutional Neural Networks (CNN) created for this purpose are used to classify the objects in the picture. CNN consists of many layers on which nonlinear operations are performed. The input photo is analyzed using a moving matrix called a filter that extracts features from the photo creating feature maps. Output from feature maps is downsampling by so-called pooling layer to reduce computational complexity for subsequent layers Finally, the data goes to a fully-connected layer where it is classified into one of the categories [1]. Architecture of CNN is shown in Fig. 3.

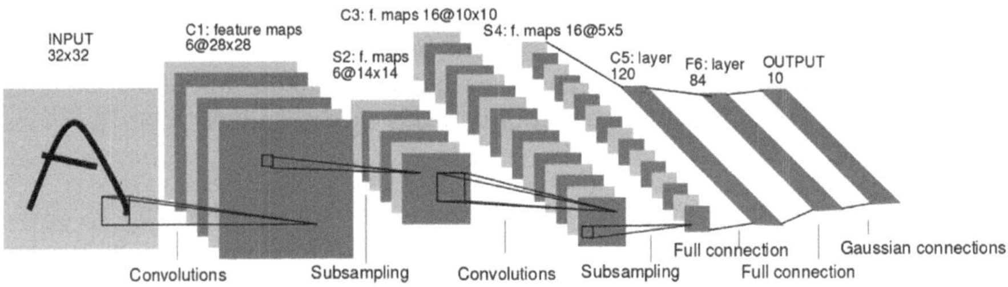


**Fig. 3.** Architecture of Convolutional Neural Network [7]

The problem with this method is the variable length of the output layer due to the variable occurrence of objects of interest in the images The solution seems to be to choose different regions of the photo and classify the object using CNN. However this leads to the need to select a large number of regions, which significantly increases computational complexity [6]. For this reason, algorithms such as Faster R-CNN or YOLO are used for vehicles detection.

**R-CNN** is an improved version of CNN called Regional-based CNN. Instead of choosing a large number of Regions of Interest it was limited to around 2000 on which a fixed-length feature vectors extraction is performed using CNN. Finally, each region is classified [6] as shown in Fig. 4.

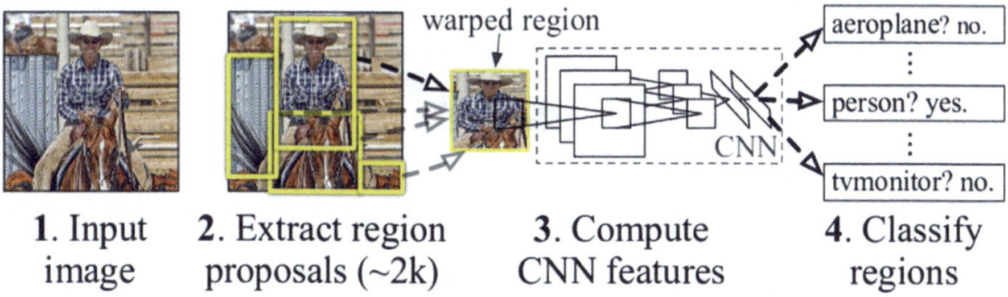


**Fig. 4.** Architecture of R-CNN [6]

Training of R-CNN is time and resource consuming process [5] because it is necessary to classify 2000 regions for each image [6]. In addition, the detection time for one photo on typical computer system is much longer than 40 ms that is required to operate continuously on typical 25 Hz video stream. To overcome these drawbacks, an improved **Fast R-CNN** algorithm has been developed. This is a similar solution only with the difference that instead of feeding CNN RoI (Region of Interests), we feed CNN with a photo to generate a convolutional feature map. Based on convolutional feature map region of proposals are identified and reshape into fixed size squares by RoI pooling layer from which fixed-length feature vector is extracted and processed by fully connected layer. This results in two output layers: softmax probabilities and per-class bounding-box regression offset [5]. Architecture of Fast R-CNN is shown in Fig. 5.

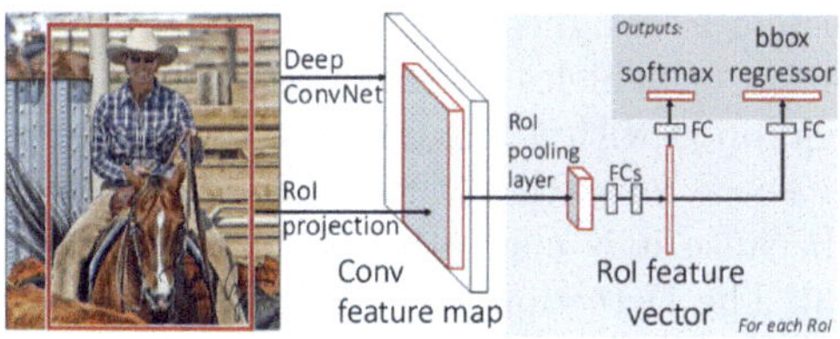


**Fig. 5.** Architecture of Fast R-CNN [5]

Even faster algorithm than R-CNN and Fast R-CNN is **Faster R-CNN**. Instead of using selective search to find the region proposals another neural network is used (Fig. 6). According to Shaoqing Ren et al. the Faster R-CNN is ten times faster than R-CNN, therefore it can be used in real-time applications [14] .

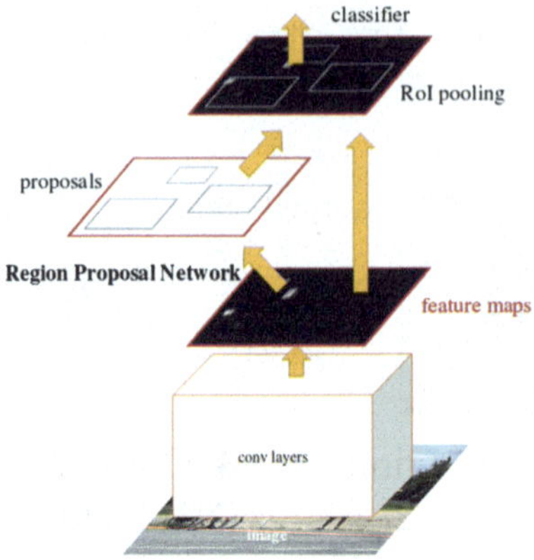


**Fig. 6.** Architecture of Faster R-CNN [14]

Another real time object detection algorithm is **YOLO**. It process parts of image with high probability of containing the object. Unlike regional-based methods process entire image during training and test. Bounding boxes and their class probabilities are predicted by single convolutional network. Image is split into $S \times S$ grid and for each grid a bounding boxes, confidence of these boxes and class probabilities are predicted. For object localization within image the bounding boxes with value above the threshold are selected [12] as it is shown in Fig. 7.

# 4 Performance and Operation

## 4.1 System Operation

The Roadsense+ system is designed for monitoring sections of the roads called scenes, and registration of vehicles committing various traffic violations. Each subsystem can be equipped with a maximum of four camera modules and analyze the state of the scene for each of them. For vehicle detection, we used the adopted YOLOv3 because it is currently the fastest network that gives satisfactory accuracy [13]. The Detection Region of Interest has been limited to one lane, regardless of scene visibility, which reduces computational complexity.

Vehicle Detector node is responsible for vehicles detection, it is listens for messages containing frames from cameras operating in the system. Detection made by this node are then passed to subsequent nodes that analyze the vehicle's features (license plate, model, make, type, color). Based on these features and the history of locations of the vehicle on the scene, a decision is made as to whether

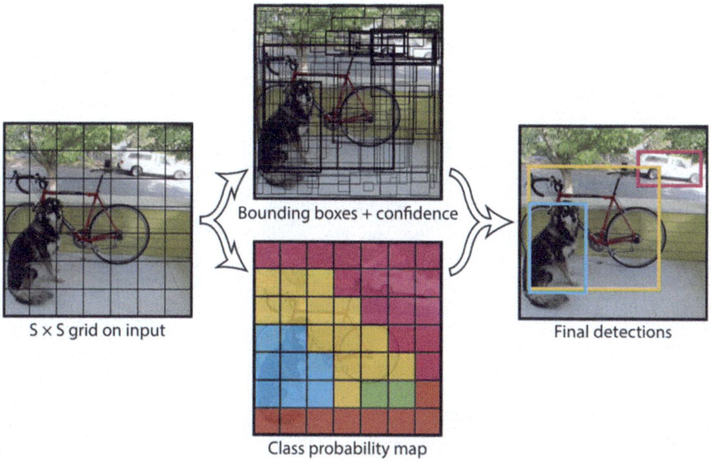


**Fig. 7.** YOLO detection model [12]

the vehicle has committed a traffic violation. If so, a photo of the vehicle is taken and violation evidence is created.

### 4.2 Processing for Various Video Stream Configurations

System requirements indicate that vehicle detection should be done at frequency higher than 10 Hz. Computer performance allows for 20 [Hz] of single video stream processing for the implemented method. Due to use of four cameras the resulting performance drops to 5 [Hz]. To solve this problem a separate graphics card for each camera may be used. Such solution would increase costs and the energy demand of the system. Instead it was decided to combine frames from individual cameras as shown in Fig. 8 and process such image by one detector. In such case it was necessary to train new network (YOLOv3-PIAP) for specific input images. The original YOLOv3 network detected non-existent objects at the places where pictures from individual cameras are joined.

### 4.3 Experiments

**Experiments Conditions and Environment.** One of the aims of the experiments was to study performance of YOLOv3-PIAP in comparison to YOLOv3 in vehicles detection and classification. Due to tendency of YOLOv3 to detect objects between frames joints it was necessary to filter detection results to make comparison possible. Detection is rejected if width or height of bounding box is lower than 30 pixels. The experimental research were carried out with a video recorded in close-to-target environment as shown in Fig. 8. Video was recorded during one day between 10:00 and 14:00 which resulted in test data-set of 8160

**Fig. 8.** Processing of streams from multiple cameras

vehicles - 7581 cars, 496 trucks and 83 bikes. The experiment consisted of streaming previously recorded video into Roadsense+ system. To simulate system working conditions recording was streamed at once up to four times.

**Quality Rates.** In order to effectively monitor a controlled section of the road, it is necessary to correctly detect and classify objects on this section. This means that the number of correctly detected vehicles should be as high as possible, and the number of non-vehicle objects classified as vehicles as low as possible. We used the following quality measures to evaluate detection systems that use different networks [10] :

– True Negative Rate

$$TNR = \frac{TN}{TN + FP} \cdot 100[\%] \tag{1}$$

where : $TN$ - True Negative stands for number of non-vehicle samples not classified as vehicle, $FP$ - False Positive is number of non-vehicle samples classified as vehicle.

– *Recall*

$$Recall = \frac{TP}{TP + FN} \cdot 100[\%] \tag{2}$$

where : $TP$ - True Positive is number of vehicle samples classified correctly, $FN$ - False Negative stands for number of vehicle samples not classified as vehicle.

– *Precision*

$$Precision = \frac{TP}{TP + FP} \cdot 100[\%] \tag{3}$$

- *Accuracy*

$$Accuracy = \frac{TP + TN}{TP + TN + FP + FN} \cdot 100[\%] \tag{4}$$

- $F_1$ score

$$F_1 = 2 \cdot \frac{Precision \cdot Recall}{Precision + Recall} \cdot 100[\%] \tag{5}$$

**Experiment Results.** During experiment all of nearly 216000 joined frames were processed and vehicles were present on 68.3 [%]. However for evaluation we excluded from consideration detections made during incomplete presence of the vehicle on scene, that mean:

- until vehicle top is invisible during entering,
- when vehicle's license plate is invisible during exceeding.

Examples of detections consider as True Positive are shown in Fig. 9.

**Fig. 9.** Examples of True Positive detections

Processing frequency never drops below 15 [Hz] during whole experiment. Reject of small object detections results in similar values of $TNR$ for both networks, examples of $FP$ detetions are shown in Fig. 10. Other quality rates reached significant higher values in case of YOLOv3-PIAP network. Results of experiment are shown in Table 1.

**Fig. 10.** Examples of False Positive detections

**Table 1.** Results of experiment

| | TNR [%] | Prc. [%] | Rec. [%] | Acc. [%] | $F_1$ [%] |
|---|---|---|---|---|---|
| YOLOv3 | | | | | |
| Car | 96.99 | 87.92 | 97.92 | 82.63 | 89.63 |
| Truck | 97.2 | 89.52 | 97.5 | 83.65 | 90.05 |
| Bike | 95.55 | 86.83 | 96.56 | 81.17 | 88.2 |
| YOLOv3-PIAP | | | | | |
| Car | 97.54 | 95.53 | 98.51 | 94.36 | 96.39 |
| Truck | 97.69 | 94.76 | 98.11 | 92.5 | 95.22 |
| Bike | 96.14 | 94.1 | 97.45 | 92.82 | 95.08 |

## 5 Conclusions

This article presents different methods to achieve vehicle detection for infrastructure monitoring. Applied solutions was tested with real data in different weather and light conditions. Achieved results shows that system is able to operate and provide reliable data in real-time. Road safety systems can use such approach to provide actual situation monitoring and behaviour analysis. Chosen solutions allowed for optimization of resources and cost-effective implementation. Further work includes tests with large sets of data and fine-tuning of different behaviour detectors in multiple conditions. Such a large scale testing is planned before final system deployment.

**Acknowledgements.** Project called "Opracowanie innowacyjnego, modułowego systemu Roadsense+ zwiększającego bezpieczeństwo ruchu drogowego", was co-financed by the European Union from the European Regional Development Fund, realized within the first priority axis called "Wykorzystanie działalności badawczo-rozwojowej w gospodarce", action 1.2 called "Działalność badawczo-rozwojowa przedsiębiorstw" by Mazovian Unit of EU Programmes Implementation in Warsaw. Grant agreement no. RPMA.01.02.00-14-6215/16-00.

## References

1. Albawi, S., Abed Mohammed, T., Alzawi, S.: Understanding of a convolutional neural network, August 2017
2. Bhaskar, P.K., Yong, S.: Image processing based vehicle detection and tracking method. In: 2014 International Conference on Computer and Information Sciences (ICCOINS), pp. 1–5, June 2014
3. Dong, E., Han, B., Yu, X., Du, S.: Moving targets detection based on improved single gaussian background model. In: 2018 IEEE International Conference on Mechatronics and Automation (ICMA), pp. 1179–1183, August 2018
4. Gazzah, S., Mhalla, A., Amara, N.E.B.: Vehicle detection on a video traffic scene: review and new perspectives. In: 2016 7th International Conference on Sciences

of Electronics, Technologies of Information and Telecommunications (SETIT), pp. 448–454, December 2016
5. Girshick, R.: Fast r-cnn (2015)
6. Girshick, R., Donahue, J., Darrell, T., Malik, J.: Rich feature hierarchies for accurate object detection and semantic segmentation (2013)
7. Lecun, Y., Bottou, L., Bengio, Y., Haffner, P.: Gradient-based learning applied to document recognition. In: Proceedings of the IEEE, pp. 2278–2324 (1998)
8. Lin, Y., Fang, M., Shihong, D.: An object reconstruction algorithm for moving vehicle detection based on three-frame differencing. In: 2015 IEEE 12th Intl Conf on Ubiquitous Intelligence and Computing and 2015 IEEE 12th Intl Conf on Autonomic and Trusted Computing and 2015 IEEE 15th Intl Conf on Scalable Computing and Communications and Its Associated Workshops (UIC-ATC-ScalCom), pp. 1864–1868, August 2015
9. Liu, Y., Tian, B., Chen, S., Zhu, F., Wang, K.: A survey of vision-based vehicle detection and tracking techniques in its. In: Proceedings of 2013 IEEE International Conference on Vehicular Electronics and Safety, pp. 72–77, July 2013
10. Olson, D.L., Delen, D.: Advanced Data Mining Techniques, 1st edn. Springer Publishing Company Incorporated, Heidelberg (2008)
11. Wan, Q., Wang, Y.: Background subtraction based on adaptive non-parametric model. In: 2008 7th World Congress on Intelligent Control and Automation, pp. 5960–5965, June 2008
12. Redmon, J., Divvala, S., Girshick, R., Farhadi, A.: You only look once: Unified, real-time object detection. In: 2016 IEEE Conference on Computer Vision and Pattern Recognition (CVPR), pp. 779–788, June 2016
13. Redmon, J., Farhadi, A.: Yolov3: An incremental improvement. arXiv (2018)
14. Ren, S., He, K., Girshick, R., Sun, J.: Faster r-cnn: Towards real-time object detection with region proposal networks (2015)
15. Sengar, S.S., Mukhopadhyay, S.: A novel method for moving object detection based on block based frame differencing. In: 2016 3rd International Conference on Recent Advances in Information Technology (RAIT), pp. 467–472, March 2016
16. Shobha, B.S., Deepu, R.: A review on video based vehicle detection, recognition and tracking. In: 2018 3rd International Conference on Computational Systems and Information Technology for Sustainable Solutions (CSITSS), pp. 183–186, December 2018
17. Tang, Y., Zhang, C., Gu, R., Li, P., Yang, B.: Vehicle detection and recognition for intelligent traffic surveillance system. Multimedia Tools Appl. **76**(4), 5817–5832 (2017)
18. Tian, B., Morris, B.T., Tang, M., Liu, Y., Yao, Y., Gou, C., Shen, D., Tang, S.: Hierarchical and networked vehicle surveillance in its: a survey. IEEE Trans. Intell. Transp. Syst. **16**(2), 557–580 (2015)
19. Wen, X., Shao, L., Fang, W., Xue, Y.: Efficient feature selection and classification for vehicle detection. IEEE Trans. Circuits Syst. Video Technol. **25**(3), 508–517 (2015)

# Fractional Order $PI^{\alpha}D^{\beta}$ Controller for the Inverted Pendulum

Krzysztof Oprzędkiewicz(✉), Maciej Rosół, and Jakub Żegleń

AGH University, al. A. Mickiewicza 30, 30 059 Kraków, Poland
kop@agh.edu.pl

**Abstract.** In the paper the Fractional Order PID (FOPID) control for inverted pendulum system is addressed. The fractional order parts of the both controllers are approximated using ORA approximation, the FOPID-s were tuned using simulations. As the cost function the integral from sum of square control and position was considered. Results of simulations show that the use of FOPID controllers allows to obtain the better control performance in the sense of considered cost function than use of classic PID controllers, however the control performance in the FOPID control system is determined also by parameters of ORA approximation used to describe fractional parts of controllers.

**Keywords:** Inverted pendulum · Fractional order control · ORA approximation

## 1 Introduction

The FOPID control of the inverted pendulum system has been presented by many Authors, for example in papers: [1,4,6].

## 2 Preliminaries

### 2.1 Basic Ideas

Presentation of elementary ideas is started with a definition of a fractional order, integro-differential operator given for example by [2]:

**Definition 1.** *(The elementary fractional order operator) The fractional order integro-differential operator is defined as follows:*

$$ {}_aD_t^{\alpha}f(t) = \begin{cases} \frac{d^{\alpha}f(t)}{dt^{\alpha}} & \alpha > 0 \\ 1 & \alpha = 0 \\ \int_a^t f(\tau)(d\tau)^{-\alpha} & \alpha < 0 \end{cases} . \quad (1) $$

*where $a$ and $t$ denote time limits of operator calculation, $\alpha \in \mathbb{R}$ denotes the fractional order of the operation.*


R. Szewczyk et al. (Eds.): AUTOMATION 2020, AISC 1140, pp. 170–181, 2020.
https://doi.org/10.1007/978-3-030-40971-5_16

Next an idea of Gamma Euler function is recalled (see for example [3]):

**Definition 2.** *The Gamma function*

$$\Gamma(x) = \int_{0}^{\infty} t^{x-1} e^{-t} dt. \tag{2}$$

An idea of Mittag-Leffler function needs to be given next. It is a non-integer order generalization of exponential function $e^{\lambda t}$ and it plays crucial role in solution of fractional order state equation. The one parameter Mittag-Leffler function is defined as follows:

**Definition 3.** *The one parameter Mittag-Leffler function*

$$E_{\alpha}(x) = \sum_{k=0}^{\infty} \frac{x^k}{\Gamma(k\alpha + 1)}. \tag{3}$$

The fractional order operator can be described by different definitions, given by: The Grünwald-Letnikov (the GL definition), Riemann-Liouville (RL definition) and Caputo (C definition). In this paper only the Caputo definition will be used.

**Definition 4.** *The Caputo definition of the FO operator*

$${}_0^C D_t^{\alpha} f(t) = \frac{1}{\Gamma(N-\alpha)} \int_{0}^{\infty} \frac{f^{(N)}(\tau)}{(t-\tau)^{\alpha+1-N}} d\tau. \tag{4}$$

For the C definition the Laplace transform is defined as follows:

**Definition 5.** *(The Laplace transform for Caputo operator)*

$$\begin{gathered} \mathcal{L}({}_0^C D_t^{\alpha} f(t)) = s^{\alpha} F(s), \quad \alpha < 0 \\ \mathcal{L}({}_0^C D_t^{\alpha} f(t)) = s^{\alpha} F(s) - \sum_{k=0}^{n-1} s^{\alpha-k-1} {}_0D_t^k f(0), \\ \alpha > 0, \quad n-1 < \alpha \leq n \in N \end{gathered} \tag{5}$$

The use of C definition (4) and (5) allows to define the fractional order transfer function, analogically to integer order case:

$$G_1(s) = \frac{b_L s^{\beta_L} + ... + b_1 s^{\beta_1} + b_0}{a_P s^{\alpha_P} + ... + a_1 s^{\alpha_1} + a_0}, \tag{6}$$

where $\beta_L ... \beta_1$ and $\alpha_P ... \alpha_1$ are fractional orders, $b_L ... b_0$ and $a_P, ..., a_0$ are coefficients.

The method proposed by Oustaloup (see [5]) allows to approximate an elementary FO transfer function $s^\alpha$ by the finite, integer-order transfer function, close to Pade approximation:

$$s^\alpha \approx k_f \prod_{n=1}^{N} \frac{1+\frac{s}{\mu_n}}{1+\frac{s}{\nu_n}} = \frac{L_{ORA}(s)}{D_{ORA}(s)}. \tag{7}$$

In (7) $N$ is the order of the approximation, $k_f$ is the steady state gain, expected to assure the convergence the step response of the approximation to step response of the real plant in a steady state. Coefficients $\mu_n$ and $\nu_n$ are calculated as follows:

$$\begin{aligned} \mu_1 &= \omega_l \sqrt{\eta} \\ \nu_n &= \mu_n \gamma,\ n = 1,..,N \\ \mu_{n+1} &= \nu_n \eta,\ n = 1,..,N-1 \end{aligned} \tag{8}$$

where:

$$\begin{aligned} \gamma &= \left(\frac{\omega_h}{\omega_l}\right)^{\frac{\alpha}{N}} \\ \eta &= \left(\frac{\omega_h}{\omega_l}\right)^{\frac{1-\alpha}{N}} \end{aligned} \tag{9}$$

In (9) $\omega_l$ and $\omega_h$ describe the bandwidth, for which parameters are be calculated. It can be expressed in the logarithmic scale:

$$\begin{aligned} \omega_l &= 10^{-W}, \\ \omega_h &= 10^{W} \end{aligned} \tag{10}$$

where $W \in \mathbb{Z}$ describes the bandwidth.

# 3 The Inverted Pendulum System

## 3.1 Model of the Pendulum

The construction of the inverted pendulum model is not much different from its typical counterpart [8]. Consider inverted pendulum on a cart that can move only in one direction on rails with two rods attached to it. Typical inverted pendulum system is shown in Fig. 1. This system has infinite equilibrium points, but the two most important ones are:

- stable - the pendulum is directed vertically downwards
- unstable - the pendulum is directed vertically upwards.

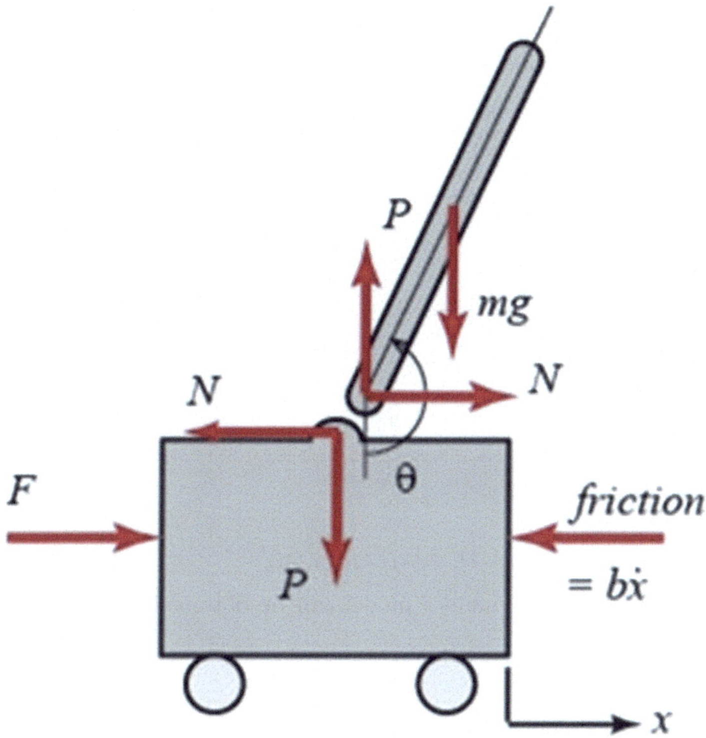


**Fig. 1.** The inverted pendulum [6].

The following variables will be used in formulas determining the mathematical model:

$M$ - mass of the cart
$m$ - mass of the pendulum
$\theta$ - pendulum angle
$g$ - standard gravity
$L$ - length of the pendulum
$I$ - moment of inertia of the pendulum
$F$ - forced applied to the cart
$x$ - cart position
$b$ - coefficient of friction

The goal of control is to stabilize the system in its unstable position, where pendulum is directed vertically upwards. Mathematical model is described by determining the kinetic and potential energy in the system. The general equations of inverted pendulum system are [8]:

$$(M + m)\ddot{x} + b\dot{x} = F + mL\dot{\theta}^2 \sin\theta - mL\ddot{\theta}\cos\theta \qquad (11)$$

$$(I + mL^2)\dot{\theta} = mgL\sin\theta - mL\dot{x}\cos\theta \tag{12}$$

Next step is to determine $\ddot{\theta}$ and $\ddot{x}$

$$\ddot{\theta} = \frac{1}{q}(FmL\cos\theta - bmL\dot{x}\cos\theta - (M+m)mLg\sin\theta + m^2L^2\dot{\theta}^2\sin\theta\cos\theta) \tag{13}$$

$$\ddot{x} = \frac{1}{q}((I + mL^2)(-F + b\dot{x} - mL\dot{\theta}^2\sin\theta) + m^2L^2g\sin\theta\cos\theta) \tag{14}$$

where:

$$q = m^2L^2\cos^2\theta - (M+m)(I + mL^2) \tag{15}$$

By entering the following constants:

$$x_1 = \theta \tag{16}$$

$$x_3 = x \tag{17}$$

and the state vector:

$$x = [x_1, x_2, x_3, x_4]^T \tag{18}$$

the following non-linear dynamics model can be obtained in the form of state equations

$$\dot{x}_1 = x_2 \tag{19}$$

$$\dot{x}_2 = \frac{1}{q}(FmL\cos x_1 - bmL\dot{x}_3\cos x_1 - (M+m)mLg\sin x_1 + m^2L^2\dot{x}_1^2\sin x_1\cos x_1 \tag{20}$$

$$\dot{x}_3 = x_4 \tag{21}$$

$$\dot{x}_4 = \frac{1}{q}((I + mL^2)(-F + b\dot{x}_3 - mL\dot{x}_1^2\sin x_1) + m^2L^2g\sin x_1\cos x_2) \tag{22}$$

where:

$$q = m^2L^2\cos^2 x_1 - (M+m)(I + mL^2) \tag{23}$$

The above model is non-linear, so it can be linearized around unstable equilibrium point $x_0 = [0 \quad 0 \quad 0 \quad 0]^T$. For small values of $x_1$ it can be assumed that $\sin x_1 \approx x_1$ and finally we obtain the following linear model:

$$\ddot{x}_1 = \frac{(M+m)g}{ML}x_1 + \frac{b}{ML}\dot{x}_3 - \frac{F}{ML} \tag{24}$$

$$\ddot{x}_3 = -\frac{mg}{M}x_1 - \frac{b}{M}\dot{x}_3 + \frac{F}{M} \tag{25}$$

If we define the state vector:

$$\boldsymbol{x} = [\dot{\theta} \quad \theta \quad \dot{x} \quad x]^T \tag{26}$$

then the nonlinear model of the pendulum is as below:

$$\dot{\boldsymbol{x}} = f(\boldsymbol{x}, F) \tag{27}$$

and the linear model takes the form:

$$\dot{\boldsymbol{x}} = A\boldsymbol{x} + Bu \tag{28}$$

where $A$ and $B$ are defined as follows:

$$A = \begin{bmatrix} 0 & \frac{(M+m)g}{ML} & \frac{b}{ML} & 0 \\ 1 & 0 & 0 & 0 \\ 0 & -\frac{mg}{M} & -\frac{b}{M} & 0 \\ 0 & 0 & 1 & 0 \end{bmatrix} \tag{29}$$

$$B = \begin{bmatrix} -\frac{1}{ML} \\ 0 \\ \frac{1}{M} \\ 0 \end{bmatrix} \tag{30}$$

### 3.2 The Integer Order Control Algorithm

The inverted pendulum can be moved from the lower position to the upper position using the swing-up controller, reacting accordingly on changes in the total value accumulated in the mechanical energy system [7,8]. This information is obtained from the feedback loop. When energy is properly added to the system, it is possible to move the pendulum in such a way, that it reaches the upper, unstable equilibrium point. Around it, the control algorithm should be switched to the second one. The task of the second regulator is to keep the pendulum around the unstable equilibrium point.

The energy $E$ of the system in the unstable equilibrium point is equal 0. This value can be calculated by the following formula:

$$E = \frac{\dot{\theta}I}{2} + mgl(\cos\theta - 1) \tag{31}$$

This allows to calculate the control signal of the swing-up controller as follows:

$$u = -\dot{\theta}\cos\theta \tag{32}$$

The controller is switched if the neighbourhood of the unstable equilibrium point is achieved. Then the typical PID controller can be applied, but in this paper the FOPID controllers are employed. They are presented in the next section. In order to properly control the pendulum around the equilibrium point, two controllers are needed. The first uses information about the position of the pendulum, while the other uses the position of the cart. Next the both control signals are summarized. The control system is shown in the Fig. 2.

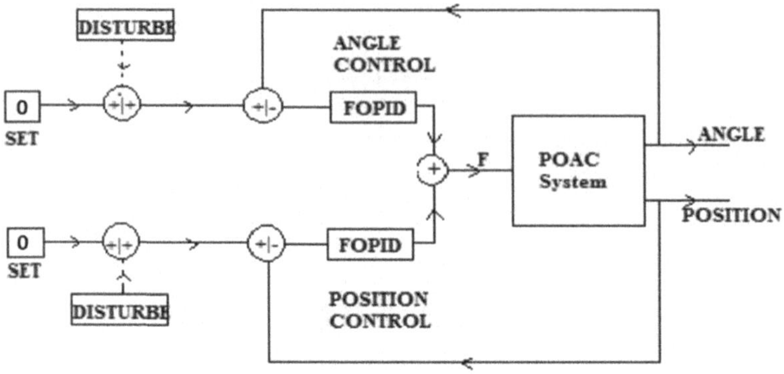


**Fig. 2.** Control system of the pendulum [1].

# 4 The FOPID Control of the Pendulum

In the control system shown in the Fig. 2 the PID controllers were replaced by FOPID-s. The first FOPID cooperates with the pendulum, the second with the cart and their signals are summarized. They are described as follows:

$$G_p(s) = K_{pp} + K_{pi}\frac{1}{s^{\alpha_p}} + K_{pd}s^{\beta_p}. \tag{33}$$

$$G_c(s) = K_{cp} + K_{ci}\frac{1}{s^{\alpha_c}} + K_{cd}s^{\beta_c}. \tag{34}$$

In (33) and (34) $K_{(p,c)(p,i,d)}$ denote coefficients of proportional, integral and derivative actions for the pendulum and cart controllers respectively, $\alpha_{p,c}$, $\beta_{p,c}$ denote orders of integration and derivative actions for the both controllers respectively. Each non integer order element was approximated using ORA approximation (7)–(9).

The main difficulty during use ORA approximation is caused by the fact that each change of order $\alpha$ or $\beta$ requires to recalculate the whole approximation (7)–(9) for each FO action of controller (together 4 approximations need to be recalculated). The control performance in the considered system is estimated using the following cost function:

$$I = \int_0^{T_f} \left(x^2(t) + u^2(t)\right) dt. \tag{35}$$

where $x(t)$ and $u(t)$ denote the cart position and control signal respectively, $T_f$ is the final time of experiment. The above cost function is a function of all parameters of the both controllers. Additionally, the control performance depends on ORA approximation parameters, e.g. its order $N$ and bandwidth $W$.

## 5 Simulations

Simulations were done using the simulink model presented in Fig. 3. The FOPID controllers were used to control both cart and pendulum position. Results of work the FOPID control system were compared to typical PID controllers.

The parameters of the ORA approximation are following: $N = 12$, $W = 2$. The "narrow" bandwidth $W = 1, 2$ guarantees the numerical stability of simulation. The final time of the experiment was equal: $T_f = 60[s]$. The parameters of the all FOPID and PID controllers are given in the Tables 1 and 2. The values of cost function (35) for both control systems are compared in the Table 3. Diagrams of positions and speeds of cart and pendulum of the both control systems are shown in Figs. 4 and 5.

**Table 1.** Pendulum controller - parameters

| | P | I | $\alpha$ | D | $\beta$ |
|---|---|---|---|---|---|
| FOPID | 20 | 0.5 | $-0.5$ | 1 | 0.5 |
| PID | 0.5 | 0.0 | $-1.0$ | 0.0 | 1 |

**Table 2.** Cart controller - parameters

| | P | I | $\alpha$ | D | $\beta$ |
|---|---|---|---|---|---|
| FOPID | 25 | 0.5 | $-0.5$ | 1 | 0.5 |
| PID | 0.5 | 0.0 | $-1.0$ | 0.0 | 1 |

**Table 3.** Cost function for PID and FOPID

| Controller | Cost function |
|---|---|
| FOPID | 1.1022 |
| PID | 1.8220 |

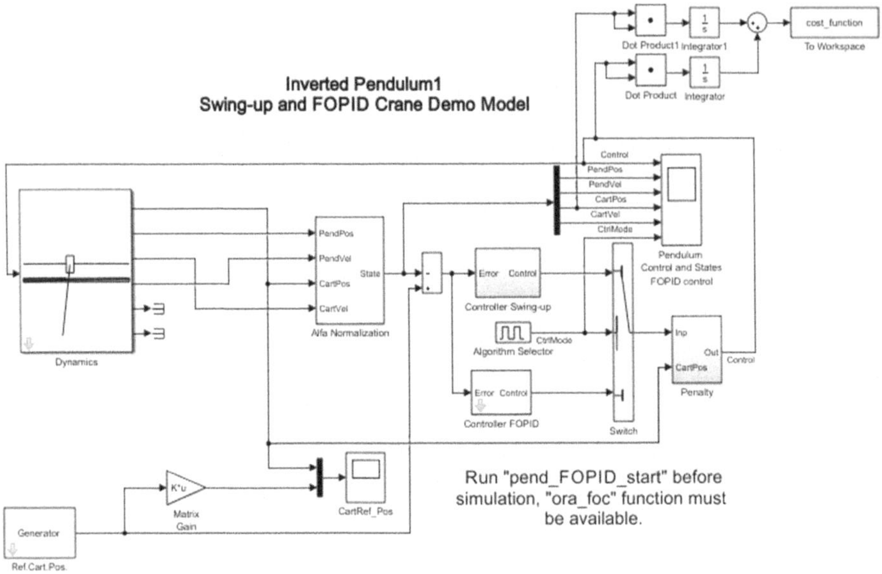


**Fig. 3.** The SIMULINK model of the control system.

**Table 4.** Cost function (35) as a function of ORA approximation parameters: $N$ and $W$.

| N | 4 | 6 | 8 | 10 | 12 | 14 | 16 |
|---|---|---|---|---|---|---|---|
| W = 1 | 1.4388 | 1.3593 | 2.4883 | 1.2044 | 2.2554 | 1.1594 | 1.4337 |
| W = 2 | 1.3127 | 1.6505 | 1.1095 | 1.1662 | 1.1022 | 1.2812 | 1.4573 |

Next the control performance as a function of ORA approximation parameters was examined. Results are presented in the Table 4 and illustrated by Fig. 6.

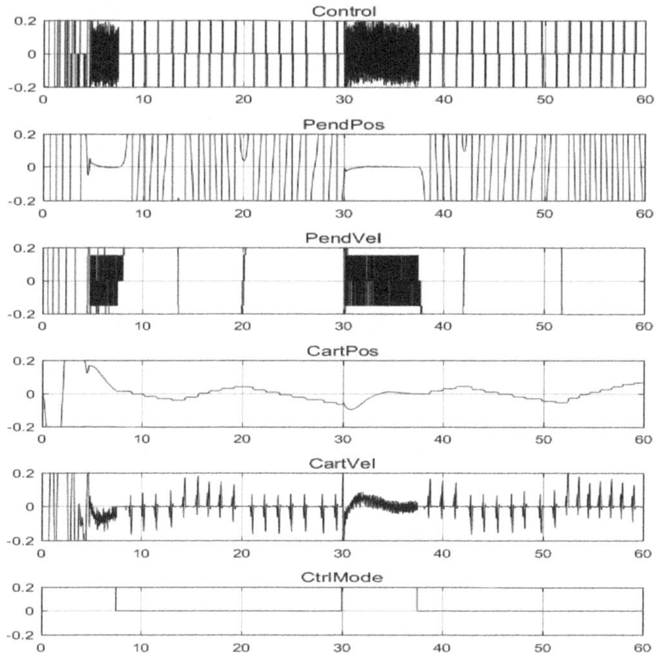


**Fig. 4.** FOPID control - variables

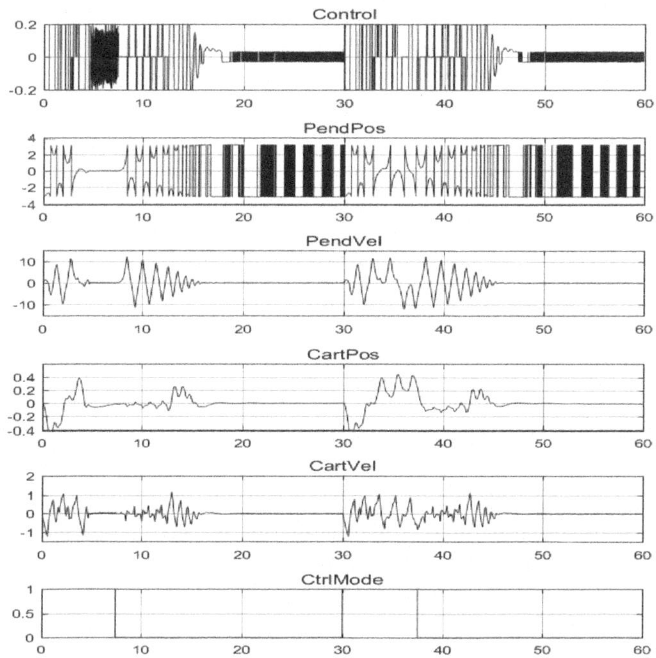


**Fig. 5.** PID control - variables

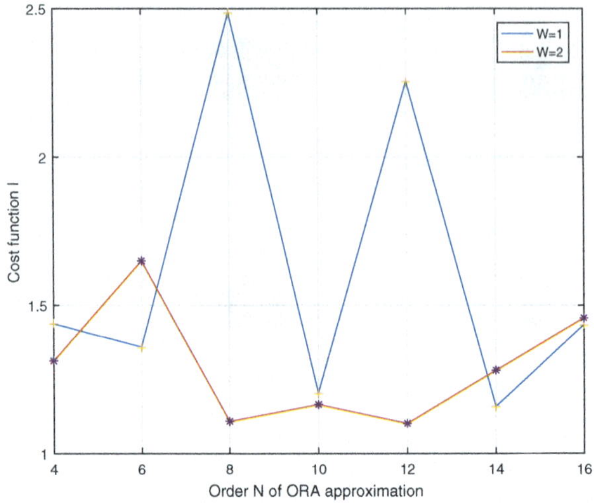


**Fig. 6.** Cost function (35) as a function of ORA approximation parameters: $N$ and $W$.

# 6 Conclusions

The main conclusion from the paper is that the proposed simple FOPID controller is able to assure the better control performance in the sense of considered cost function than the classic PID controller. Next, the control performance is determined not only by controllers parameters, but it also strongly depends on ORA approximation parameters.

The further investigations will cover tests using real experimental system and deeper analysis of the considered control system. For example stability of the system with FOPID needs to be analyzed as well as effective tuning methods for the both FOPID controllers. This problem seems to be interesting due to relatively high number of parameters necessary to assign. The biologically inspired optimization methods are planned to implement. Finally the proposed control system is planned to PLC implementation.

**Acknowledgment.** This paper was sponsored by AGH project no 16.16.120.773.

# References

1. Agarwal, H., Singh, A.P., Srivastava, P.: Fractional order controller design for inverted Pendulum on a Cart System (POAC). WSEAS Transaction on Systems and Control, vol. 10 (2015). E-ISSN: 2224-2856
2. Kaczorek, T.: Selected Problems in Fractional Systems Theory. Springer, Heidelberg (2011)
3. Kaczorek, T., Rogowski, K.: Fractional Linear Systems and Electrical Circuits. Bialystok University of Technology (2014)

4. Mishra, S.K., Chandra, D.: Stabilization and tracking control of inverted pendulum using fractional order PID controllers. J. Eng. vol. 2014 (2014). Hindawi Publishing Corporation, Article ID 752918, 9 pages
5. Oustaloup, A., Levron, F., Mathieu, B., Nanot, F.M.: Frequency-band complex noninteger differentiator: characterization and synthesis. IEEE Trans. Circuits Syst. I: Fundam. Theory Appl. **47**(1), 25–39 (2000)
6. Paliwal, S., Pathak, V.K.: Analysis & control of inverted pendulum system using PID controller. J. Eng. Res. Appl. **7**(5), 01–04 (2018). ISSN: 2248-9622. http://www.ijera.com/papers/Vol7_issue5/Part-4/A0705040104.pdf
7. Lam, J.: Control of an Inverted Pendulum (2018). https://pdfs.semanticscholar.org/b6b2/f079bb6011a28ccd5d28d08df00f57ef5697.pdf
8. Żegleń, J.: Advanced control methods of an inverted pendulum on the card implemented in PLC (Zaawansowane metody sterowania modelem laboratoryjnym wahadła na wózku implementowane w sterowniku PLC) Master Thesis at AGH University under supervision M. Rosół (2018)

# Robotics

# Using a WebSocket Server and the JSON Format to Transmit Data for the Purposes of Implementing the Brain-Computer Technology

Szczepan Paszkiel(✉)

Department of Biomedical Engineering, Opole University of Technology,
Faculty of Electrical Engineering, Automatic Control and Informatics,
Proszkowska 76, 45–271 Opole, Poland
s.paszkiel@po.opole.pl

**Abstract.** This article describes the Emotiv Cortex software and a Raspberry PI device comprising an original solution in the field of control based on the BCI brain-computer interface technology. For the purposes of the conducted research, a Raspberry PI with a Wi-Fi module was used. The Raspberry PI was connected to a universal board, which enabled easy linking of other modules, to be controlled by the persons operating the solution. The brain-computer interface developed by Emotiv (EPOC+ NeuroHeadset) communicates with the Emotiv Cortex software, the messages are sent to the Raspberry PI 2B and then can be used to control a device operating in a smart home system.

**Keywords:** WebSocket · JSON · BCI · Transmit data

## 1 Introduction

Controlling various technical devices using BCI technology is implemented by many units around the world [2,3]. The available literature already describes solutions based on implementing Emotiv technology in smart homes. This is described in, among others, an article titled: Home Automation using Emotiv: Controlling TV by Brainwaves, written by Marquos Zaki, Ali Alquraini, Tarek Sheltami [8]. This design utilizes an operation model similar to the solution proposed in this article, which is the Emotiv Cortex software and a Raspberry PI device. The difference however is that the solution presented in this article utilizes a newer model of an Emotiv device, the Emotiv EPOC+ NeuroHeadset with a higher EEG signal sampling frequency and a cheaper version of the Raspberry PI, in the light of increased operational efficiency, while minimizing the implementation costs. Another solution is a project named: A Brain Computer Interface for Smart Home Control developed by Wei Tuck Lee, Humaira Nisar,

---

S. Paszkiel—PhD. Eng., Assistant Professor.


R. Szewczyk et al. (Eds.): AUTOMATION 2020, AISC 1140, pp. 185–194, 2020.
https://doi.org/10.1007/978-3-030-40971-5_17

Aamir S. Malik, Kim Ho Yeap [6]. It is, however, based on a completely different idea involving control using a small 8051 processor. The concept was enhanced with a graphical interface presenting a virtual home, which can be controlled using a mouse. The graphical interface is also a kind of feedback imaging in the form of changing simulation states. The system enables controlling: multimedia, lighting, central heating system, etc. For the purposes of the research carried out at the Laboratory for NeuroInformatics and Decision Systems of the Opole University of Technology, the Raspberry PI was connected to a universal board, which enabled easy linking of other modules, to be controlled by the persons operating the solution. The principle of operation is shown in Fig. 1. The brain-computer interface developed by Emotiv communicates with Cortex. The messages are then sent to a Raspberry PI and used to control the device. The entire concept behind the application was divided into three blocks: data collection, transport and control.

**Fig. 1.** Operating principle of the proposed solution

## 2 Communication

Communication with the Emotiv system is possible through the provided Emotiv Cortex software. The Cortex software enables communication between external applications and the Emotiv device through a WebSocket server. Emotiv provides an API, which enables communication with the server. The Cortex protocol comprises three parts: WebSocket, JSON, JSON-RPC. The WebSocket module ensures real-time communication with the Cortex service. JSON is a data transmission (sending and receiving) format. JSON-RPC is a JSON method used to send and receive the results. The WebSocket client created by the Cortex software communicates over port: 54321 at: wss://Emotivcortex.com:54321. JSON-RPC is used to monitor queries through adding JSON keys. The response is a returned value in the JSON format with fields such as: jsonrpc key and the values depend on whether the sent message was correct or not. Communication messages with the Cortex software are received synchronously. As a standard, the session authorization process is not required, however, if the application requires access to high-resolution EEG data [2], application authorization is necessary. Authorization is executed by using the Authorize method.

Within the proposed solution, a session represents a constant period of collecting data from the Emotiv neuroheadset. The sessions status can be: Open, Active or Closed. A new session starts with a status of: Open or Active and automatically changes its status to Closed, when the connection with the neuroheadset is interrupted or the headset is deactivated. The conducted research and simulations indicate that a delay in changing the status to Closed is approximately one minute if the USB BT module is being used. When the session status is set to Open, it is only possible to subscribe to: movement, mental command, facial expression, information about device power, information about the device, low-resolution data. When the session status is set to Active, the options depend on the activated software license. It is possible to subscribe to: raw EEG data [7], high-resolution data and both of these variants. When the session is closed, there is no subscription possibility.

All of the potential transitions between the sessions are shown in Fig. 2. As can be seen, each of the session state changes is preceded by an appropriate action presented as a transition line description.

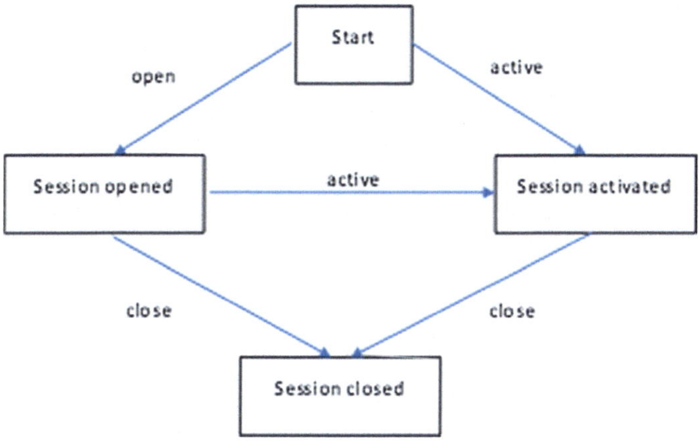


**Fig. 2.** Session diagram for Emotiv Cortex

A single connection enables creating a session with one device operating within the brain-computer technology. For example, application "A" creates sessions "As" with device "INSIGHT-1", and if application A wants to create another session with the headset, it is necessary to close sessions "As" prior to creating another one. Whereas application "B" can create sessions with a neuroheadset "INSIGHT-1". The conclusion is as follows: a device based on the brain-computer technology can have numerous sessions but with different applications. Multisessions, according to the aforementioned example, are possible only for a device operating based on the BCI technology and not for applications. As indicated by the conducted tests, Emotiv enables recording the sessions. After the session status changes to Open, the data is sent and recorded on the Emotiv server. In order to receive data for a given session, one needs to subscribe to a certain data type. Subscription is possible through the "subscribe" method. After calling the subscription method, we need to define the type of information we are interested in. Markers are reference points for streams, which are used to link external events with EEG data. The format of the markers in EEG stream consists of a pair of two channels: IEE_MARKER and IEE_MARKER_HARDWARE. IEE_MARKER is responsible for the current marker value. If none of the markers is triggered within a specified time, value 0 is added to the file. Otherwise, the value of the byte transmitted through the USB/COM port will be set to the sample. IEE_MARKER_HARDWARE is the index of the device, which was used to trigger it. The value depends on the device sequence (USB, serial, keyboard). As indicated by the conducted simulation, the first device will receive an index value of 1, the next one 2, etc.

## 3 Solution Implementation Concept

Figure 3 shows the idea and concept behind implementing smart home control via a device based on the brain-computer technology made by Emotiv and the Cortex software. The assumption in the course of the research was designing such a system, which would enable high-level communication with the neuroheadset to the greatest possible abstraction extent. The entire system is made up of equipment comprising: Emotiv EPOC+ Neuroheadset and a Raspberry PI [1]. The only software used within the solution implementation concept is available under the name Cortex and would have to be saved in the Raspberry.

**Fig. 3.** Concept of smart home control implementation

A default server named Cortex, made available by Emotiv, would have to be installed in order to connect external software. An account would have to be created and Cortex software installed for the purposes of the installation. After running the software, the user is shown the login window. After logging in and connecting the neuroheadset via Bluetooth, Cortex software window appears, which is shown in Fig. 4 (left). The window also contains: BCI device battery charge level, percentage of connected electrodes, connection type, its name and BCI neuroheadset ID. The transitional state of electrode connection at the set-up stage is shown in Cortex graphically, in the right window in Fig. 4.

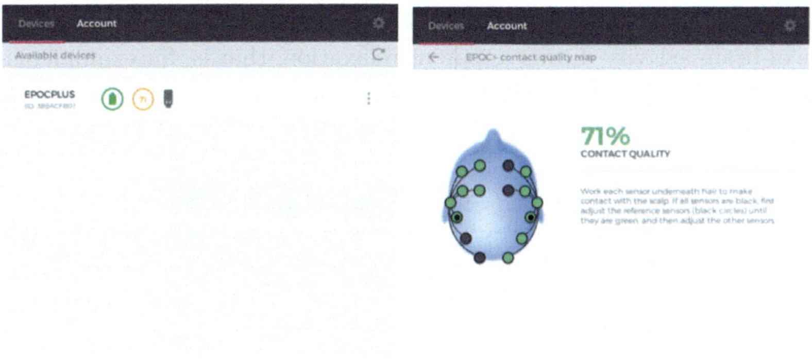


**Fig. 4.** Cortex software window after logging in and the transitional state of electrode connection

In order to prove the concept regarding the communication via the aforementioned mechanisms, an attempt was made at connecting with the Cortex server using software developed in the JS language. Figure 5 shows the communication outcome. The first textbox, "Location", contains an entered address together with the port, described in the first part of this article. The textbox "Message" contains an entered sample message, predefined by Emotiv as an example. The Log field contains three messages: two of these described a change in the state, i.e., inform that: the application was connected correctly, it received a response regarding the user logging out and logging out from the application.

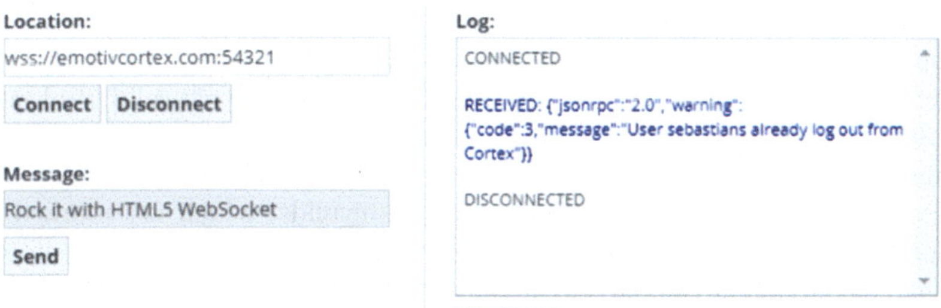


**Fig. 5.** Example of communication with a Cortex server

## 4 EEG Signal Artefacts as a Control Signal Source

To testing sample artefacts, which could be used to control a smart home involved using additional software (Emotiv Xavier TestBench) owned by the Laboratory for neuroinformatics and decision systems at the Opole University of Technology, which enables visualization of brain activity during certain activities, including the diversification between various types of brain waves. Figure 6 shows sample graphs of brain activity during apparent relaxation, that is, for example: lack of intentional eye movement, swallowing, teeth gritting, etc.

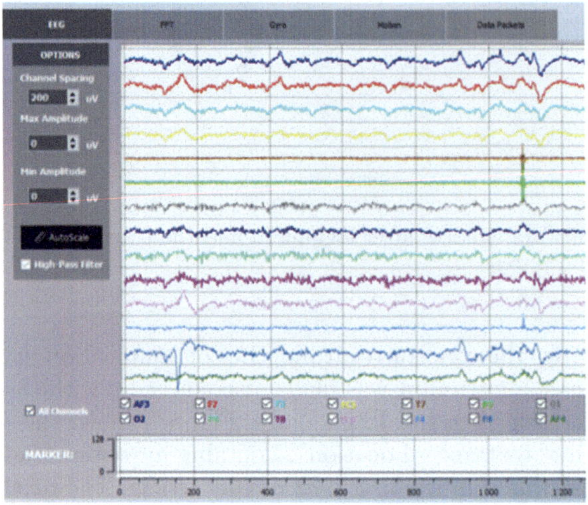


**Fig. 6.** Oscillations of EEG waves from Emotiv Xavier TestBench software, showing the reference state (without intentional artefacts)

Figure 7 shows EEG signals associated with the activity of brain waves correlated with hand movements.

The next analysed EEG signal artefacts, which the research focused on were the ones linked with intentional eyelid blinking. In the case of eyelid blinking, we can observe periodic changes in the EEG signal activity for each of the electrodes coming from the Emotiv EPOC+ NeuroHeadset. Figure 8 shows changes of the EEG signal oscillations for three blinks of the eyelids, which are marked with red squares. The changes can be observed to a greater or lesser extent on all electrodes (apart from the currently not working ones - the deactivated).

Another tested activity was teeth gritting. As shown in Fig. 9, teeth gritting generates a significant number of artefacts. The artefacts are generated on most electrodes throughout the entire length when saving the EEG signal. As indicated by the tests, EEG signal increment occurs when the pressure force on the teeth increases, whereas a sharp decline is observed when the clenched teeth are released.

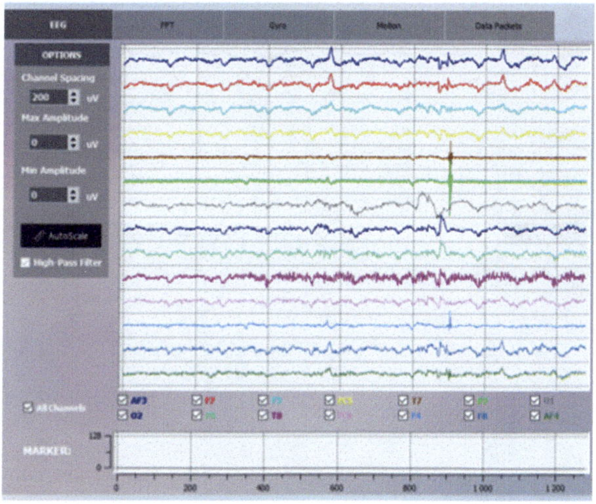


**Fig. 7.** Oscillations of EEG waves from Emotiv Xavier TestBench software, showing brain activity during hand movement

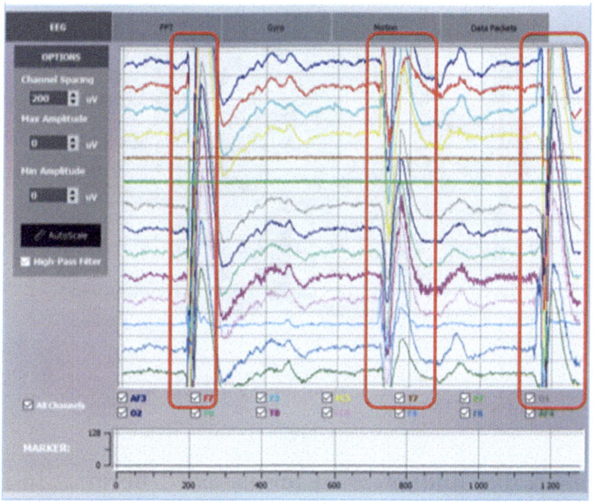


**Fig. 8.** Oscillations of EEG waves from Emotiv Xavier TestBench software, showing brain activity during eyelid blinking

# 5 Hardware and Software Concept of Solution Implementation

The hardware concept was designed in terms of ease of programming, greatest possibility of using packet analysis and ease of code development. The Raspberry PI 2B [5], shown in Fig. 10 was used as the base of a smart home controller.

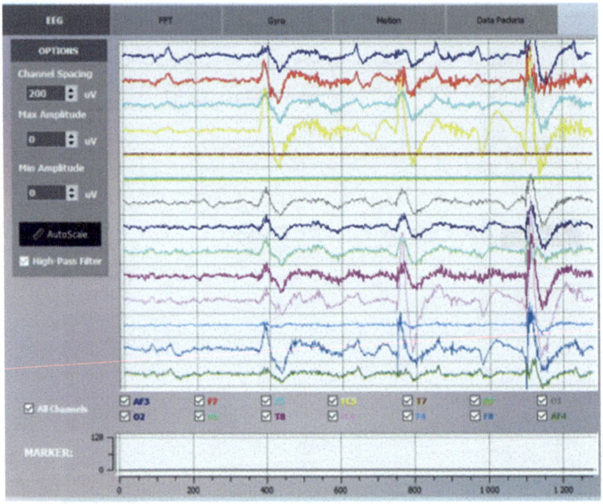


**Fig. 9.** Oscillations of EEG waves from Emotiv Xavier TestBench software, showing brain activity during teeth gritting

**Fig. 10.** Raspberry PI 2B

Linux Raspbian distribution, with pre-installed libraries to operate GPIO named Wiring-PI, available for the Python and C/C++ languages, was used as the operating system. Owing to the high GPIO control speed and very vast control possibilities, it was decided to limit the research to digital GPIO control, without peripherals, which can be ultimately connected. It should be noted that it is recommended to use galvanic isolation for the Raspberry PI GPIO [9]. In order to enable the communication discussed in the first section of the article, it is necessary to have an Internet connection. For the purposes of communication, it is possible to use an Ethernet connection or connect a Wi-Fi card via a USB 2.0 port. However, given the need to ensure security and stability, it is advisable to set up an Ethernet connection. In the course of the research, a segment of the software was developed in Python. Python enables easy data parsing and utilizing the JSON format, discussed in the first part of the research paper. An important aspect is to send queries via WebSocket. The libraries below: asyncio and websockets were selected for this purpose. The entire control process requires the process of learning the neutral state or basic commands, such as: push, pull, etc.

## 6 Conclusions

Python language was used to develop the software used within the conducted experiments. A Raspberry PI 2B was used in order to enable controlling home automation elements since it is a fully functional computer with an installed Linux system. In a longer time perspective, such a configuration enables developing the project and make the platform independent from a commercial system, for example: MS Windows. An important advantage of the presented structure is its high mobility and the option to utilize numerous ready solutions as an expansion of the control process. The development opportunities determined in the course of the research work include, among others: implementation of encryption and adding the possibility of remotely controlling the home from anywhere in the world using a port forwarding setting or adding a DDNS. Another development option for the presented solution is adding wireless devices controlling the elements of lighting, heating or multimedia within a smart home system.

## References

1. Bolaños, F., LeDue, J.M., Murphy, T.H.: Cost effective Raspberry PI - based radio frequency identification tagging of mice suitable for automated in vivo imaging. J. Neurosci. Methods **276**(30), 79–83 (2017). https://doi.org/10.1016/j.jneumeth.2016.11.011
2. Cegielska, A., Olszewski, M.: Nieinwazyjny interfejs mózg-komputer do zastosowańtechnicznych. Pomiary Automatyka Robotyka **3**(19), 5–14 (2015)
3. Gorska, M., Olszewski, M.: Interfejs mózg-komputer w zadaniu sterowania robotem mobilnym. Pomiary Automatyka Robotyka **3**(19), 15–24 (2015)
4. Ghaemia, A., Rashedia, E., Mohammad, Pourrahimib A., Kamandara, M., Rahdaric, F.: Automatic channel selection in EEG signals for classification of left or right hand movement in Brain Computer Interfaces using improved binary gravitation search algorithm. Biomed. Signal Process. Control **33**, 109–118 (2017). https://doi.org/10.1016/j.bspc.2016.11.018
5. Kuziek, J.W.P., Shienh, A., Mathewson, K.E.: Transitioning EEG experiments away from the laboratory using a Raspberry Pi 2. J. Neurosci. Methods **277**(1), 75–82 (2017). https://doi.org/10.1016/j.jneumeth.2016.11.013
6. Lee, W.T., Nisar, H., Malik, A.S., Yeap, K.H.: A brain computer interface for smart home control. In: International Symposium on Consumer Electronics (ISCE), pp. 35–36 (2013). https://doi.org/10.1109/ISCE.2013.6570240
7. Libenson, M.H.: Practical Approach to Electroencephalography. Elsevier Health Sciences, Edinburgh (2012)

8. Marquos, Z., Alquraini, A., Sheltami, T.: Home automation using emotiv: controlling TV by brainwaves. J. Ubiquit. Syst. Pervasive Netw. **10**(1), 27–32 (2018). https://doi.org/10.5383/JUSPN.10.01.004
9. Paszkiel, S.: Using the Raspberry PI2 module and the brain-computer technology for controlling a mobile vehicle. In: Szewczyk, R., Zieliński, C., Kaliczyńska, M. (eds.) Automation 2019, Progress in Automation, Robotics and Measurement Techniques. Advances in Intelligent Systems and Computing, pp. 356–366. Springer, Switzerland (2020). https://doi.org/10.1007/978-3-030-13273-6_34

# Leveraging Object Recognition in Reliable Vehicle Localization from Monocular Images

Tomasz Nowak(✉), Michał R. Nowicki, Krzysztof Ćwian, and Piotr Skrzypczyński

Institute of Control, Robotics, and Information Engineering, Poznań University of Technology, ul. Piotrowo 3A, 60-965 Poznań, Poland
{tomasz.nowak,michal.r.nowicki,krzysztof.cwian, piotr.skrzypczynski}@put.poznan.pl

**Abstract.** We present the processing pipeline of a monocular vision system that successfully performs the task of detecting, identifying and localizing a city bus electric charger station. This task is essential to the operation of an advanced driver assistance system that helps the driver to dock the long vehicle at the charging station. The focus is on the role of machine learning techniques in developing a robust detection and classification procedure that allows our system to localize the camera with respect to the charger even from long distances. We demonstrate that the learned detection procedure improves robustness of the vision techniques for monocular localization, while the geometric relations estimated by our system can be used to improve the learning results.

## 1 Introduction

Driver assistance technologies become increasingly applied in urban public transport due to a shortage of skilled personnel and high costs of training. In electric buses, that also rapidly become popular, one of the main problems is to automatize docking at the electric charging stations. This operation is complicated because it requires the driver assistance system to start guidance of the long vehicle at a distance of 30 to 40 m from the charging station, which has to be detected, recognized, and localized automatically. For that purpose passive vision is applied, because the laser scanners used for general localization (e.g. with respect to building walls) [1] and for terrain perception [11] produce sparse point clouds that do not guarantee to capture shapes of small objects from long distances. A monocular camera configuration is chosen according to the requirements of the bus manufacturer, who wanted to have a single sensor that can be easily integrated with buses that differ with respect to their body details, particularly in the roof area.

Localization with monocular vision, while being cost-effective and easily deployable on commercial vehicles is known for its problems with initialization


R. Szewczyk et al. (Eds.): AUTOMATION 2020, AISC 1140, pp. 195–205, 2020.
https://doi.org/10.1007/978-3-030-40971-5_18

and estimation of the proper scale due to the lack of 3-D information in the individual images [8]. Therefore, in this paper, we present a new approach to localization with monocular vision that takes into account the specific requirements of the city bus application, and combines the classic techniques of landmark-based localization and 3-D vision with the modern, learning-based object detection and classification module. This module makes it possible to learn the appearance of the electric chargers used by the buses (even if they differ between particular bus operators), and to consider the charger and its post (pylon) as a known landmark. Although the knowledge about the charger's size and location makes it possible to compute the camera's pose from individual images using the $n$-point algorithm [4], the detection of the feature points on the image remains a problem, as the charger itself can be very small on the acquired images. Whereas we are interested only in 2-D localization (i.e. position $x, y$ and orientation $\theta$) of the camera with respect to the known charging station position, the computations have to consider that the charger is located at some elevation above the camera (attached to the bus roof), and that the vehicle can approach the charging station from different angles, which results in different appearance of the charger. We can use only the bounding box of the detected object to determine its size, the shape mask that allows our system to determine the size with higher accuracy, or we can detect the characteristic points that are directly used by the $n$-point algorithm. The contribution of our work can be summarized as:

- novel procedure for localization with learned landmark characteristics,
- evaluation of different approaches to the extraction of the object's characteristics from the images,
- learning procedures for the convolutional neural network (CNN) that result in reliable charger detection and extraction of the geometric characteristics,
- proposal of a new loss function in the CNN learning procedure that considers the distance to the object computed by the localization procedure.

## 2 Related Work

Although self-driving cars and other autonomous vehicles localize themselves relying on the Global Positioning System (GPS), the GPS signal is degraded in many environments due to the presence of objects that obscure the view of the satellites. This is a common problem in urban scenarios with tall buildings, which limits the application of GPS for accurate localization of vehicles [15]. Even in open spaces achieving sub-meter position accuracy requires differential GPS (DGPS), which involves the transmission of correction signals from a ground station. The relatively high cost of DGPS maintenance and the insufficient reliability of the satellite-based systems in the urban environment motivate us to look for a solution that uses only local features and the most affordable sensor – monocular camera.

The accuracy of monocular vision for outdoor localization with respect to the DGPS performance was shown by Royer *et al.* [13] back in 2005. Since that time the monocular localization methods have developed significantly, allowing practical vehicle localization with a single camera, as demonstrated by the ORB-SLAM results [8] on the KITTI dataset [2]. However, the visual odometry or simultaneous localization and mapping approaches need to observe a significant number of salient features to work properly, and are incremental, i.e. they have to start from a known position of the camera.

Our task calls for a different approach: we have a known object that is observed on an arbitrary (possibly non-static) background. The position of this object (the charger) is known, while the bus starts to localize from an unknown position. Therefore, we consider the localization of the bus-mounted camera with respect to the fixed coordinate system of the charger. While docking with the visual position feedback is similar to the well-studied visual servoing problem, the existing visual servoing methods, such as the ViSP library [7] can handle only localization with respect to known objects that appear big enough on the images, while the charger observed from a long distance consists of just a dozen of pixels in the image. This situation requires an object detection algorithm that reliably distinguishes objects of the given class from the background clutter in spite of the large changes in their appearance. Fortunately, powerful object detection and classification methods are available nowadays due to the rapid progress in deep neural networks. Architectures based on the convolutional neural networks (CNN) are commonly used in autonomous vehicles, e.g. for detection of traffic lights [6] and other elements of the road infrastructure. We use the Faster R-CNN [3] and Mask R-CNN [5] architectures, extending our previous work [10] that applied only the Faster R-CNN to detect bounding boxes around the chargers, focusing on the explainable learning [14] approach that allowed us to reduce false detections of the chargers almost to zero. Here we use the same approach to build the training dataset, but we focus on the localization task. The Faster R-CNN was also used to detect and localize landmark objects in the recent work [9], but we are the first, who show that the modified R-CNN-family networks can be applied as direct feature extractors for monocular localization with a single landmark.

## 3 Charger Detection

Our initial approach to charger detection, described in [10], was based on the modified Faster R-CNN network [12]. The first step of processing is feature extraction. In the R-CNN-family architectures, it is usually done using one of the pretrained networks specialized in image classification. Accordingly, we used the Inception ResNet v2 architecture pretrained on the popular and large COCO dataset. The output of this step consists of 512 feature maps of size $75 \times 100$ "pixels". Then for each "pixel", the Region Proposal Network (RPN) generates anchors (early-stage box proposals), and for each anchor determines the possibility of containing an object belonging to one of the known classes. Outputs of this step are 300 box proposals.

The final processing stage in Faster R-CNN consists of two predictor heads: the classifier predicts which one of the known classes represents each box proposal, while the regressor refines the bounding box location and size (Fig. 1).

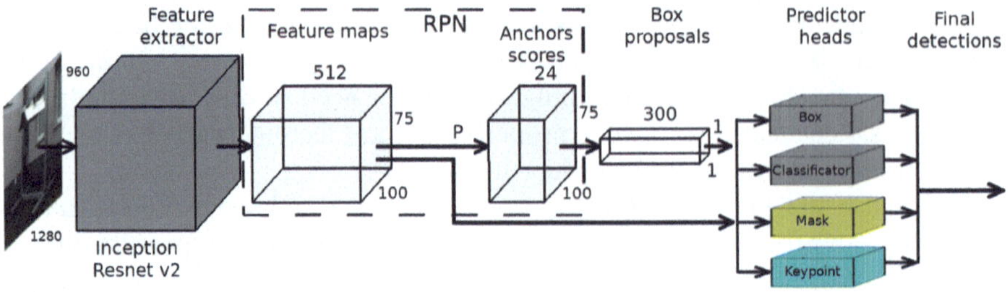


**Fig. 1.** Simplified representation of the Faster and Mask R-CNN networks. The mask and the key point predictor heads are added to achieve additional output information

Eventually, the outputs are bounding box coordinates, the class label of the detected object, and a confidence score. An important improvement we made in [10] was the visualization of attention heat maps (not shown in Fig. 1), which made it possible to select the negative examples (e.g. glazed building facades, other vehicles) augmenting the training dataset. This procedure radically reduced the number of false charger detections and may be used as a tool when fielding the system with the given bus operator and environment type.

**Fig. 2.** Visualisation of extracting the image fragment with charger roof

This baseline approach provided us with a bounding box of the detected charger, that often was oversized and sometimes undersized with respect to the actual size of the detected object. As our localization procedure uses the size of the detected object projection in the image to compute the distance, we were interested in obtaining more accurate boundaries of the charger. This is possible using the Mask R-CNN, which is based on the Faster R-CNN architecture. The main difference is in the final step, where except the classifier and regressor heads, there is a mask predictor head, which predicts the exact segmentation mask of the detected object. As the R-CNN architecture makes it possible to add other heads designed for particular purposes, we finally modified the Mask R-CNN by adding regressor network that detects the characteristic points (feature points) of the detected object. To train network for feature points detection, we need to provide additional metadata about localization of these points on images in the training dataset. Actual key points prediction takes place at the end of the pipeline, parallel to mask prediction. The head consists of 8 convolutional layers followed by one fully connected.

The localization task requires the image size big enough to ensure the best accuracy at a reasonable processing time. Images from the camera used in our experiments have the resolution of $5472 \times 3648$ pixels, and the typical size of images processed by the network is $960 \times 720$ pixels. Hence, 42 consecutive passes through the network (white grid on Fig. 2) are required to scan the whole image in search of the object. As this approach is inefficient, the object detection pipeline is divided into two parts. Firstly, we detect the entire charging post on a downsampled image, then, having a rough localization of the charger on the image, we perform actual detection of the charger's head (called a "roof", due to its shape). This approach to detection allows us to provide the localization system in real-time with the information extracted from images.

# 4 Charger Localization

## 4.1 Using the Bounding Box

Using a monocular camera it is possible to determine the distance to an object providing that the dimensions of this object are known in advance. This is our case, as a CAD model of the electric bus charger is available. The distance can be computed directly from the proportion between the known dimensions of the object, and the dimensions of this object's projection on the image. The size of the charger on the image can be obtained directly from the width and height of the bounding box produced by Faster R-CNN (Fig. 3).

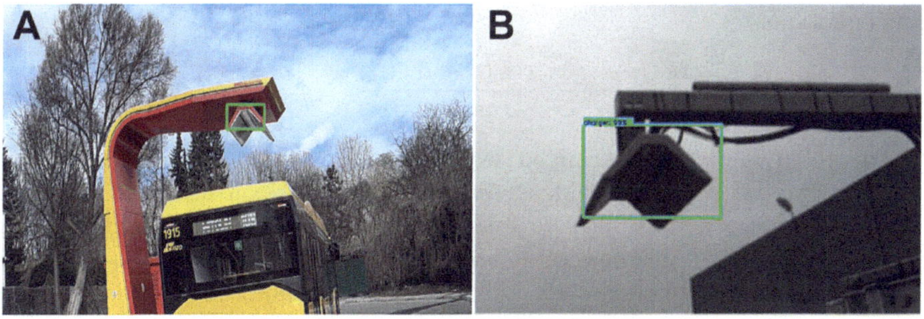


**Fig. 3.** Examples of Faster R-CNN detection: correct bounding box (A), and oversized bounding box (B)

Figure 4 shows the geometric construction that is used to compute the distance having only the dimensions of the bounding box. The coordinates $x, y$

and $\theta$ define the position and orientation of the charger roof with respect to the camera coordinate system.

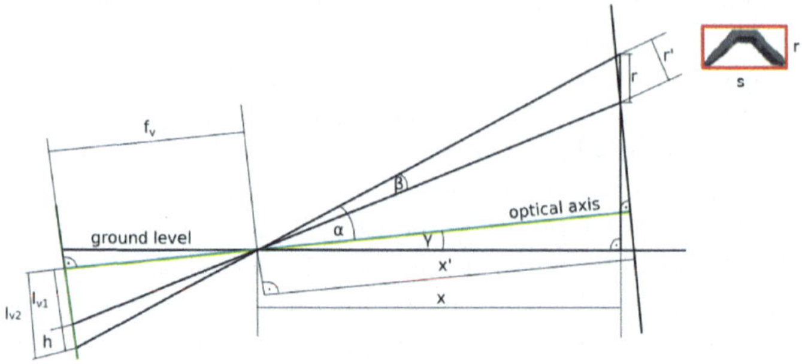


**Fig. 4.** Visualization of distance calculation in $x$ axis

To compute the distance between the camera and the charger we need to know the bounding box size, the camera intrinsic parameters (from calibration), and the charger dimensions. Pixels' locations on an image are described in $(u, v)$ coordinate system. Denoting $(f_u, f_v)$ the camera focal length, $(p_u, p_v)$ the camera image center, $r$ and $s$ the charger height and width, $h$ and $w$ the charger projection height and width on the image in pixels, $I_{v1}$ and $I_{v2}$ the bottom and top edge of the charger projection relative to the image center, $I_u$ the localization of charger center on image along $u$ axis, we compute the $\alpha$ and $\beta$ angles using equations:

$$\alpha = \arctan\left(\frac{I_{v2}}{f_v}\right), \quad \beta = \alpha - \arctan\left(\frac{I_{v1}}{f_v}\right). \tag{1}$$

Then, having a known camera pitch angle ($\gamma$) we can estimate the translation from equations:

$$x = \frac{f_v \cdot r}{h} \cdot \frac{\cos(\alpha + \gamma)}{\cos\alpha} \cdot \frac{\cos(\alpha - \beta + \gamma)}{\cos(\alpha - \beta)}, \quad y = \frac{x \cdot (I_u - p_u)}{f_u}. \tag{2}$$

The pitch angle is determined with respect to the gravity vector using accelerometers of the Inertial Measurements Unit (IMU) attached to the camera. There is also a possibility to estimate the $\theta$ angle using the geometrical correspondences, but this method is prone to errors due to the small size of the charger projection. Unfortunately, this procedure cannot compute a reliable $\theta$ estimate, as the rectangular bounding box does not provide the necessary information about the object's shape.

### 4.2 Using the Mask Shape

Bounding boxes do not provide any additional information about detection to improve estimation, and they are often oversized with respect to the real size of the object's projection on the image (Fig. 3B). To circumvent these problems we have decided to use the Mask R-CNN network. Information about the charger silhouette (Fig. 5A) allows us to determine the $\theta$ angle, but the computations are prone to errors, because small changes in the detected shape (e.g. due to lighting fluctuations) induce large changes in the estimated angle. The raw mask output from the network is a grayscale patch having the size of the detection bounding box. The intensity of each pixel represents the confidence whether this pixel belongs to the charger or not. By thresholding this patch, we can extract the charger silhouette. We experimented with the extraction of the edges and characteristic points from the segmentation mask, but the variety of the possible poses and open-world setting made it very hard to solve in a classic way. Hence, we extract only four charger points which define the width and height of the object, and we use the same method to compute the translation as described in Subsect. 4.1.

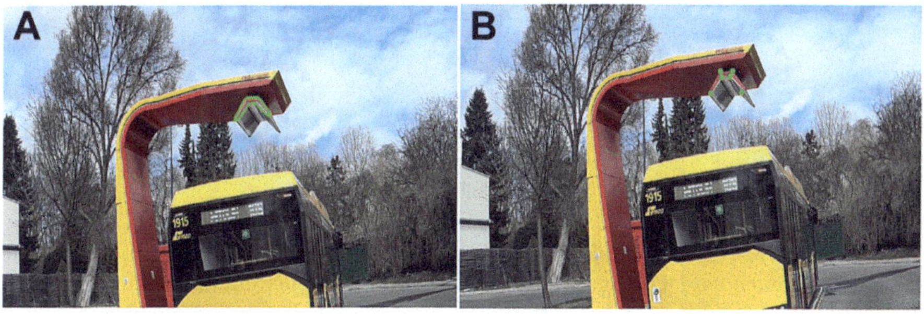


**Fig. 5.** Examples of Mask R-CNN detection (A), and key points detection (B)

### 4.3 Using the Point Features

We enriched the Mask R-CNN model with a key points prediction head. Having four charger corners (Fig. 5B) on image, localization of the corresponding points on the real charger model, and camera intrinsic parameters matrix (from calibration), we can use the iterative SolvePnP algorithm implemented in OpenCV library to estimate the six-degrees-of-freedom pose of the charger. This algorithm iteratively minimizes the reprojection error defined as sum of squared distances ($d$) between point localization on image and object model points projected on image:

$$[\mathbf{R}|\mathbf{t}]^* = \underset{\mathbf{R},\mathbf{t}}{\operatorname{argmin}} \sum_{i=1}^{4} d\left(Pc_i, \mathbf{K}[\mathbf{R}|\mathbf{t}]Pw_i\right)^2, \tag{3}$$

where $Pc_i$ are coordinates of the $i$-th point's projection to the image, $\mathbf{K}$ is a camera intrinsic parameters matrix, $[\mathbf{R}|\mathbf{t}]$ is a rigid transformation matrix (rotation and translation, $[\mathbf{R}|\mathbf{t}]^*$ stands for the estimate), and $Pw_i$ are coordinates of the $i$-th real object's point corresponding to $Pc_i$. This function uses Levenberg-Marquardt optimization and computes not only the $x$ and $y$ coordinates, but also the $z$ coordinate, and the roll, pitch and yaw angles. The $z$ coordinate is ignored, as it is a constant difference between the camera elevation and the (known) elevation of the charger post. Similarly, the roll and pitch angles of the camera are known from the accelerometers. The yaw angle is the $\theta$ heading that is used by our system. Currently, we use only four points at the front of the charger, but to achieve better accuracy, we can provide more points. Especially the addition of points from the back of the charger should improve stability and further reduce the $\theta$ estimation error. Figure 6 gives a comparison between the mask-based (from Mask R-CNN) and the keypoint-based (using SolvePnP) charger localization methods.

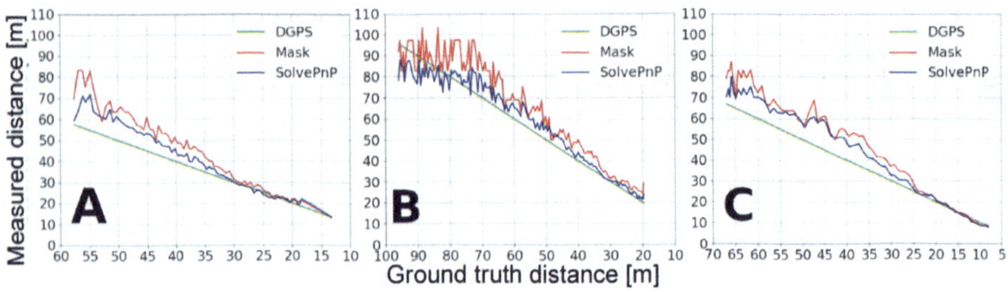


**Fig. 6.** Comparison of the mask-based method and the SolvePnP-based method. The estimated distance vs. DGPS ground truth for three sequences: a short one (A) and a longer one (B) recorded at dusk, and another one recorded at full daylight (C)

## 5 Improving Learning with the Localization Results

Neural networks learn their target behavior using a loss function that indicates how far the output of the network is from the desired one. Thus, a proper loss function is essential for the training of a deep neural network. The loss function that is used in the implementation of Mask R-CNN consists of six components. The first two ($L_{RPNloc}$ and $L_{RPNobj}$) come from the RPN network and are responsible for box proposals localization and objectness classification. The next three components penalise final box classification ($L_{class}$), bounding box and key points localization ($L_{loc}$), and mask segmentation ($L_{mask}$) errors. The last component is a regularization loss ($L_{reg}$).

$$L_{total} = L_{RPNloc} + L_{RPNobj} + L_{class} + 4 \cdot L_{loc} + 2 \cdot L_{mask} + L_{reg}. \quad (4)$$

However, the ultimate goal of our network is to provide information for the localization procedure, which is not reflected by the standard loss function. Thus, to further improve the estimation accuracy, we modified the loss function used during training the model. We calculate the six-d.o.f. pose using keypoints detected by the network and the SolvePnP algorithm. Then, we calculate the distance to the object. The difference between the estimated distance and the ground truth from DGPS (accuracy of 2 cm) is added to the box classifier localization loss. With the proposed modifications, the localization loss consist of the bounding box ($L_{bbox}$), key points ($L_{kp}$), and distance ($L_{dist}$) components:

$$L_{loc} = L_{bbox} + L_{kp} + L_{dist}. \quad (5)$$

Figure 7 demonstrates the difference in accuracy between the original Mask R-CNN network and the same network trained with the additional loss component.

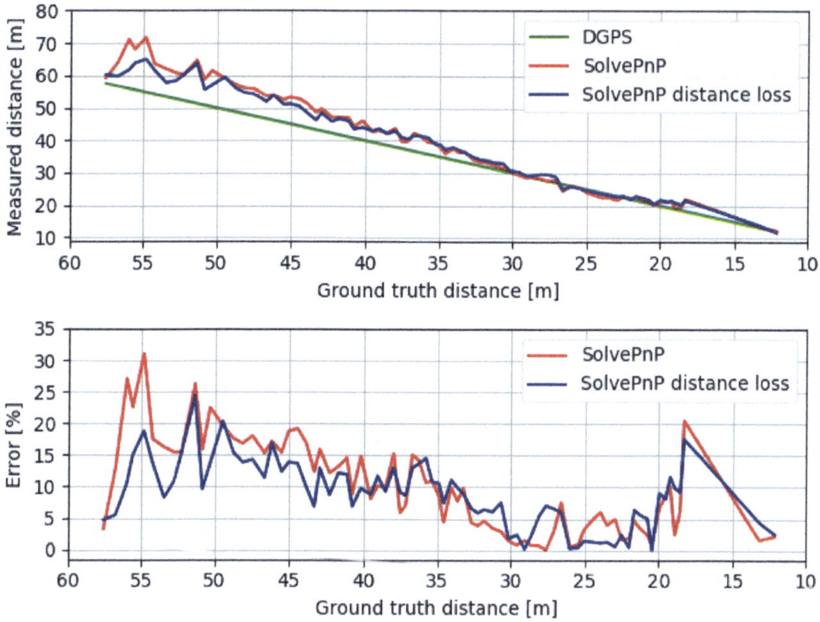


**Fig. 7.** Comparison of model trained with standard loss function and loss function with distance

Finally, the accuracy of the considered methods was tested on three video sequences recorded in different lighting conditions. In this experiment, the distance to the charging station differs from 95 to 13 m. We do not plot results for the $\theta$ computations, because of the lack of ground truth for the camera orientation during the experiment (it is not provided by the single DGPS receiver). The distance errors are defined as:

$$AvgErr[\%] = \frac{\sum_{i=1}^{n} \frac{|e_i - gt_i|}{gt_i}}{n} \cdot 100, \quad AvgErr[m] = \frac{\sum_{i=1}^{n} |e_i - gt_i|}{n}, \quad (6)$$

where $n$ is the number of images, $gt_i$ is the ground truth distance, and $e_i$ is the estimated distance.

All examined methods are compared in Table 1. In the considered sequences, the distance estimated from the bounding box and the mask-based method were bigger than the ground truth distance. This explains the fact that the average error for the Faster R-CNN method (bounding box) is smaller than for the Mask R-CNN version. The charger projection dimensions extracted from a mask are always smaller than the same parameters extracted from the bounding box. The methods that used simple proportions to compute the distance resulted in worse accuracy of the distance estimation and do not provide reliable information about the orientation ($\theta$ angle). In contrary, the method based on the SolvePnP algorithm provides full six-d.o.f. transformations and presents smaller distance estimation errors.

**Table 1.** Comparison of average distance measurement errors for different methods

| Method | Faster | Mask | SolvePnP | SolvePnP distance Loss |
|---|---|---|---|---|
| Average error [%] | 10.72 | 13.41 | 9.66 | 7.92 |
| Average error [m] | 4.95 | 6.26 | 4.70 | 3.87 |

## 6 Conclusions

We presented a simple yet efficient approach to localization of the electric charger stations for electric buses using a monocular camera. The obtained $x, y$ values allow us to position the charger with respect to the bus-fixed camera and to plan the necessary maneuvers starting from a large enough distance. The larger than expected uncertainty in the $\theta$ computation should not be a problem during real docking, as the chargers are always oriented along the road axis, thus the orientation can be predicted from the bus motion (dead reckoning), and then used to discard unreliable $\theta$ estimates.

**Acknowledgement.** This work was funded by the National Centre for Research and Development grant POIR.04.01.02-00-0081/17.

## References

1. Ćwian, K.: Feature-Based Laser Simultaneous Localization and Mapping for Automotive Applications. Agencja Wydawnicza Impuls, Kraków (2019)
2. Geiger, A., Lenz, P., Urtasun, R.: Are we ready for autonomous driving? the KITTI vision benchmark suite. In: IEEE Conference on Computer Vision and Pattern Recognition (CVPR), Rhode Island, pp. 3354–3361 (2012)
3. Girshick, R., Donahue, J., Darrell, T., Malik, J.: Rich feature hierarchies for accurate object detection and semantic segmentation. In: IEEE Conference on Computer Vision and Pattern Recognition (CVPR), Columbus, pp. 580–587 (2014)

4. Hartley, R.I., Zisserman, A.: Multiple View Geometry in Computer Vision. Cambridge University Press, Cambridge (2004)
5. He, K., Gkioxari, G., Dollár, P., Girshick, R.: Mask R-CNN. In: IEEE International Conference on Computer Vision (ICCV), Venice, pp. 2980–2988 (2017)
6. Kim, J., Cho, H., Hwangbo, M., Choi, J., Canny, J., Kwon, Y.P.: Deep traffic light detection for self-driving cars from a large-scale dataset. In: IEEE International Conference on Intelligent Transportation Systems, Maui, pp. 280–285 (2018)
7. Marchand, E., Spindler, F., Chaumette, F.: ViSP for visual servoing: a generic software platform with a wide class of robot control skills. IEEE Robot. Autom. Mag. **12**(4), 40–52 (2005)
8. Mur-Artal, R., Montiel, J.M.M., Tardós, J.D.: ORB-SLAM: a versatile and accurate monocular SLAM system. IEEE Trans. Robot. **31**(5), 1147–1163 (2015)
9. Nilwong, S., Hossain, D., Kaneko, S., Capi, G.: Deep learning-based landmark detection for mobile robot outdoor localization. Machines **7**(2), 25 (2019)
10. Nowak, T., Nowicki, M., Ćwian, K., Skrzypczyński, P.: How to improve object detection in a driver assistance system applying explainable deep learning. In: IEEE Intelligent Vehicles Symposium, Paris, pp. 226–231 (2019)
11. Nowicki, M., Nowak, T., Skrzypczyński, P.: Laser-based localization and terrain mapping for driver assistance in a city bus. In: Szewczyk, R., et al. (eds.) Automation 2019 Progress in Automation, Robotics and Measurement Technique, AISC 920, pp. 502–512. Springer (2019)
12. Ren, S., He, K., Girshick, R., Sun, J.: Faster R-CNN: Towards real-time object detection with region proposal networks. In: Advances in Neural Information Processing Systems (NIPS), pp. 91–99 (2015)
13. Royer, E., Lhuillier, M., Dhome, M., Chateau T.: Localization in urban environments: monocular vision compared to a differential GPS sensor. In: IEEE Conference on Computer Vision and Pattern Recognition CVPR, San Diego (2005)
14. Zhang, Q., Zhu, S.: Visual interpretability for deep learning: a survey, frontiers of information technology & electronic. Engineering **19**(1), 27–39 (2018)
15. Zhu, N., Marais, J., Betaille, D., Berbineau, M.: GNSS position integrity in urban environments: a review of literature. IEEE Trans. Intell. Transp. Syst. **19**(9), 2762–2778 (2018)

# Motion Planning of the Cooperative Robot with Visual Markers

Michał Spławski[2], Rafał Staszak[1], Filip Jarecki[2], Jakub Chudziński[2], Piotr Kaczmarek[1,2], Paweł Drapikowski[1,2], and Dominik Belter[1,2(✉)]

[1] Institute of Control, Robotics and Information Engineering, Poznan University of Technology, Poznan, Poland
dominik.belter@put.poznan.pl
[2] A.S. Adrian Stern, Wilczak 45/47, 61-623 Poznan, Poland
http://www.lrm.put.poznan.pl, https://rob-tech.pl/en/

**Abstract.** In recent years numerous robotic solutions have been developed to improve production processes. However, most of the robotic arms are applied in large factories where they perform repetitive tasks supporting mass production. The complexity of programming and adapting robots for changing production is still a barrier for the application of robot arms in flexible and small-scale production. In this paper, we present the system which allows programming the robot using visual cameras and a set of visual markers. We develop a method for flexible motion planning for the cooperative robot to quickly define the trajectory of the robot in the 3D space. We present the architecture of the system, calibration method and the performance of the system. We also show how to significantly improve the detection range of the visual markers using Convolutional Neural Networks for object detection.

**Keywords:** Motion planning · Cooperative robot · Visual programming

## 1 Introduction

The industrial robot can be programmed manually using scripts and teaching pendant. Robot programming is a time-consuming task and requires expert knowledge. Teaching by doing reduces the time needed to program the robot but the trained trajectory can not be changed without re-programming the robot. In this paper, we consider the problem of an industrial robot working in a small factory. The production changes frequently and the application of industrial robots requires continuous re-programming and tunning. In this paper, we propose a system that can be used to program the robot using visual markers attached to the planar object which can be later drilled or milled.

Recently, new ways of robot programming/control have been presented. The robot can be controlled using voice commands [17,18]. In this case, the robot recognizes a defined set of orders and executes the motion commands. This


R. Szewczyk et al. (Eds.): AUTOMATION 2020, AISC 1140, pp. 206–215, 2020.
https://doi.org/10.1007/978-3-030-40971-5_19

kind of interface is problematic in industrial and noisy environments [17]. The motion of the robot can also be programmed using human posture or hand tracking devices. In this case, the robot mimics the motion of the human operator. The most popular interfaces for industrial robots are devices that track hand gestures [20] or augmented reality human-robot interaction devices [10]. Also, multi-degrees of freedom walking robots are teleoperated by hand gestures and hand tracking devices [4].

In the paper, we consider motion planning and programming a robot which is equipped with drilling or milling device. In the considered scenario, the robot is working in a small workshop, e.g. in the carpenter workshop. When the carpenter wants to drill holes in the wooden planar pieces, he attaches the visual markers to the material. Each marker defines a property of the robot's motion (operation speed) and properties related to the tool like hole diameter or the drill depth. The robot uses the visual camera to detect markers and to plan its motion. When the first trajectory is executed the robot repeats the programmed trajectory for all the remaining pieces. In this paper, we show how to integrate the visual system for tracking markers with the motion planning library and calibrate the system. We also show the application of the deep neural network to improve the detection range of the visual markers.

## 2 Related Work

The most popular artificial visual markers are ArUco [5,6], AprilTags [14], QR and Maxicode codes [15]. Each type of tag contains black and white fields which are used to store information. It is also possible to detect the marker and estimate its pose in the 3D space. The properties of the AprilTags in the context of robot localization were investigated in [21]. Also Babinec et al. [3] investigated the maximal detection range, localization error, and system properties with the ArUco markers. Recently, an improved method of AprilTags detection has been proposed [19]. The new image thresholding, segmentation, and quad detections

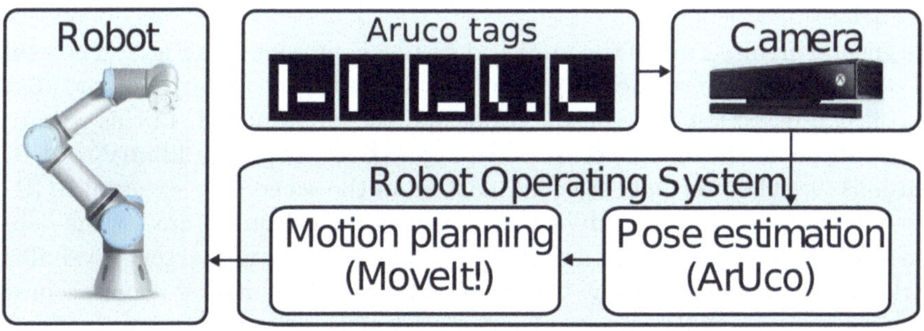


**Fig. 1.** Architecture of the proposed motion planning system

improve the detection rate, detection range, and reduces the false positives detection rate. The accuracy of the marker-based localization system was improved by using multiple fiducial markers (AprilTags), and optimization with RANSAC to remove outliers [13].

Visual markers were used to localize, calibrate many mobile robots but also to track ants and investigate their social behaviour [12]. A new type of visual markers was proposed by Li et al. to localize a mobile robot in the warehouse [11]. The standard QR-codes were used on the mobile LabBot robot to localize the robot in the home environment [16]. The Kalman filter was applied to fuse information from multiple sources. The uncertainty model of pose measurements with respect to the visual markers was proposed and used in the Kalman Filter to localize the robot. 2D barcodes (DataMatrix) were used as artificial landmarks to localize humanoid robot Nao [7]. This solution was proven to be efficient for robot localization in the indoor environment. This solution is also fast and the localization system can work on the onboard controller of the robot [7].

QR-codes were also applied for calibration and localization of the mobile industrial arm [2]. With the QR-code the robot could localize itself with respect to the workstation and find the robot's camera pose. The obtained precision of the robot is below 4 mm. Also, the cooperative industrial arm UR10 mounted on the mobile platform MiR100 was calibrated using QR-code tags [1]. This strategy reduced the position error when the robot reached the workstation.

The main contribution of this article is the system which allows programming the motion of the robot using visual markers. We present standard components of the system which are integrated and calibrated to give the robot a new capability to plan its motion semi-autonomously in the environment prepared by the user, with the feedback from the visual system. We also show how to significantly improve the detection range of the visual markers by utilizing a deep neural network object detection framework.

## 3 Motion Planning with Visual Markers

### 3.1 General Architecture

The general architecture of the proposed system is presented in Fig. 1. The external stationary RGB-D camera $C_K$ observes the workspace of the robot and allows for collision detection between the robot and the environment. In this research, we use the Asus Xtion Pro sensor. We applied the OctoMap library to detect collisions [9]. The user puts selected markers in the scene. The external RGB-D camera can be used to identify a tag but the standard procedure for the detection of the marker fails in the detection of small and distant markers. Even if the markers are detected properly, the estimated pose of the marker is not accurate. Thus we also use RGB camera $C_W$ mounted at the tip of the arm $T$ (Fig. 2). This camera is used to scan the environment and precisely estimate the pose of the markers in 3D space.

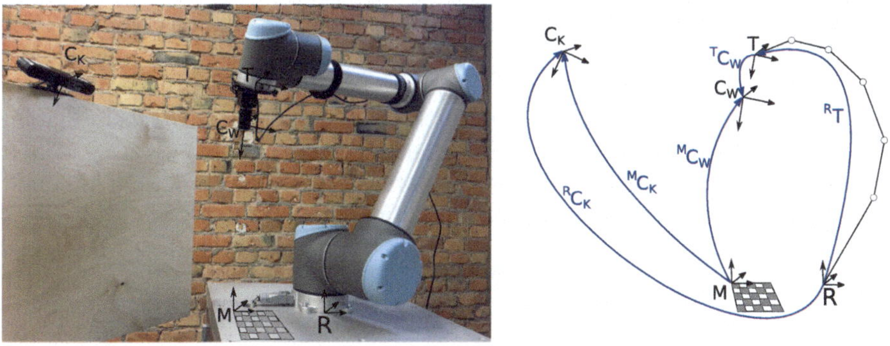


**Fig. 2.** Kinematic structure of the system and spatial relations between robot $R$, robot's wrist $T$, RGB camera on the wrist $C_W$, external RGB-D camera $C_K$, and marker $M$ coordinate systems

To detect and estimate poses of visual markers we use ArUco implementation in the OpenCV library. The detection function returns ${}^{M}\mathbf{C}_W$ poses of the markers and identifiers stored in each visual marker. We use original ArUco dictionary markers with sizes $4 \times 4$ cm. Each marker from this dictionary has $7 \times 7$ data matrix. The main data is stored in $5 \times 5$ array. The border columns and rows are black to improve the detection rate (additionally, we add white outer border columns and rows to improve marker detecting on a black surface).

The detected set of ArUco markers is used to plan the motion of the robot. During planning, we take into account the identifiers of the markers and actions associated with the identifiers. We use the MoveIt library to plan the motion of the robot in the Euclidean space. We plan the whole trajectory at once. It means that the robot starts in the default position, performs all actions defined by the markers and ends the motion in the default position. If the robot cannot perform the whole trajectory we return information about the success rate. Finally, the robot executes the planned trajectory. Trajectory execution is also managed by the MoveIt library on the Robot Operating System (ROS).

### 3.2 System Calibration

At the beginning of the experiment, we calibrate the perception system of the industrial robot. We compute the internal parameters of the RGB-D and RGB camera using the camera calibration toolbox[1]. We also perform external calibration of the camera which estimates the transformations between the robot's base and the coordinate system fixed at RGB-D camera $C_K$ and the camera attached to the robot's wrist $C_W$.

The kinematic structure of the proposed system is presented in Fig. 2. We attached the RGB camera $C_W$ to the tip of the robot. The transformation between the camera $C_W$ and the robot's base is defined as ${}^{R}\mathbf{C}_W = {}^{R}\mathbf{T} \cdot {}^{T}\mathbf{C}_W$.

[1] www.vision.caltech.edu/bouguetj/calib_doc.

We can determine the pose of the tool $T$ with respect to the base of the robot using the forward kinematic model of the robot. We also use a static transformation ${}^{R}\mathbf{C}_{K}$ between the camera $C_K$ and the robot's base. During calibration, the camera estimates the pose of the visual marker. Finally, the transformation ${}^{M}\mathbf{C}_{W}$ between the calibration pattern (marker) $M$ and the camera $C_W$ is used to find the transformation between the tool and the camera ${}^{T}\mathbf{C}_{W}$. The found transformation should fulfill the given condition:

$$ {}^{T}\mathbf{C}_{W} = {}^{M}\mathbf{C}_{W}^{-1} \cdot {}^{M}\mathbf{R} \cdot {}^{R}\mathbf{T}. \tag{1} $$

Similarly, we perform the external calibration for the RGB-D camera $C_K$, but in this case, we have to move the marker and change transformation ${}^{M}\mathbf{C}_{K}$. The found transformation ${}^{R}\mathbf{C}_{K}$ should fulfill the given condition:

$$ {}^{R}\mathbf{C}_{K} = {}^{R}\mathbf{M} \cdot {}^{M}\mathbf{C}_{K}. \tag{2} $$

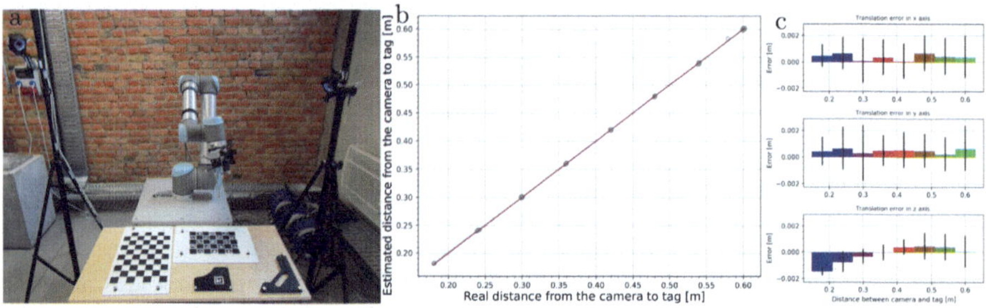


**Fig. 3.** Experimental set with the UR10 robot and Optitrack reference tracking system (a) used to verify the calibration results: obtained relation between the real and estimated distance between the camera and the tag (b), and the translational error (c)

The experimental set used to calibrate the system is presented in Fig. 3a. We also use the Optitrack reference system to compare the obtained results with the ground truth data. Thus we attach Optitrack markers to characteristic points of the robot, visual markers, and the RGB camera $C_W$.

### 3.3 Motion Planning

Currently, each marker stores information about the distance of the tool to the center of the marker. In the future, the identifier of the marker can also determine other properties like the speed and the type of the tool. In the current implementation, we move the tool along the linear trajectory between markers. This can be used to mill the material between two markers or move the tool above the surface of the product. The marker can also be associated with the dxf file and used for milling 2.5D shapes on the planar surfaces. Currently, we parse dxf file to define a 3D path for the robot in the 3D space.

The markers found by the vision system of the robot define the waypoints for the robot's tool. We plan the motion of the robot in a three-dimensional Cartesian coordinate system using the MoveIt library. The path is generated using the inverse kinematic model of the robot. The MoveIt library is also used to execute the planned trajectory.

### 3.4 Tag Detection Enhancement with Mask R-CNN

In the standard application of the proposed system, we move the RGB camera $C_W$ above the surface of the object to detect and estimate the initial position of markers. Then, we move the robot to set the distance between the camera and the tag to 0.35 m. From this distance, we estimate precisely the 3D pose of the marker which is later used to plan the motion of the robot. In this section, we show the independent module which allows detecting the markers on the RGB image from the significantly larger distance using, for example, the external RGB-D library. This approach reduces the time needed to initially scan the environment.

Improper lighting conditions and high distance may influence the performance of the tag detection. In practical applications, these problems have been addressed in order to ensure reliability in a given environment. When the tag presented on the scene is not detected the system suspends its action. The process can be resumed manually by providing a correct position of the tag, but it contradicts the main concept of the autonomous execution of manipulation tasks.

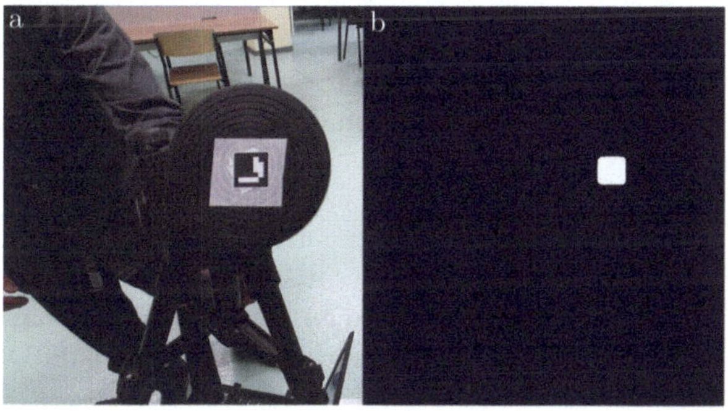


**Fig. 4.** The image generated to train the neural network and the mask representing the detected ArUco marker

In the paper, we propose a method of minimizing the influence of the problem mentioned above. We used and trained Mask R-CNN neural architecture [8] as it has shown a significant surge of detection performance for image data. We prepared a sample dataset of images depicting tags at different distances to the camera (Fig. 4). Moreover, we manually created binary masks that represent the location of the tag on an image. The number of images in the dataset has been increased through data augmentation.

## 4 Results

We start the experiment with the calibration of the robot and the cameras. The calibration results are presented in Fig. 3. In Fig. 3b we show the estimated distance between the camera $C_W$ compared to the real distance measured by the Optitrack reference system. We also fit a line to the obtained data to show the accuracy of the calibration. The translational errors along each axis are presented in Fig. 3c. For the distance equal to 0.35 m, the standard deviation is around 2 mm but the mean value computed for a series of measurements is less than 1 mm along each axis. It means that the proposed system can be used in small factories or workshops for drilling and milling 2.5D shapes. The obtained accuracy is significantly better than the accuracy of the human operating hand tools.

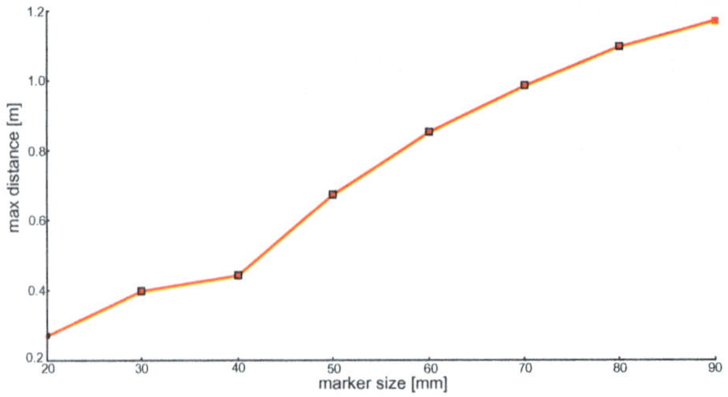


**Fig. 5.** Maximum detection distance between the camera and the ArUco tag

We also verified the maximal detection distance between the camera and the tag. Because the detection rate depends on the size of the marker we tested various tag sizes. The minimal marker size is 20 × 20 mm and the maximal marker size is 90 × 90 mm. During the experiment, we keep the marker surface perpendicular to the camera axis. The results of the experiments are presented in Fig. 5. The maximal detection distance for the 20 × 20 mm markers is 0.27 m. The maximal marker size considered for this application is 90 × 90 mm. The maximal detection distance for this marker is 1.17 m.

We managed to carry out a successful detection of markers beyond the prior limited range using Mask R-CNN. We made it possible to detect the markers virtually in every example provided by our dataset which is, in fact, a major improvement compared to the previous results. The final maximal detection distance reached 1.87 m for a 20 × 20 mm marker (Fig. 6). Thus, we improved the detection range 6.93 times by applying the neural network-based object detector. Mask R-CNN provides a binary mask for every detected instance of a marker on the scene. It allows determining a bounding box which wraps the marker and creates an excerpt of the image for further analysis.

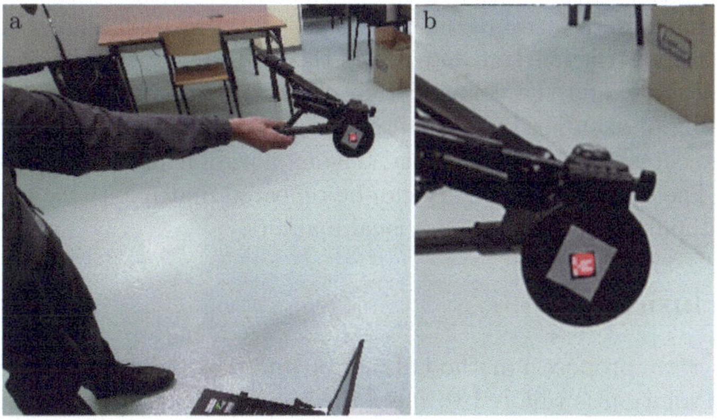


**Fig. 6.** Detected marker located at distance of 1.8 m(a), marker close-up (b)

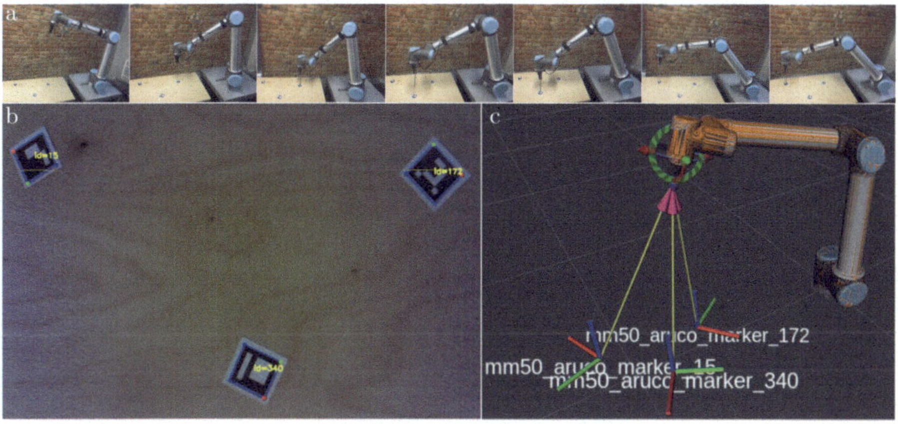


**Fig. 7.** Robot executing a path planned by the controller (a) – example camera view (b) and visualization of the detected markers in the RViz (c). The trajectory of the robot depends on the identifiers of the tags on the scene.

Finally, we performed the experiments on the real Universal Robot robot arm equipped with the ELP 5–50 mm varifocal lens 1080P USB camera. The experimental set is presented in Fig. 7. The ArUco markers are randomly attached to the planar surface. The visual system detects markers. Then, the robot plans its motion taking into account identifiers of the markers and their poses. In the experiment, the sequence of the motions depends on the identifiers of the markers. We sort the identifiers and related poses of the marker in ascending order.

The planned sequence is presented in Fig. 7. Before the experiment, the robot scans the environment to detects all markers. Then, it goes to the position which allows better estimating the pose of the marker (around 0.35 m from the sensor).

In our experiments, the error of the marker pose estimation is smaller than 1 mm. In Fig. 7b we show the markers detected in the RGB image. The 3D visualization of the robot state and the configuration of the markers is presented in Fig. 7. If the position of each marker is determined, the robot plans the motion of the tool and executes the final trajectory. The exploration procedure is performed only once. The obtained trajectory can be repeated multiple times for the same product without the need of using visual markers.

## 5 Conclusions

In this paper, we propose a method of motion and task planning for an industrial robot. Our solution is tailored to meet the specific requirements of small-scale industry, which assumes both simplicity and flexibility of the production process. The potential lack of a qualified workforce in the upcoming years may result in a huge demand for solutions that simplify the process of programming and reconfiguring industrial robots. We proved that it is possible to create such a solution which involves neither specialized hardware nor any significant changes at a specific workstation.

We show the system which consists of three main software modules. The first module is responsible for marker detection and poses estimation. The second node plans the path of the robot using 3D poses of the visual markers. The third module is used offline to determine the geometric relations between the main components of the system (robot and cameras). Moreover, we have managed to improve the performance and reliability of tags detection by applying Mask R-CNN neural architecture despite unsatisfactory lightning conditions and a large distance to the camera.

In the future, we are going to perform experiments with milling tools. It also includes the adjustment of the robot's trajectory to the properties of the material and properties of the used tool.

**Acknowledgments.** This work was supported by the European Regional Development Fund (ERDF) Wielkopolska Regional Operational Programme for 2014–2020, Measure 1.2.

## References

1. Andersen, R.E., Hansen, E.B., Cerny, D., Madsen, S., Pulendralingam, B., Bøgh, S., Chrysostomou, D.: Integration of a skill-based collaborative mobile robot in a smart cyber-physical environment. Proc. Manuf. **11**, 114–123 (2017)
2. Andersen, R.S., Damgaard, J.S., Madsen, O., Moeslund, T.B.: Fast calibration of industrial mobile robots to workstations using QR codes. In: IEEE International Symposium on Robotics, pp. 1–6 (2013)
3. Babinec, A., Jurišica, L., Hubinský, P., Duchoň, F.: Visual localization of mobile robot using artificial markers. Proc. Eng. **96**(6), 1–9 (2014)
4. Cieślak, W., Rodykow, S., Belter, D.: Teleoperation of a six-legged walking robot using a hand tracking interface. In: Silva, M.E., et al. (eds.) Human-Centric Robotics, pp. 527–536. World Scientific, Singapore (2017)

5. Garrido-Jurado, S., Muñoz-Salinas, R., Madrid-Cuevas, F., Medina-Carnicer, R.: Generation of fiducial marker dictionaries using mixed integer linear programming. Pattern Recogn. **51**, 481–491 (2016)
6. Garrido-Jurado, S., Muñoz-Salinas, R., Madrid-Cuevas, F., Marín-Jiménez, M.: Automatic generation and detection of highly reliable fiducial markers under occlusion. Pattern Recogn. **47**(6), 2280–2292 (2014)
7. George, L., Mazel, A.: Humanoid robot indoor navigation based on 2D bar codes: application to the NAO robot. In: IEEE-RAS International Conference on Humanoid Robots (Humanoids), pp. 329–335 (2013)
8. He, K., Gkioxari, G., Dollar, P., Girshick, R.: Mask R-CNN. In: IEEE International Conference on Computer Vision, pp. 2980–2988 (2017)
9. Hornung, A., Wurm, K.M., Bennewitz, M., Stachniss, C., Burgard, W.: OctoMap: an efficient probabilistic 3d mapping framework based on octrees. Auton. Robots **34**(3), 189–206 (2013)
10. Krupke, D., Starke, S., Einig, L., Zhang, J., Steinicke, F.: Prototyping of immersive HRI scenarios. In: Silva , M. E., et al. (eds.) Human-Centric Robotics. World Scientific, Singapore (2017)
11. Li, Y., Zhu, S., Yu, Y., Wang, Z.: An improved graph-based visual localization system for indoor mobile robot using newly designed markers. Int. J. Adv. Robot. Syst. **15**(2), 1–15 (2018)
12. Mersch, D.P., Crespi, A., Keller, L.: Tracking individuals shows spatial fidelity is a key regulator of ant social organization. Science **340**(6136), 1090–1093 (2013)
13. Nissler, C., Büttner, S., Marton, Z.-C., Beckmann, L., Thomas, U.: Evaluation and improvement of global pose estimation with multiple AprilTags for industrial manipulators. In: IEEE International Conference on Emerging Technologies and Factory Automation (ETFA), pp. 1–8 (2016)
14. Olson, E.: Apriltag: a robust and flexible visual fiducial system. In: IEEE International Conference on Robotics and Automation, pp. 3400–3407 (2011)
15. Ouaviani, E., Pavan, A., Bottazzi, M., Brunelli, E., Caselli, F., Guerrero, M.: A common image processing framework for 2d barcode reading. In: 7th International Conference on Image Processing and its Applications, Conference Publishing, vol. 2, no. 465, pp. 652–655 (1999)
16. Rostkowska, M., Skrzypczyński, P.: A practical application of QR-codes for mobile robot localization in home environment. In: Silva, M.E., et al. (eds.) Human-Centric Robotics, pp. 311–318. World Scientific, Singapore (2017)
17. Silaghi, H., Rohde, U., Spoial, V., Silaghi, A., Gergely, E., Nagy, Z.: Voice command of an industrial robot in a noisy environment. In: International Symposium on Fundamentals of Electrical Engineering (ISFEE), pp. 1–5 (2014)
18. Pires, J.N.: Experiments on commanding an industrial robot using the human voice. Ind. Robot: Int. J. **32**(6), 5–11 (2005)
19. Wang, J., Olson, E.: AprilTag 2: efficient and robust fiducial detection. In: 2016 IEEE/RSJ International Conference on Intelligent Robots and Systems, pp. 4193–4198 (2016)
20. Wilson, F., Cueva, C., Hugo, S., Torres, M., Kern, M.J.: 7 DOF industrial robot controlled by hand gestures using microsoft kinect v2. In: IEEE 3rd Colombian Conference on Automatic Control, pp. 1–6 (2017)
21. Xavier, R.S., da Silva, B.M.F., Gonzalves, L.M.G.: Accuracy analysis of augmented reality markers for visual mapping and localization. In: Workshop of Computer Vision, pp. 73–77 (2017)

# The Infrastructure of RFID-Based Fast Moving Consumer Goods System Using Cloud

Bartosz Pawłowicz, Mateusz Salach(✉), and Bartosz Trybus

Rzeszów University of Technology, al. Powstańcow Warszawy 12,
35-959 Rzeszów, Poland
{barpaw, m.salach, btrybus}@prz.edu.pl

**Abstract.** The paper presents the application of RFID identification in the field of fast moving consumer goods (FMCG). The concept of a distributed system including household appliances, local embedded controllers and mobile platforms for product identification is presented, and RFID data processing is discussed. An important element of the solution is the use of cloud solutions, such as a central database, middleware connecting household appliances with other system modules and cloud services. The prototype of this solution has been developed using RFID read/write devices and the popular Raspberry Pi platform as a communication module for a home fridge with Microsoft Azure cloud services.

**Keywords:** RFID · Fast Moving Consumer Goods · Cloud computing · Internet of Things

## 1 Introduction

Identification of various objects plays an important role in today's world. They may be encountered in multiple areas of life, such as everyday shopping, maintaining warehouse records or advanced identification processes in industry. Identification systems may be either contact or contactless. The first group involve magnetic and chip cards. Barcodes and RFID (*Radio Frequency Identification*) technology are examples of contactless solutions.

The RFID systems are more and more frequently used in automated processes in various areas [1]. For example, multiple applications are carried out in the range of the Internet of Things (IoT) [2]. Reliable and safe identification of moving objects is also a significant research area [2, 4]. Modern applications involve maintaining public transport or transport of goods, city traffic management and monitoring systems or collecting road toll [5, 6]. The need for advanced object identification raises new research challenges, so existing standards are constantly updated and new standards are developed, especially for the RFID systems using HF and UHF frequency ranges [5].

Some researchers focus on Fast Moving Consumer Goods (FMCG) in supply chains [3, 17]. This application area is also the topic of this paper. The purpose is to present an overall software architecture of a FMCG solution, involving data transfer from RFID transponders, local management controllers running on embedded platforms and cloud-based services. The concepts are illustrated with a prototype IoT


R. Szewczyk et al. (Eds.): AUTOMATION 2020, AISC 1140, pp. 216–226, 2020.
https://doi.org/10.1007/978-3-030-40971-5_20

solution for management of food products. The prototype extends the well-known concept of a "smart" home refrigerator with RFID-based product management and cloud connectivity.

## 2 RFID Usage in FMCG Supply Chains

Currently, the most popular solution for consumer goods identification is a barcode. A barcode contains a constant amount of digits represented by vertical lines with different width which is related to a specific digit. EAN-13 (European Article Number) is an example of a standardized barcode format. The first three digits describe the country prefix. The second block including four digits is related to the producer code. The third block which contains five digits identifies a product. The last number is a checksum. Barcode is a good solution in stores where the cashier can manually rotate the product so that the scanner can read the code. However, it is a low efficient solution for warehouses and delivery systems, because data can only be read by placing the scanner directly on the barcode sticker.

To avoid such problems, a wireless RFID identification system can be applied. Such solution can significantly speed up the process of product identification in warehouses, stores and during product delivery [7–10]. As RFID systems are becoming more and more cost-effective, this can have important practical implications in the near future [11, 12].

A contactless RFID identification system involves a reader-writer device (RWD) with an antenna and the identified product equipped with an RFID transponder chip (or tag). The transponder can hold all the product data as the barcode and even more if necessary, up to several kilobytes. Moreover, some new chips provide supplementary autonomous functions, such as measuring physical quantities (humidity, temperature, light intensity, pressure, acceleration, gas, etc. [13–15]). The information can be read from the transponder by the RFID reader which is generating a radio signal which powers up the RFID transponder (Fig. 1a). After the transponder is charged with the signal from the reader, a microchip (μP) in the transponder starts operation and sends back the data to the reader device. In addition, the RWD can also write or update the data performing a double function in the system (transmitter/receiver) and enabling data transmission in two directions [16].

RWDs and transponders constitute the hardware part of the system. In the case of FMCG supply chain, products are equipped with RFID transponders (instead of barcode stickers). RFID transponders can also be placed on collective packaging (or pallet) to improve the transportation tracking (Fig. 1b). The data can be also used by household appliances such as a refrigerator or a washing machine (Fig. 1c). For example, a fridge equipped with an RWD can gather specific information from products that are stored inside. Thus, ordering products may include not only their availability in the fridge but also the expiry date.

The software part serves for both direct controlling of individual digital devices as well as for managing the whole system. The use of such solution is additionally interesting due to the fact that RFID data can be read using a mobile device.

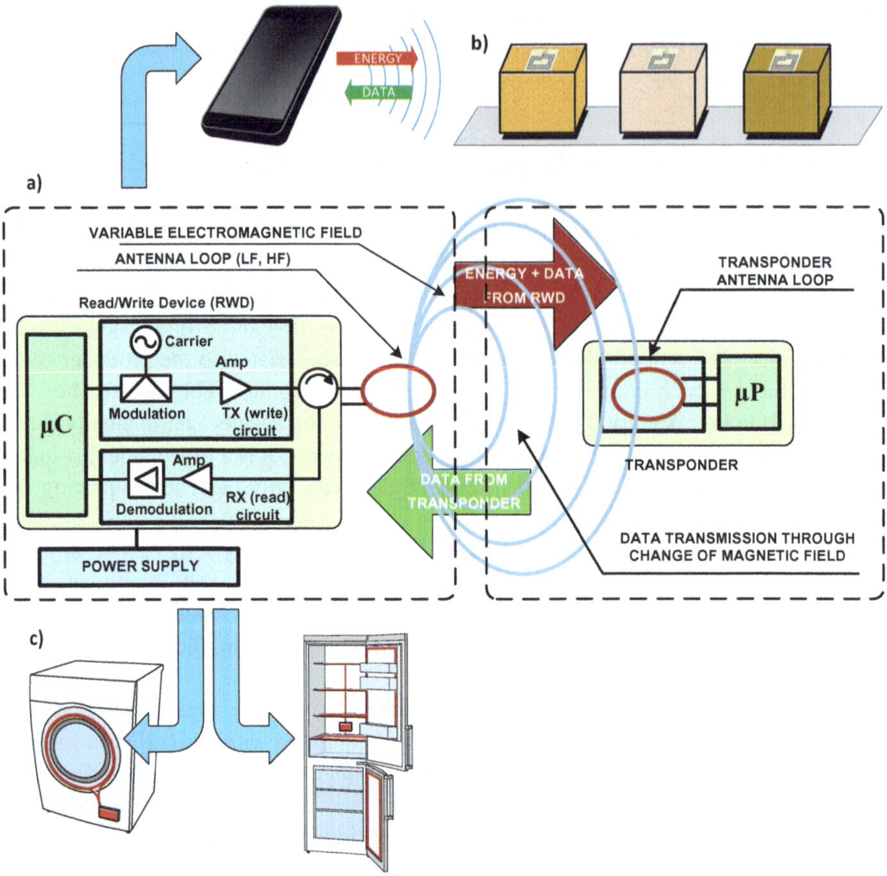


**Fig. 1.** FMCG identification system: (a) reader-writer device operation with an RFID transponder, (b) tracking of product packaging, (c) household appliances with RFID identification.

## 3 Processing of RFID Data During Production and Transportation of Goods

An RFID transponder can be applied to store data on each stage of the product life cycle for various scenarios. For example, the RFID transponders may be used on the production line to hold some specific data, such as manufacturing date, serial number, version, etc. The manufacturer can also introduce some reserved information for internal needs. Later, the data fetched from the RFID transponders can be sent to a production management system. A different bunch of data will be used for transportation purposes [18].

As said, the data can be read from an RFID transponder by using a dedicated RWD device. Interestingly, a smartphone can also play the role of such a device, due to the use of NFC (Near Field Communication), a wireless communication technology for short-range data exchange. Since NFC share the same frequency as RFID, mutual communication using appropriate protocols is possible between RFID and NFC devices. Thanks to this, infrastructural requirements can be significantly reduced, especially during transportation of goods.

In Fast Moving Consumer Goods supply chain, the speed of product transfer plays an important role. RFID-based solutions can lower the time needed for managing and tracking products. When a product is ready to be delivered to a store or other location, an employee uses an RWD to store data containing target information into RFID transponders located on the product packaging (Fig. 2). By utilizing a cloud database storage, management and control of consumer goods flow can be shared with other participants. For instance, the information that a product is ready for transfer is stored to the cloud. A delivery company is notified about the fact by the cloud service and assigns a vehicle (a truck, lorry, train, maybe a drone) for transportation. After leaving the warehouse, the product packaging is monitored by an RFID read/write device built in the vehicle used for transportation. When the delivery is completed, an information about reaching the destination is written to the product RFID transponder. The data can be visualized with life tracking of consumer goods.

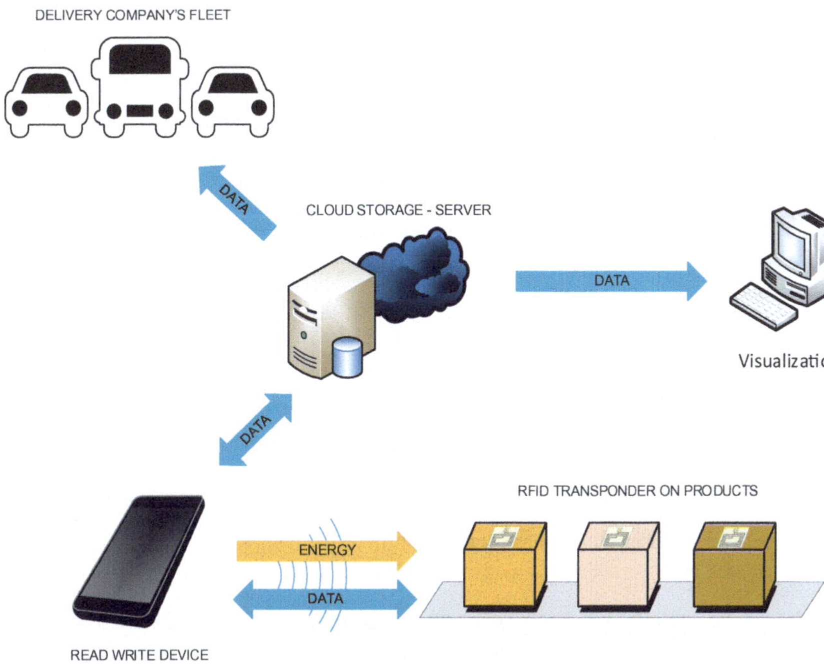


**Fig. 2.** Product delivery system using a cloud and RFID techniques.

By using an RWD device anyone can modify the product data what may be treated as a security risk in some scenarios. This threat may be reduced by using encryption of data stored in the RFID transponder's memory. For example, data update may be performed only when a dedicated authorization key is used. A simplified algorithm for secure writing of data to an RFID transponder involving an authentication key is shown in Fig. 3. The RFID system waits until the RWD device is within range. If the device is in range, the system checks the authentication key. If the key is wrong, the update is prohibited and the connection is dropped. Otherwise, when the key is correct, the update starts. After the data is written, the success is signalled and the connection is closed.

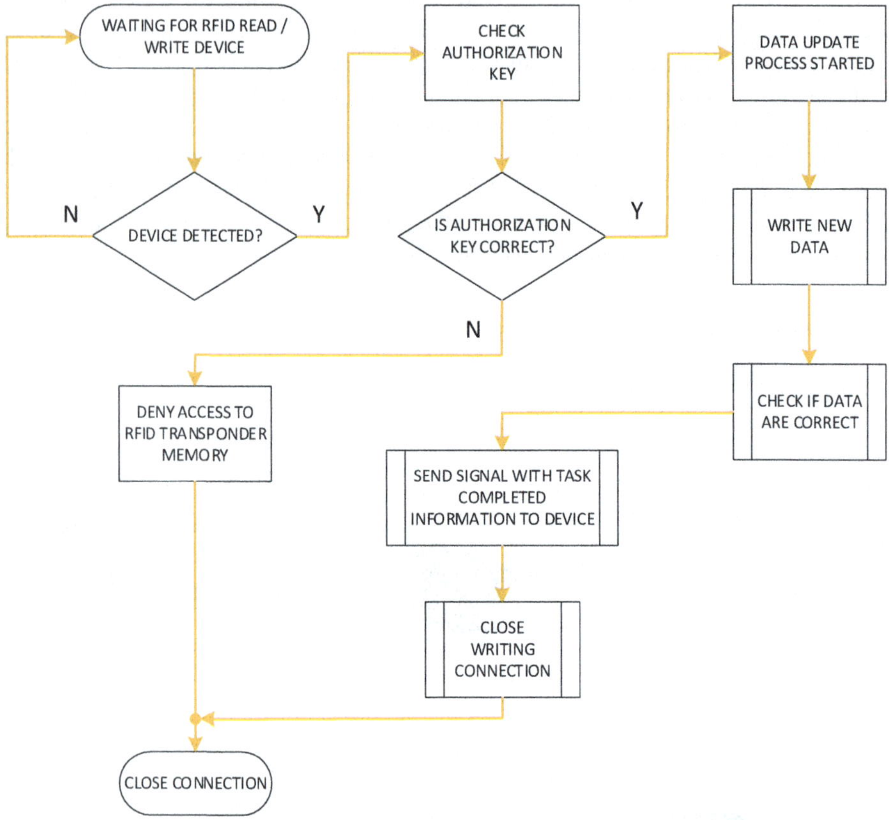


**Fig. 3.** Secure data writing in an RFID system.

It should be stated, that such scenario is possible when the system allows data feedback. This may be achieved using e.g. a smartphone and an RWD device. In such a case, the RWD can retrieve information from the smartphone and send back other data, for example to visualise them on the screen. Another application of two-way data exchange can be a production line, when a mobile device can send an identification code for a new product and receive feedback information with data related to another step of production or redirection to a target warehouse.

## 4 Prototype Food Product Management System with Cloud Connectivity

The concepts of identifying products based on RFID using cloud services have been tested by making a laboratory prototype of a food management system. This was conceived as a new extension to the well-known idea of intelligent home appliances with the ability to monitor product quantity and expiration date with the possibility of performing additional tasks such as remote ordering.

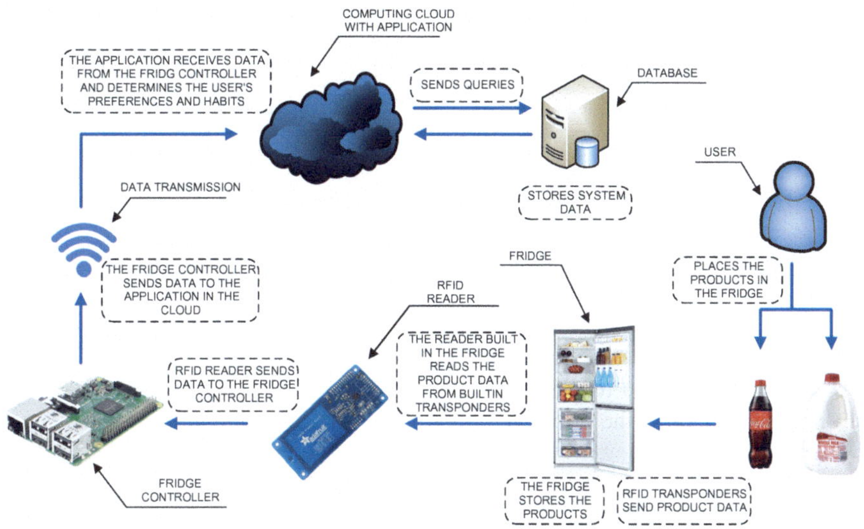


**Fig. 4.** Concept of an RFID Refrigerator with cloud system.

Components of the system are shown in Fig. 4. The fridge is equipped with an RFID RWD device and a local embedded controller. When the user places food products in the fridge, RFID data stored by their manufacturers are read and sent to the controller. Using its networking capabilities, the controller communicates with a cloud application. The communication involves sending information about the products in the fridge. A cloud database is used to store the data, allowing to perform an analysis to adjust the system to user preferences and habits.

As it can be deducted from Fig. 4, the fridge controller plays an important part in the system, being the integrating element between the RFID transponders and upper level modules. The prototype solution involves the popular Raspberry Pi (RPi) microcomputer board. Adafruit PN532 is used as the RFID RWD device. Data exchange between RPi and RWD is done using I2C protocol. The RPi 3 model is equipped with a built-in Wi-Fi module necessary for communication with the cloud.

The controller application for RPi written in Python maintains products in the fridge and performs connectivity with the cloud. Structure of the application showing the core activities is presented in Fig. 5. They include monitoring state of the fridge, handling the situation when the users opens the door and ordering a product.

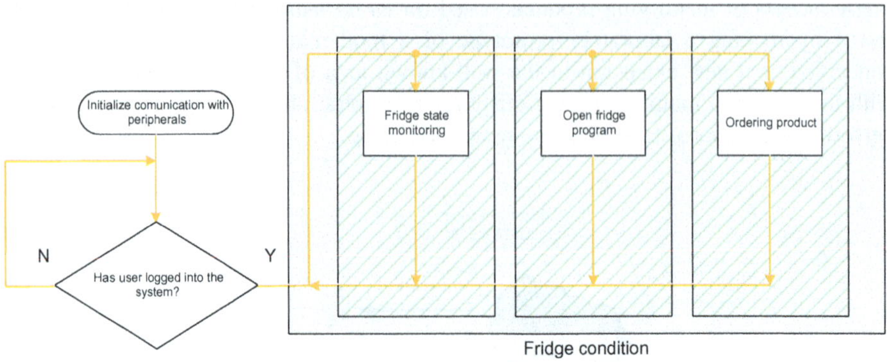


**Fig. 5.** Core activities of the fridge controller.

When the system detects that the door is open, it monitors user actions. In particular, the user can:

- take a product from the refrigerator
- put a new product into refrigerator
- do nothing and close the door.

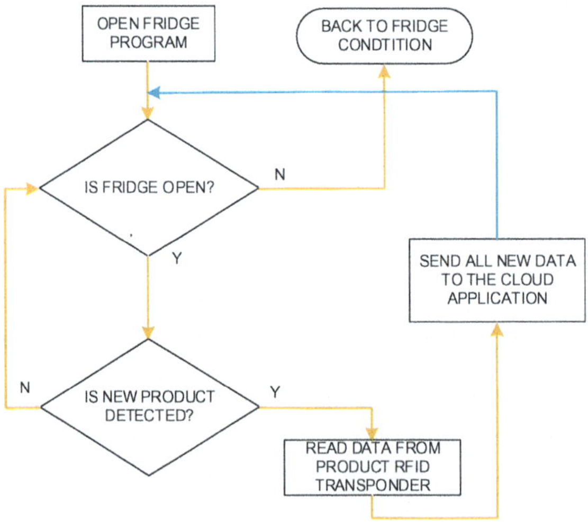


**Fig. 6.** Open fridge door program.

Figure 6 presents a part of the program handling the open door scenario. If the door is open, the controller tries to detect the products by invoking the RFID RWD device to get data from their RFID transponders. If the user decides not to do anything and closes the door, the RWD is turned off and no action is performed. In case of adding new product or taking it from refrigerator the program is redirected to the discovery algorithm shown in Fig. 7.

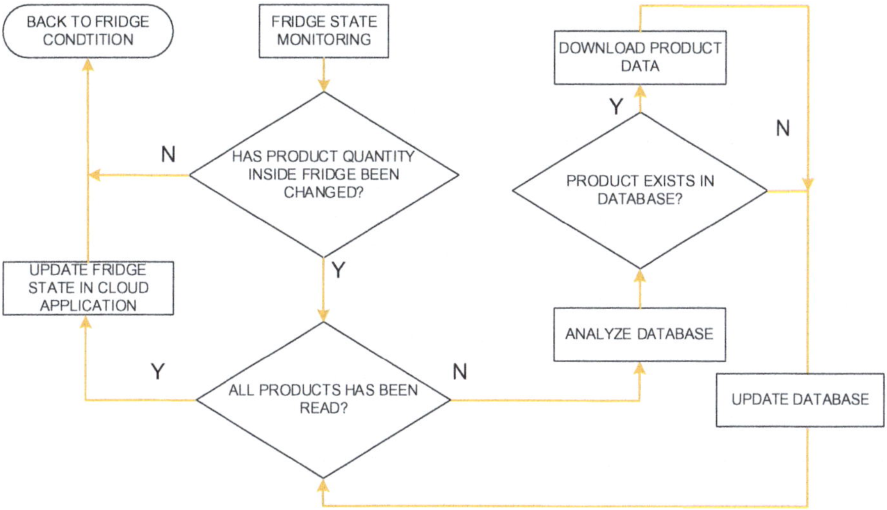


**Fig. 7.** Product discovery by the fridge controller.

The discovery is done after the product passes the built-in RFID RWD and data is read from its RFID transponder. At this point, the program is querying a local database of products to get information about the product. If the product does not yet exist in the database (a new product), the data downloaded from the RFID is added to the database. If all the products are recognized, the system is signalling an update to the cloud. If the product already exists in the database, its data is fetched to maintain this situation which may be assumed as taking the product out from the refrigerator and then returning it back.

It should be noted that the RWD device in the fridge can put extra data in the transponder. For example, it may mark a particular product with an unique identification number to allow distinguishing two or more products of the same kind. Additional information about the product usage can also be stored in the transponder and used later when the product is taken to another fridge (e.g. from home to work).

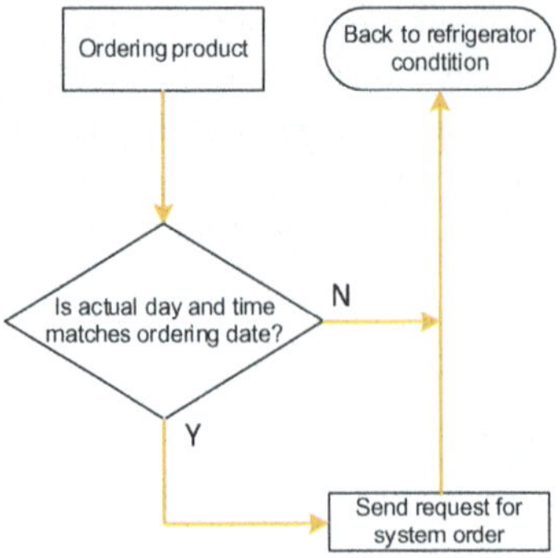


**Fig. 8.** Scheduling product orders.

The system allows to order products from a supplier. The order list is created based on the product usage. If a product is exhausted, the system will automatically add it to the list of ordered products. The system may also add a product to the order list if its expiration date has passed. However, the user can edit the list, before the order is actually sent to the supplier. This is done at a scheduled time and day (Fig. 8) or upon request.

The cloud part of the RFID-based fridge goods management involves a cloud application, database storage and services. The application exposes an function interface for the fridge controller to store product data in the cloud as well as to get extra information from manufacturers or authorities. For example, the user may be informed that the product is withdrawn from the market for some reasons or it has been declared unsafe for health.

The prototype uses the Microsoft Azure as the cloud platform with Cosmos DB NoSQL data storage services. The application has been created with .NET Core framework and provides an API to communicate with the fridge controller.

**Fig. 9.** Cloud-based virtual fridge with RFID identification.

In addition, the cloud application creates a "virtual fridge", i.e. the user can order products from everywhere using data stored in the RFID transponder on the product packaging (Fig. 9). The virtual fridge involves a mobile application to read the RFID data with a NFC-enabled smartphone. The mobile application connects the user with the cloud account via Microsoft Azure Identity mechanism. The virtual fridge may be then filled with new records with product information. The records are added by the mobile application when the user scans RFID data of the required product no matter where the user is located. Then the products may be ordered and delivered to the real fridge in the system.

## 5 Conclusions

RFID identification brings significant benefits in the area of FMCG supply chains, provided that the required hardware infrastructure is available. Some of the requirements can be reduced using NFC-capable mobile devices, such as smartphones instead of dedicated RWD devices. However, RFID solutions cannot work without an appropriate software architecture. Such software involve modules running on different levels of the system, starting from an embedded controller up to the cloud-based applications. The use of the cloud increases the size of the solution, but also provides additional unique features, including remote access, scalable data storage and central monitoring. In the case of the presented system prototype, the cloud services will enable tuning the ordering system to the user's habits and preferences. This may be achieved with Azure Machine Learning consuming the collected data. Azure IoT Hub and business analysis services can also be a point of interest.

**Acknowledgment.** This project is financed by the Minister of Science and Higher Education of the Republic of Poland within the "Regional Initiative of Excellence" program for years 2019–2022. Project number 027/RID/2018/19, amount granted 11 999 900 PLN.

## References

1. Ustundag, A. (ed.): The Value of RFID. Benefits vs. Costs. 1st edn. Springer, London (2013)
2. Giusto, D., Iera, A., Morabito, G., Atzori, L. (eds.): The Internet of Things, 20th Tyrrhenian Workshop on Digital Communications. Springer, New York (2010)
3. Traub, K. (ed.): The GS1 EPCglobal Architecture Framework. Version 1.6. GS1 (2014)
4. Jankowski-Mihułowicz, P., Węglarski, M.: Determination of 3-dimentional interrogation zone in anticollision RFID systems with inductive coupling by using monte carlo method. Acta Phys. Pol. A **121**(4), 936–940 (2012)
5. Dobkin, D.M.: The RF in RFID: UHF RFID in Practice. 2nd edn. Newnes (2012)
6. Pawłowicz, B., Salach, M., Trybus, B.: Infrastructure of RFID-Based Smart City Traffic Control System: Progress in Automation, Robotics and Measurement Techniques, Automation (2019)
7. IDTechEx: Continued growth as market for RFID exceeds $10bn milestone, pp. 38–39. ID World Magazine, December 2015
8. Diorio, C.: New intelligence for RFID resurgence at item level, pp: 44–45. ID World Magazine. December 2014
9. Brown, D.E.: RFID Implementation. McGraw-Hill, New York City (2007)
10. Das, R., Harrop, P.: RFID Forecasts, Players and Opportunities 2014–2024. Report. IDTechEx (2014)

11. Volpato Filho, O., Piva, F.: NFC-enabled decentralized checkout system. In: 2014 IEEE Brasil RFID, pp. 35–37 (2014)
12. Tyagi, S., Ansari, A.Q., Khan, M.A.: RFID Data Management. INTECH Open Access Publisher (2010). http://cdn.intechweb.org/pdfs/8488.pdf
13. Abad, E., Mazzolai, B., Juarros, A., Gómez, D., Mondini, A., Sayhan, I., Krenkow, A., Becker, Th.: Fabrication process for a flexible tag microlab. In: Proceedings of SPIE, vol. 6589, p. 65890O (2007)
14. Oprea, A., Courbat, J., Bârsan, N., Briand, D., de Rooij, N.F., Weimar, U.: Temperature, humidity and gas sensors integrated on plastic foil for low power applications. Sensors Actuators B Chem. **140**(1), 227–232 (2009)
15. Yang, L., Rida, A., Wu, T., Basat, S., Tentzeris, M.M.: Integration of sensors and inkjet-printed RFID tags on paper-based substrates for UHF cognitive intelligence applications. In: IEEE Antennas and Propagation Society International Symposium, Honolulu, USA, 9–15 June 2007, pp. 1193–1196 (2007)
16. Finkenzeller, K.: RFID Handbook – Fundamentals and Applications in Contactless Smart Cards and Identification. Wiley, Hoboken (2003)
17. Zhigang, Z.: Applying RFID to reduce bullwhip effect in a FMCG supply chain. In: 2011 International Conference on E-Business and E-Government (ICEE), Shanghai, China, pp. 1–4 (2011)
18. Wrycza, P., Hille, A., Anderseck, B.: 1.5 Billion events a week - a statistical assessment for smart transport items in German FMCG supply chains. In: Smart SysTech 2014; European Conference on Smart Objects, Systems and Technologies, Dortmund, Germany, 1–2 July 2014 (2014)

# The Use of Facial Expressions Identified from the Level of the EEG Signal for Controlling a Mobile Vehicle Based on a State Machine

Szczepan Paszkiel(✉)

Faculty of Electrical Engineering, Automatic Control and Informatics, Department of Biomedical Engineering, Opole University of Technology, Opole, Poland
s.paszkiel@po.edu.pl

**Abstract.** The development of brain-computer technology resulted in affordable Brain-Computer Interface peripherals. The purpose of the article is to choose BCI device and the way of passing commands to remote vehicle. Next part of the work is preparation of physical device, which can be controlled through Wi-Fi from Personal Computer. The vehicle is expected to make movements like forward, backward and rotation. Finally, the whole process will be orchestrated by software on PC, which passes commands from BCI device to the vehicle. Thus the robot is controlled with facial expressions, intercepted by BCI device.

**Keywords:** Control · Mobile robot · Facial expressions · Brain-computer interfaces

## 1 Introduction

Nowadays in many implementations can be found examples facial expressions identified from the EEG signal for controlling the mobile robots, including, inter alia, by means of LabVIEW [13] and proprietary solutions program [14]. However, this process has not yet been described on the base state machines as presented in this article. The development of brain-computer technology resulted in affordable Brain Computer Interface peripherals. Interpretation of intercepted EEG waves and passing movement commands based on predefined scheme.

The most important part of the article is design and implementation of system realizing stated goals. The software is expected to have modular design. Therefore, each component can be tested independently and system is easier to maintain. This software can be extended to steer wheelchair. Proposed solution can be valuable for people with disabilities – who have limited movement capabilities. The solution can be also used for entertainment, i.e. simulator can be extended to be a game.

The brain-computer interface allows direct communication between the brain and an appropriate external device [11]. The device used in the conducted research was created in the laboratories of Emotiv System Inc. The software delivered by the manufacturer reads and interprets conscious and unconscious emotions, head


R. Szewczyk et al. (Eds.): AUTOMATION 2020, AISC 1140, pp. 227–238, 2020.
https://doi.org/10.1007/978-3-030-40971-5_21

movement and even facial expressions. The Emotiv EPOC+ Neuro Headset consists of 14 electrodes + 2 reference electrodes, which read brain wave signals. It operates on the principle of an electroencephalograph. Electrodes placed on the head measure electrical potential changes, which appear as a result of neuron activity in the cerebral cortex. A signal is formed as a result of brain activity associated with a specific action [10]. Based on the measurement of this signal and assigning it to an appropriate activity, it is then possible to control a given device. As the conducted analyses show, it is currently impossible to read human thoughts.

## 2 Current Research and BCI Device

In the past the one way to communicate with computer was typing of commands on terminal. Today there is a variety of input methods available, such as modern desktop controlled with a mouse, touch panels, motion sensors and voice input. Now there is a continuous investigation of Brain Computer Interfaces technologies – device control with the use of brain waves [2]. The main challenges of the research are efficient filtering of artifacts caused by wireless networks, electromagnetic field and other interference emitted by many devices around us. It was already proposed to control vehicle with Brain–Computer Interface devices [6]. Such solution might be helpful for disabled people. Due to population aging, there will be in future further development of BCI technology.

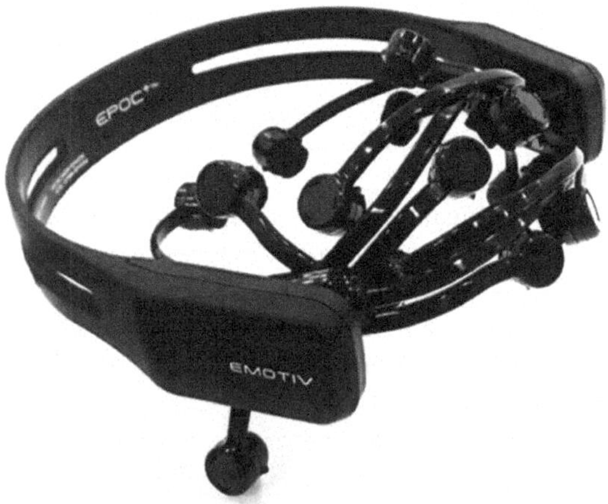


**Fig. 1.** Emotiv EPOC+ Neuro Headset.

Clinical electroencephalography is one of several methods of data acquisition from human brain. It was introduced by Hans Berger, a German psychiatrist in the 1930s. It is a noninvasive method consisting in detection and registration of electrical activity of the brain using electrodes attached to the scalp which register changes of electric potential on the skin surface coming from the activity of cerebral neurons and after their amplification they form a record – an encephalogram [4]. The value of the potential registered by consecutive electrodes can be described by the Eq. (1).

$$V_n = V_{EEGn} + V_{CMS} \tag{1}$$

where: $V_n$ – potential value on electrodes, $V_{EEGn}$ – potential connected with electrical activity of the brain. $V_{CMS}$ – common signal on all electrodes, also connected with interference from the network. Presently, EEG is most often used by neurologists to differentiate functional from organic brain diseases, to diagnose sleep disorders, headaches, dizziness, to monitor brain activity during heart operations. EEG offers high temporal resolution which is not possible with MRI. Obtaining electrode resistance not exceeding 10 kΩ at the start of the investigation is an essential condition for obtaining a good quality EEG. It depends, to a large extent, on proper preparation of the scalp, which before the electrodes are attached, should be carefully degreased and the superficial calloused layer of the epidermis removed. The disadvantage of using an electroencephalograph in practice is, among others, limitation of resolution by equipment capabilities and the need to use a computer to view and analyze data.

In this work the BCI device of choice is Emotiv EPOC+ Neuro Headset (Fig. 1) – wireless EEG headset with 14 channels capturing brain activity. The device realizes noninvasive method of signal acquisition. The device sends to PC obtained signal through Bluetooth 4.0 Low Energy. The hardware is bundled with proprietary software containing detection suite. Brain activity is recognized by algorithm based on machine learning. Access with SDK is described below.

## 3 Applied Technology

The proposed robot is continuous track vehicle based on Genuino Uno platform (Fig. 2). The main components of the robot are described below.

Two motors let the vehicle move forward, backward and rotate. They are connected to the H Bridge. An H Bridge is an electronic circuit that switches the polarity of a voltage applied to a load. These circuits are often used in robotics and other applications to allow DC motors to run forwards or backwards. The motors are controlled through H-Bridge, which applies voltage to motors in either direction. Therefore, the vehicle can move not only forward, but also backward and can perform rotation.

The used H-Bridge is discrete component DRV8835 from Texas Instruments. The component provides pins described in Table 1. M−, M+, MA+, MA−, MB+, MB− are not connected to Genuino, the rest of pins are. The Wi-Fi module used in the robot is ESP-12E, which is designed to be used by hobbyists. The communication with the module is realized through UART, i.e. with pins RX and TX. Essentially, the ESP

**Fig. 2.** The proposed vehicle.

**Table 1.** Pins for H-Bridge DRV8835

| Pin | Description |
|---|---|
| M−, M+ | Motor power (2–11 V) |
| MA+ , MA−, MB+ , MB- | Power for motors A and B |
| ENA, ENB | Enable motor A or B (3.3 V) |
| PHA, PHB | Phase for motors (Pulse Width Modulation) |
| VCC | Input power 3.3 V |
| GND | Ground |

module is a computer in itself – by default it has software which can connect to or host WPA protected networks. The module can even host DHCP server. Genuino Uno is a micro-controller board for hobbyists. It is based on ATmega328P. The device is programmed through USB serial connection. The main role of the vehicle is to move with included motors. Therefore, it is important to analyze possible movements by the device. The multiple combinations are possible by assigning to each motor one of the states in H Bridge: stop, forward and backward. The possible states are described in Table 2.

**Table 2.** Movements of robot as a function of motor movement direction

| Left motor state | Right motor state | Obtained movement | |
|---|---|---|---|
| Stop | Stop | Stop | |
| Forward | Forward |  | Forward |
| Backward | Backward |  | Backward |
| Forward | Backward | 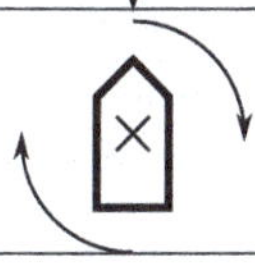 | Clockwise rotation |
| Backward | Forward | Counterclockwise rotation | |
| Stop | Forward |  | Rotate left, forward |
| Stop | Backward |  | Rotate left, backward |
| Forward | Stop |  | Rotate right, forward |
| Backward | Stop |  | Rotate right, backward |

## 4 Mapping of User Input into Vehicle Movements

The BCI input will be mapped into robot movements. Although robot movement could be in simple one-to-one relation to subset of BCI commands, it was decided to introduce command schema based on *finite state machine*. This makes the robot control more robust. Each state movement is caused by *BCI input*. Some transitions take into account time elapsed since last state switch. The input consists of *facial expressions* – headset user face state. The software will interpret following facial movements, which are captured by Emotiv EPOC+ Neuro Headset: Closed/open eyes When eyes are closed, there appears *posterior rhythm*; Left/right smirk Movement is reflected as *muscle artifact* on the EEG; Left/right eye movement Movement is reflected as *jagged eye movement artifact* on the EEG; Neutral No facial expression was detected.

According to the previous subsection, it can be interpreted from EEG whether eyes are closed. Therefore, blinking can be detected as a sequence of eye opening and closing. The state machine for detection of 3 blinks is in the Fig. 3. Symbols "eye open" and "eye closed" represent detection of open eyes and closed eyes respectively. Dashed transitions represent return to initial state, which are activated after 1.5 s since last change of the state. Therefore, accidental eye blinking doesn't trigger state change. Assume that the current state is $A$. For instance, consider the following sequence: user twice closes and opens eyes. Thus the current state is $O_3$. After 5 s user again blinks. However, timeout occurred, so the current state is $O_1$, not $B$. The dashed transitions are not ordinary in the sense that when they are activated, the BCI input is again processed by the state machine. Thus, when the current state is $O_2$ and user send input to switch state from $A$ with smiling – the facial state will be consumed by state $A$, no $O_2$.

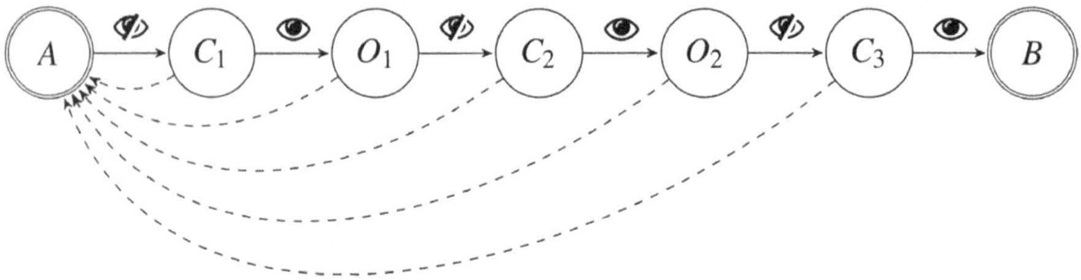


**Fig. 3.** State machine of blink handling with close count equal 3.

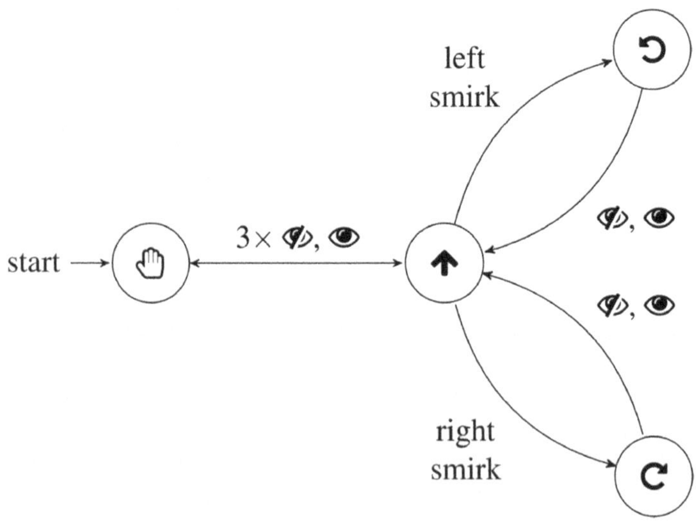


**Fig. 4.** State machine of BCI input map into robot movements.

Blinking sequences will be replaced with transitions labeled "$n\times$ closed eye, open eye", where $n$ is a natural number denoting close count. Therefore, state machine from 3 will be denoted with label "3× closed eye, open eye". The proposed state machine is in the Fig. 4. Symbols "hand", "arrow up", "counterclockwise rotation" and "clockwise rotation" denote vehicle movements from Sect. 3: *Stop*; *Forward*; *Rotate Left, Forward* and *Rotate Right, Forward* respectively. The command scheme allows to obtain four robot movements, because the state machine has four vertices and forms *strongly connected directed graph*. The transitions from forward to rotations require at least 1.5 s since state switch. Thus robot will move forward at least 1.5 s. There is also a variation of the command scheme, where left and right smirk are replaced with left and right eye movement.

## 5 System Design

This section presents proposed software solution to the stated objectives. The design is presented with support of class diagrams [7]. Class diagrams were prepared with "C# profile", i.e. they are extended with stereotypes describing specific constructs for C#. For instance, stereotype "«C# property»" assigned to attribute means, that the attribute is accessed through public set and access (get) methods. This section briefly discusses third-party solutions and their integration with system. Majority of subsections consider software at client side. The software running on Windows is built on *.NET Framework 4.5* and implemented in *C# 6.0* [8]. The software was developed with *Visual Studio 2015 Enterprise*. .NET Framework provides GUI and I/O library including TCP/IP connection. Graphical User Interface is built on *Windows Presentation Foundation* [9], which is part of *.NET Framework 4.5*. Crucial part of interaction with WPF is *data binding*, which lets the view (ex. window) monitor state of *data context* (data behind the view). The mechanism is based on *observer pattern* [3]. Each class implementing INotifyPropertyChanged is *observable*, that is: it notifies subscribed observers about change of contained state. On the other hand, view is an *observer*: when notified about change of *data context*, it refreshes its contents. To utilize this mechanism, Microsoft proposed Model-View-View Model pattern, which is variation of Model-View-Controller pattern. It consists of three parts. Model Class or classes containing business logic, such as data access or calculations. It's optimized for usage as a library, isn't aware of graphical user interface. View Model Adapts the model to form more usable from human perspective, but doesn't manipulate any graphical components directly. The view model class must be observable.

View Contains graphical components representing state of view model. Updates itself through mentioned *data binding*. View should have almost no logic, rather it should be placed in view model. Class SomeModel is hypothetical implementation of business logic. Class SomeViewModel adapts the logic of SomeModel for user interface. Class SomeView presents data from SomeViewModel. SomeViewModel passes calls to model and updates own properties. Update of Value raises PropertyChanged event. From this point it is assumed, that each instance of INotifyPropertyChanged notifies about any change of its properties. It's worth to notice visibility of objects in this pattern: model doesn't see view model and view model is independent of view. View model manipulates graphical components of view only indirectly through *data binding*. This system contains *unit tests* of logic mapping user input into vehicle movement [1]. The *NUnit* unit testing framework is used to check software correctness empirically through Arrange-Act-Assert logic: Arrange Preparation of objects for testing, including memory allocation; Act Execution of action to be tested. Method is feed with data from arrangement; Assert Comparison of obtained output with expected output. Each unit test checks correctness of *particular method* in *particular class*. In order to provide separation of classes, the unit tests utilize *NSubstitute* library. The tool creates mocks of components required in arrange step, based on predefined scenario.

Particular components depend on some other components, for instance robot steering View Model requires the vehicle driver. Although that initialization could be done manually in code, this system decided to use dependency injection container (*DI container*), which resolves the dependencies automatically. The DI container of choice is *Simple Injector*. First stage of its lifetime is setup: to each interface in system is assigned particular implementation. Next system requests from DI container implementation of given interface. There is delivered official SDK which allows custom software to communicate with Emotiv EPOC+ Neuro Headset. The possible operations include raw EEG fetch and use of algorithms delivered by Emotiv. The built in algorithms include "mental command" and "facial expressions" detection. The SDK is distributed as C library in DLL file. To the SDK is attached library wrapper, which makes SDK routines available to software developed in .NET Framework. However, the provided wrapper has following drawbacks: The library is "thin wrapper", i.e. it's structure is very close to the organization of native SDK library, thus it doesn't utilize many C# mechanisms. Therefore the library is a bit awkward in usage. The library provides functions which are not required for this research, such as support for multiple headsets connected at once. Hence it requires the developer to take care of multiple headsets. Software on Genuino is developed with C++ using manufacturer SDK. Utilizes third party library which communicates with Wi-Fi module.

This section covers system design and division of tasks among classes. It should be noticed, that over each important component of the system is defined layer of abstraction – C# interface. Thus other components depend on class interface, not particular implementation [3]. This has important consequences: classes for particular parts of the system are independent and unit testing is possible in isolation of other components.

Communication between PC and vehicle is based on TCP/IP socket. Computer sends to the device commands describing which engine should be turned on, including their direction and speed. Communication protocol is based on plain ASCII. Each command consists of 4 bytes followed by *Carriage Return* and *Line Feed*. Table 3. presents detailed scheme of protocol. For instance, command [AF22] orders left engine to move forward with speed 34 (13% of full speed). The role of vehicle is only to parse and execute incoming commands.

**Table 3.** Command contents divided into bytes

| Byte position | Description | Contents |
|---|---|---|
| 1 | Which engine is affected | A (left), B (right) |
| 2 | Direction of movement | F (forward), B (backward), S (stop) |
| 3, 4 | Speed in range 0–255 | Hexadecimal number |
| 5, 6 | End of command | *Carriage Return* and *Line Feed* |

In order to drive the robot from PC, the software must send commands to it. This is done by XRobotDriver, which is configured by robot IP address. Enumeration EngineState denotes direction of engine movement. Commands are send by setting LeftEngine and RightEngine properties value. Moreover, SlowMode lets the vehicle to move slower. Property Connected is used to check whether connection with robot exists. Often instead of movement of particular motor this research works will consider states, which are captured in enumerator RobotEngineStates. The class RobotStates includes methods which convert to and from the enumerator. Software intercepts input from Emotiv EPOC+ Neuro Headset through SDK. Because the program will utilize fraction of SDK, the library will be accessed through *facade* [3]. The class Device-Driver provides simple interface to the SDK. One instance of DeviceDriver can handle only one headset. Indeed, DeviceDriver accesses the SDK through singleton from the wrapper. The design of interfaces follows *interface segregation principle* [5]. The roles of particular interfaces are following: IBrainCommandObservable is implemented by object, which raises notifications about detected changes of facial expressions. The observers subscribe through Face-StateChanged and are feed with instances of FaceStateChangedEvent-Args – which describe last state of facial expression.

IDeviceStatus is implemented by object describing state of Emotiv EPOC+ Neuro Headset, such as battery charge level or wireless signal status. It should be noticed, that the object should be observable. Therefore, the object can be queried for current device state and it can notify about status change. The IDeviceStatus includes method Connect, which tells the object to connect to the headset. Method ProcessEvents is expected to be run regularly in order to process events from SDK, such as detected facial expression. The system will perform mapping of user input into vehicle movements according to command schema described above. The mapping is performed by instance of the class BrainFaceRobotPilot, which takes as an input messages from Emotiv EPOC+ Neuro Headset (objects implementing interfaces IDeviceStatus and IBrainCommandObservable). The output is reflected as change of the robot state (object implementing IRobotDriver). As an additional input it takes instance of IStopwatch, which is generalization of the concept System.Diagnostics.Stopwatch. Although during normal work of the system the mapper will use just SystemStopwatch, the generalization will be crucial during automated testing of the module. The class BrainFaceRobotPilot is based on *state design pattern* [3].

Graphical User Interface is built with Windows Presentation Foundation, the design of interaction with the library follows MVVM pattern. The view will be class implementing property change propagation. Because all view-models share the same method notifying about state change, it was extracted as class VoidViewModel. Therefore, all view-models derive from class VoidViewModel.

The vehicle simulator takes as an input vehicle commands (exposes interface IRobotDriver), as an output generates visualization of robot movements on a plane. Therefore there arises natural separation of module into two parts: Simulator logic Class RobotVirtualSandbox calculates robot position, direction and velocity. The input are vehicle commands. Output consists of robot angle, position on a plane and velocity – which is accessible through the class instance properties. Visual representation Class RobotSandboxViewModel represents RobotVirtual-Sandbox and notifies RobotSandboxPage about robot position change. The module described in this part of article don't raise robot commands. Robot during movement is in one of the states described above, which are determined by current robot motors configuration. Each state is modeled as instance of the interface IRobotState – this is realization of the state design pattern [3].

Although the vehicle can be controlled in software through IRobotDriver, it is required that the robot can be manipulated with GUI. The design follows MVVM from section above. The pilot is expected to show to the user current state of robot movement. The challenge posed by the requirements is the fact, that the robot state can be changed outside of the pilot.

The first task is to choose *model*. In this case it is instance of IRobotDriver. However, the model does not suite the requirements of the user interface: The model is manipulated by settings of particular motors. The user will be passing commands to the pilot as states. Interface of the model IRobotDriver doesn't provide any means of notifying about its state change. However, the pilot is expected to show running motors even when the vehicle is controlled outside of the pilot. This can happen when pilot is turned on and the robot is manipulated through Emotiv EPOC+ Neuro Headset.

The "View-Model" part of MVVM pattern will adapt the model to the just mentioned requirements. The class RobotPilotViewModel has properties such as LeftBackward, which will be used to update GUI components to reflect current robot state. Moreover, because the class derives from VoidViewModel, the view will be notified by the viewmodel about change of motors state. There are methods TurnModeOff and TurnModeOn, which are based on RobotEngineStates enumeration. In order to intercept calls to the instance of IRobotDriver, the architecture will utilize *proxy* design pattern [3]. There will be instance of RobotDriverInterceptor, which aggregates the original instance of IRobotDriver. The rest of the system will communicate with the driver through the proxy. Proxy notifies RobotDriverObserver about changes, which notifies the observers – including RobotPilotViewModel (*observer* pattern [3]). The view-model propagates change further through INotifyProperty-Changed. Of course, classes RobotDriverInterceptor and RobotDriverObserver could be merged into one class – however it was split as a workaround of DI container limitations. The page doesn't have any fields except reference to the view-model. Vehicle pilot and mapping of user input can manipulate exactly one instance of IRobotDriver. However, this research works supports multiple vehicles, including real vehicle and simulated vehicle. This problem can be solved with *composite* design pattern [3]. The main purpose of composite pattern is to treat single object and objects collection uniformly. The class RobotDriverAggregator exposes interface of IRobotDriver and aggregates instances of

IRobotDriver. Each call is redirected to the aggregated robot drivers. Each query for motor state is redirected to the first passed.

In terms of the composite design pattern, each driver is a *leaf* (for instance XRobotDriver) and RobotDriverAggregator is a *composite*. The composite can be passed wherever IRobotDriver is expected. Both leaf and composite are *components*.

## 6 Summary

The research works was successfully implemented as a software realizing stated goals. The whole problem was first divided into separate components, which had clearly defined points of interaction through notion of interface. Then the components were implemented according to the specification. Usage of utility libraries to create mock-up objects allowed to test components in isolation. Therefore, it was possible to check, whether component realizing mapping of user input into vehicle movements gives expected output for appropriate input. One of the important implementation details was utilization of state design pattern – general description of classes realizing state machine. Although end user does not see, how the modules were implemented, this particular solution makes them more maintainable. The Emotiv EPOC+ Neuro Headset has limited accuracy, what limits application of the realized research works. This shows that the Brain-Computer interface technology needs further improvement. Today the technology is not sufficiently accurate to control automobiles or airplanes. The author of this work believes, that in the future the technology might be applied in situations, where hands are busy – for instance in order to get the phone. The other application could be entertainment. For instance, the simulator in this research works could be extended with barriers and enemies to obtain simple game utilizing the headset. It is possible to extend the application to control the wheelchair – what could be potentially valuable for people with disabilities [12].

## Bibliography


1. Beck, K.: Test Driven Development: By Example. Addison-Wesley Longman Publishing Co., Inc., Boston (2002)
2. Ebrahimi, T., Vesin, J.M., Garcia, G.: Brain-computer interface in multimedia communication. IEEE Signal Process. Mag. **20**(1), 14–24 (2003)
3. Gamma, E., Helm, R., Johnson, R., Vlissides, J.: Design Patterns: Elements of Reusable Object-oriented Software. Addison-Wesley Longman Publishing Co., Inc., Boston (1995)
4. Libenson, M.H.: Practical Approach to Electroencephalography. Elsevier Health Sciences, Philadelphia (2012)
5. Martin, R.C.: Agile Software Development: Principles, Patterns, and Practices. Prentice Hall PTR, Upper Saddle River (2003)
6. Paszkiel, S., Błachowicz, A.: Zastosowanie BCI do sterowania robotem mobilnym. Pomiary Automatyka Robotyka **16**(182), 270–274 (2012)
7. Pilone, D.: UML 2.0 Pocket Reference. O'Reilly Media Inc, Sebastopol (2006)
8. Troelsen, A., Japikse, P.: C# 6.0 and the.NET 4.6 Framework, 7th edn. Apress, Berkely (2015)

9. Wolpaw, J.R., Birbaumer, N., McFarland, D.J., Pfurtscheller, G., Vaughan, T.M.: Brain-computer interfaces for communication and control. Clin. Neurophysiol. **113**(6), 767–791 (2002)
10. Ghaemia, A., Rashedia, E., Mohammad Pourrahimib, A., Kamandara, M., Rahdaric, F.: Automatic channel selection in EEG signals for classification of left or right hand movement in brain computer interfaces using improved binary gravitation search algorithm. Biomed. Signal Process. Control **33**, 109–118 (2017). https://doi.org/10.1016/j.bspc.2016.11.018
11. Wei-Yen, H.: Brain-computer interface connected to telemedicine and telecommunication in virtual reality applications. Telematics Inform. **34**(4), 224–238 (2017). https://doi.org/10.1016/j.tele.2016.01.003
12. Mathewson, K.E., Lleras, A., Beck, D.M., Fabiani, M., Ro, T., Gratton, G.: Pulsed out of awareness: EEG alpha oscillations represent a pulsed-inhibition of ongoing cortical processing. Front. Psychol. (2011). https://doi.org/10.3389/fpsyg.2011.00099
13. Paszkiel, Sz.: Facial expressions as an artifact in EEG signal used in the process of controlling a mobile robot with LabVIEW. Electr. Rev. **2017**(4), 156–160 (2017). https://doi.org/10.15199/48.2017.04.38
14. Paszkiel, Sz.: Using the raspberry PI2 module and the brain-computer technology for controlling a mobile vehicle. In: Szewczyk, R., Zieliński, C., Kaliczyńska, M. (eds.) Automation 2019, Progress in Automation, Robotics and Measurement Techniques. Advances in Intelligent Systems and Computing, pp. 356–366. Springer, Switzerland (2020). https://doi.org/10.1007/978-3-030-13273-6_34

# An Approach Towards High-Precision Docking of the Mobile Robots for Industrial Purposes

Piotr Falkowski[1(✉)], Michał Smater[1], Jakub Koper[1], Arkadiusz Myśliwiec[1], and Tomasz Mackiewicz[2]

[1] ŁUKASIEWICZ Research Network – Industrial Research Institute of Automation and Measurements PIAP, Warsaw, Poland
pfalkowski@piap.pl

[2] Centre for Research and Development of Modern Technologies, Września, Poland

**Abstract.** The fourth industrial revolution caused a dynamic development of production capabilities. To manufacture faster and more efficiently, production lines are being equipped with numerous sensors based on IoT technology. Thus, the machines, and especially robots working in the factories, need to operate in an even more flexible way. This does not only apply to the interchangeability of their tasks but also requires their mobility between the different workstations. Not necessarily does it have to be realised with the linear tracks. The era of Industry 4.0 brings about a real need for the implementation of autonomous mobile robots in the manufacturing sector. Therefore, dedicated docking systems need to be developed to allow improved mobility of industrial manipulators and maintain a satisfactory level of their precision. The following paper constitutes a case study exploring the viability of the implementation of industrial robots, placed on a manual trolley and an autonomous mobile robot MiR 100, for automotive training purposes. A detailed description of the design-and-test iteration process is presented herein, followed by results obtained from a real-life validation.

**Keywords:** Docking systems · Industry 4.0 · MiR 100 · Mobile industrial robots · Training in automotive

## 1 Introduction

In the era of Industry 4.0 flexibility became one of the leading trends in the design of industrial robots [1]. Possibility of changing the current task, not only at one work-station but also across several stands or machines, can increase the efficiency of the robotic set-up. Moreover, it enables dynamic changes in the production line, thereby eliminating the robot's downtime. These aspects are highly desirable in industry, and in particular in its automotive sector [2].

While using industrial robots, it is possible to easily change their function. It typically involves their relocation, exchange of the tool and the switch of the program. While the program may be shifted automatically via the PLC controller and the additional fast tool change system may be assembled to the robot's flange, only


R. Szewczyk et al. (Eds.): AUTOMATION 2020, AISC 1140, pp. 239–247, 2020.
https://doi.org/10.1007/978-3-030-40971-5_22

transferring of the manipulator to a new location remains inefficient and problematic. It is most often achieved by the demounting and assembling to another workstation manually or with the use of linear tracks. The first method is not convenient due to the need of lifting the manipulator and its precision setup at another position. Thus, it is also not applicable to heavier robots. On the other hand, tracks enable motion of almost any type of industrial machines but at the same time, their motion is limited to just a single direction linear displacement. Moreover, these devices tend to be expensive.

Another possible solution is to assemble the robots on mobile constructions and dock them to workspaces. This approach is particularly useful for the lighter manipulators. For this, either simple trolleys or advanced autonomous mobile robots may be applied as the mentioned mobile constructions [3, 4]. However, for mobile robots, lack of repeatability in their exact positioning when docking to the workstation poses a major problem [5].

Moreover, application of the manipulators on mobile robots is associated with the need for providing pneumatic and electric supply to them [1]. It may be either solved with the constant connection or with the docking connection at the workstations. The first solution may be implemented for one mobile robot or for more mobile robots which working areas do not intersect. The second method is easy to implement with the modular connectors, but it requires additional precision while docking, so as to avoid damaging electric pins.

The following paper describes an evaluation of several different approaches towards high-precision positioning of a mobile robot with the industrial robot assembled onto it. It is divided into six sections covering the case description and consecutive stages of the project's evolution. Its main objective is to evaluate, by means of computational tests and actual experiments, the idea of applying manipulators on mobile robots for standard applications such as assembling in the automotive sector.

## 2 Case Description

The setup used for the tests of the docking system consisted of two robotic arms (ABB IRB 1200-7/0.7), one assembled on a manual trolley, and the other one mounted onto an autonomous mobile robot (MiR 100; see Table 1), and four workstations arranged in a rectangular pattern (see Fig. 1). The practical task of the manipulators was to manipulate cubic objects with the edge length of 60 mm, drawn from the storage, and to transport them to the upper panel of the mobile robot's casing.

**Table 1.** Parameters of *MiR 100* mobile robot [6]

| Parameter | MiR 100 |
|---|---|
| Mass | 65 kg |
| Dimensions | 890 mm/580 mm/352 mm |
| Maximum payload | 100 kg |
| Maximum speed | 1.5 m/s (forwards)/0.3 m/s (backwards) |
| Repeatability | ±50 mm/±10 mm |
| Running time | 10 h/20 km |

As the robotic arm requires pneumatic and electric supply, it needs to be connected to their sources while being docked to the workstation. The chosen mobile robot does not require a constant supply, but it has to be charged every few hours. This is realised automatically at the charging station.

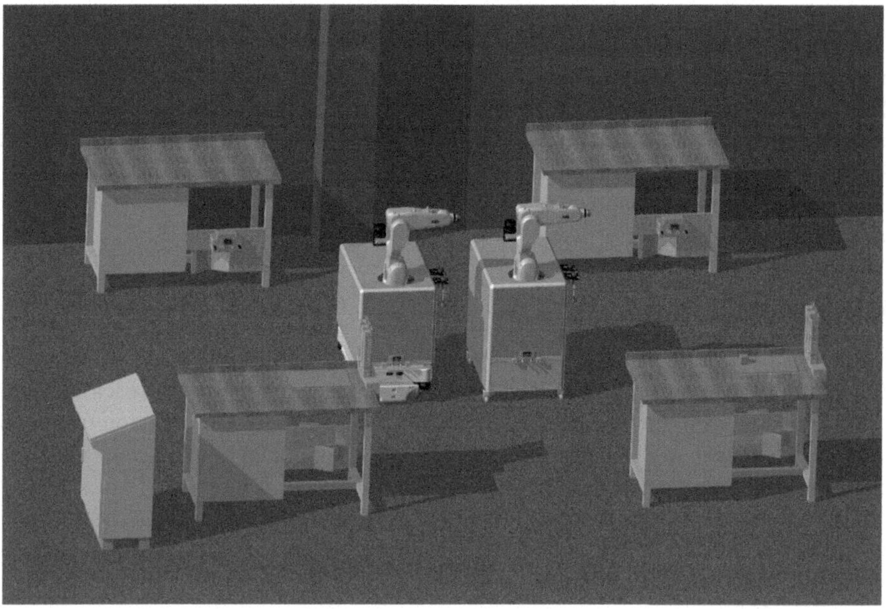

**Fig. 1.** Schematic visualization of the installation.

MiR 100 was chosen because it has the highest payload of all mobile robots considered to be suitable for the application with a small robotic arm. Its repeatability is similar to that of other mobile robots and may also be improved to ±1 mm by the addition of the MiR Precise Docking module. However, this value is still much bigger than the one characteristic for the manipulators [7]. To achieve a repeatable process of assembling of the small elements, the required tool position repeatability value of maximum ±0.5 mm is defined for this case [8]. Additionally, the ±0.02 mm pose repeatability of the ABB IRB 1200-7/0.7 manipulator must be accounted for.

To obtain a satisfactory level of precision, a dedicated guidance system needs to be implemented. Its design is optimised in terms of the possibility of its application also with the manual trolley, and the components' price. Moreover, the system is also responsible for assessing the immobility of the whole construction while working.

# 3 Docking System

## 3.1 Electric and Pneumatic Supply System

Electric and pneumatic supply system enables control over the industrial robot with the PLC and realisation of its tasks while it is docked to one of the workstations. It is based on the *Staubli CombiTac* connector system with pneumatic, ethernet, signal pins and electric power supply modules. To avoid their damage, the system is also equipped with rails positioned such that they prevent angular misalignment of the connectors. On their own, they are not, however, capable of ensuring straight docking of the mobile robot itself. The whole system was initially intended to be assembled on springs dumping the impact of the robot's hit at the final stage of docking. As the connector occurred to be robust enough and to have enough backlash, the springs were excluded from the design.

## 3.2 Guidance System

Initial tests of MiR 100 mobile robot's positioning capabilities involved the use of the different docking systems (see Table 2). As the robot was able to can reach the programmed point with the repeatability of $\pm 50$ mm based on the information from two laser scanners and a 3D camera, an additional positioning method was required. Therefore, differently shaped metal markers were tested to achieve the best accuracy. Considered V and VL markers were designed as described in the robot's specification and tested [9]. Final pose repeatability reached the level of $\pm 10$ mm for the application of the VL version. Even this result was significantly insufficient for simple manipulation tasks [6].

Following the initial tests, it was concluded that avoiding collisions of the connectors, as well as ensuring precise and repeatable operations of the industrial manipulator at the workstation required a dedicated guidance system. The designed system contained two connectors with two chamfered pins or sleeves each (see Fig. 2). Their construction with the sliding fit enabled improvement of the robotic system's positioning repeatability from $\pm 10$ mm up to $\pm 0.2$ mm. This value was deemed sufficient to perform all the planned operations.

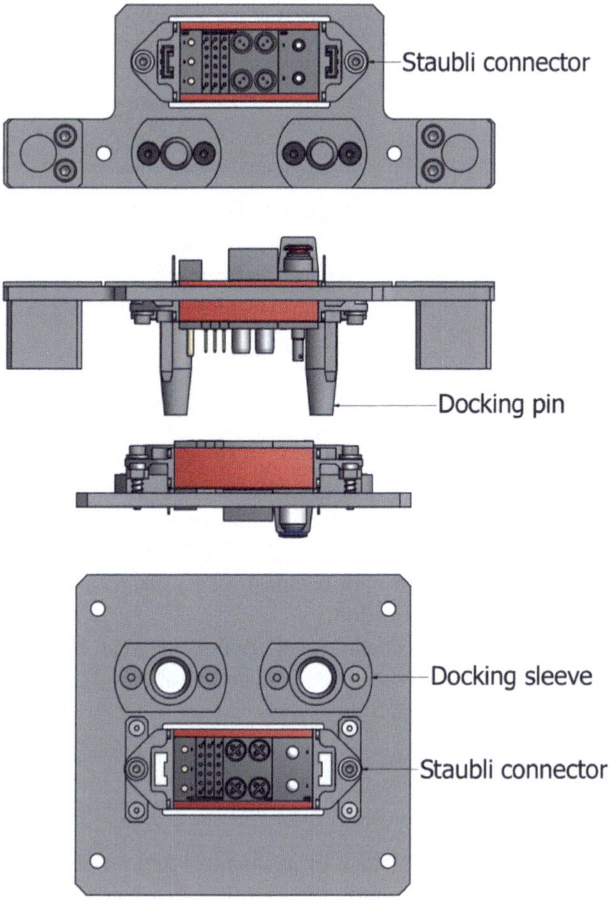


**Fig. 2.** Schematic of both parts of the docking system.

However, the mechanical guidance system was not capable of solving the holistic problem of high-precision docking. Mobile robot MiR 100 creates its motion path prior to beginning its movement and then uses only 16 basic directions (22.5° one from another) when docking to the marker [6]. Thus, the whole process was divided into two stages – docking to the marker with a frontal offset of 100 mm followed by a simple straight linear motion of 100 mm forwards. This distance eliminated over 99% of failed dockings and did not affect the robot's ability to detect markers from a pre-docking position.

The docking markers occurred to be particularly vulnerable to any changes to their exposition, and any other objects placed in close proximity to them. When lighting in the room changing, MiR 100 had major problems recognizing their shapes. Also, additional elements located within a small distance from the markers were sometimes considered to be their integral parts. Therefore, the mobile robot was not recognizing the markers as references for docking. To avoid this confusion, pre-docking position, reached by the robot in order to localize the marker and to set its further motion path, was reduced to 50 mm, which is equal the standard robot positioning repeatability (see Table 2).

**Table 2.** Comparison of the different approaches towards positioning.

| Positioning | Repeatability | Requirements for application |
|---|---|---|
| Standard robot positioning | ±50 mm | No requirements |
| Docking robot to the marker | ±10 mm | No requirements |
| Centring pins | ±0.2 mm | Prior repeatability of ±10 mm |

### 3.3 Locking System

The guidance system excluded the negative effect of potential side displacement of the mobile robot. However, it did not protect the machine from its backwards motion caused by the reaction and centrifugal forces of the operating manipulator. To resolve this issue, simple welding locking systems were added on both sides of the docking system. They were assembled with high backlash (±1 mm) so as not to overconstrain the docking system's degrees of freedom.

The locking system introduced additional friction between the docking system's parts. With its application, total measured industrial robot's TCP pose repeatability maintained within the range of ±0.5 mm.

## 4 Effectiveness

Validation trials were run at every stage of mechanical and program improvements of the setup (see Fig. 3). In the beginning, the mobile robot was equipped just with the guidance system only. However, as the results were not satisfactory the additional docking markers were attached to the workstations. Their effectiveness strongly interfered by the visual distractions. Such disruptions typically led to abortions of the trials. It was possible to resolve this issue by shifting the pre-docking position closer to the final docking position. To eliminate inaccurate dockings, such as pneumatic undocking caused by small angular or axial misalignments, the locking system was added to the system at the final stage.

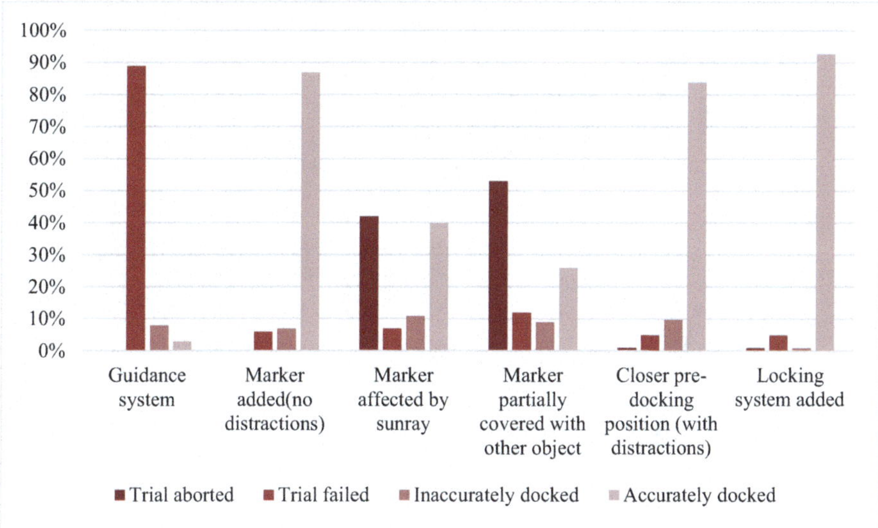


**Fig. 3.** Outcomes of the trials with the different docking methods and distractions.

## 5 Application

The designed system (see Fig. 4) was applied to the training installation for the automotive sector (see Fig. 5). It enabled docking both manual trolley and autonomous mobile robot, with the industrial manipulators mounted onto them, to the four workstations.

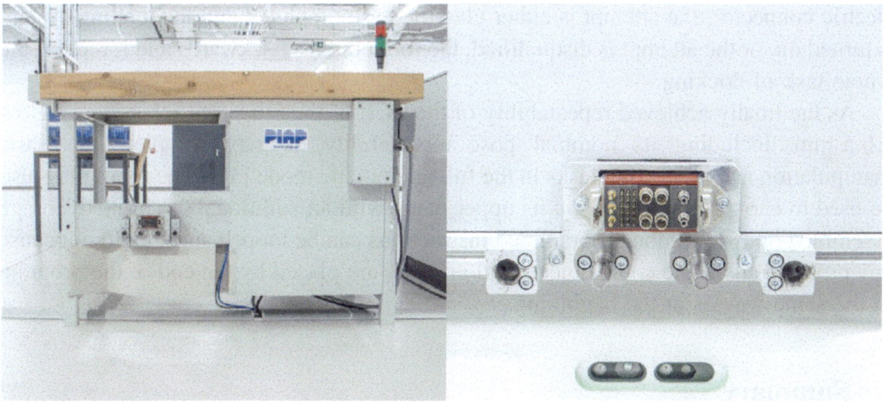

**Fig. 4.** Applied docking system.

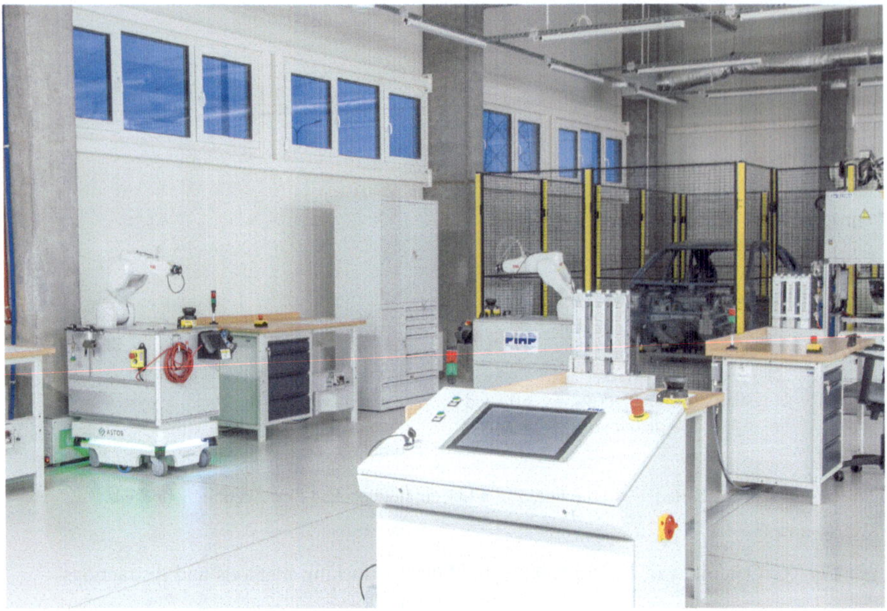

**Fig. 5.** The solution applied to the real setup – an industrial manipulator on a manual trolley standing in the centre and an industrial manipulator on a mobile robot between two workstations with docking systems.

After conducted tests and implementation of necessary improvements in the process of rapid prototyping, MiR 100 robot reached 90% effectiveness of the docking at the first attempt, that is 30 times improved efficiency compared to the initial 3% effectiveness as obtained with guidance system only. Based on the test signal at the electric connector, the attempt is either classified as completed and the locking system is turned on, or the attempt is disqualified, the robot moves backwards and it repeats the whole task of docking.

As the finally achieved repeatability of the TCP of the industrial robot stays below $\pm 0.5$ mm, including its nominal pose repeatability, it may complete the basic manipulation and palletizing tasks in the fully automatic mode [10]. The robot may also be used to carry cubic objects on its upper panel with an additional 3D printed storage assembled onto it. All the programmed instructions can be looped and should not cause any problems if all the cubic objects and grippers are placed by the end of the program in the same way as at the beginning.

## 6 Summary

Even the most effective autonomous mobile robots cannot reach pose repeatability corresponding to the standards of the industrial manipulators. It was shown here that improvements can be made through the use of special docking markers, but a satisfactory result was only achieved after an additional guiding and locking system was implemented. This approach may be used for the high-precision docking to guarantee reaching the programmed point by TCP with at least ±0.5 mm precision.

Such high accuracy may be applied for the tasks of manipulation and palletization characteristic for the automotive sector, such as tightening the screws, assembling or bonding. The flexibility of the installation also enables maximum exploitation of fewer manipulators, thereby reducing costs and increasing profits of the company [2].

## References


1. Weyer, S., et al.: Towards Industry 4.0-Standardization as the crucial challenge for highly modular, multi-vendor production systems. IFAC-PapersOnLine **48**(3), 579–584 (2015)
2. Angerer, S., et al.: Give me a hand - the potential of mobile assistive robots in automotive logistics and assembly applications. In: 2012 IEEE International Conference on Technologies for Practical Robot Applications (TePRA), pp. 111–116. IEEE (2012)
3. Virk, G.S.: Industrial mobile robots: the future. Ind. Robot **24**, 102–105 (1997)
4. Schneier, M., Bostelman, R.: Literature review of mobile robots for manufacturing. US Department of Commerce. National Institute of Standards and Technology (2015)
5. Michalos, G., et al.: Performance assessment of production systems with mobile robots. Procedia CIRP **41**, 195–200 (2016)
6. MiR's webpage. https://www.mobile-industrial-robots.com/en/products/mir100/. Accessed 03 Oct 2019, 06 Oct 2019
7. Low, H.: Industrial Robotics: Programming, Simulation and Applications. Pro literatur Verlag, Mammendorf (2006)
8. Falkowski, P., Wittels, B., Pilat, Z., Smater, M.: Capabilities of the additive manufacturing in rapid prototyping of the grippers' precision jaws. In: Szewczyk, R., Zieliński, C., Kaliczyńska, M. (eds.) Automation 2019: Progress in Automation, Robotics and Measurement Techniques, pp. 379–387. Springer, Cham (2019)
9. MiR Robot Interface 2.0 – Reference Guide. https://www.mobile-industrial-robots.com/media/2406/mir-robot-interface-20-reference-guide-v-11_en.pdf. Accessed 20 Nov 2019
10. ABB IRB-1200 robot datasheet. https://new.abb.com/products/robotics/industrial-robots/irb-1200/irb-1200-data. Accessed 26 Oct 2019

# Recent Advances in Mechanical Micro- and Nanomanipulation

Mantas Makulavicius[1(✉)], Oleksii Balitskyi[2], Rimgaudas Urbonas[1], Andrius Dzedzickis[1], Vytautas Bučinskas[1], Algirdas Petronis[1], Mykyta Kovalenko[1], and Inga Morkvenaite-Vilkonciene[1,2]

[1] Vilnius Gediminas Technical University, Vilnius J. Basanaviciaus str. 28, Vilnius, Lithuania
{m.makulavicius, inga.morkvenaite-vilkonciene}@vgtu.lt
[2] Center for Physical Sciences and Technology, Sauletekio 3, Vilnius, Lithuania

**Abstract.** In order to achieve an accurate positioning for an investigation of micro- or nano objects or to operate with them in micro or nanoscale, modern and sophisticated tools are required. Over the past years of the development of micro- and nanomanipulation tools, novel types were invented, or old ones were upgraded. Those tools such as microgrippers, microrobots for micromanipulation and piezoelectric actuators for nanomanipulation with visual monitoring using atomic force microscopy (AFM) are implemented in nowadays research. In this article different types and tools of micro- and nanomanipulation, the design of micro- or nanomanipulation devices, the environment for work with, and the development process are described. Furthermore, the comparison of different types of micromanipulation and, separately, of nanomanipulation are provided.

**Keywords:** Atomic force microscopy · Micromanipulation · Microgripper · Nanomanipulation · Nanoparticle · Piezoelectric motor

## 1 Introduction

Different micromanipulation techniques can offer new possibilities to manipulate small-size objects, such as living cells or small particles and obtain data in different ranges of speed and accuracy in microscale. Those techniques rely on microrobotics, micro-grippers, microelectromechanical systems and other [1–4]. Moreover, accurate positioning and manipulation of nanometer-scale objects requires appropriate microscopic techniques [5–8]. In the development of micro and nanomanipulation systems, physics and chemistry researchers mainly use atomic force microscopes (AFM) [9–13]. Even though the AFM manipulation techniques have several advantages, such as a wide selection of materials (not limited to leading materials), as well as high resolution and precision, they have some limitations. AFM is used for both moving and visualizing the starting and ending positions of the nanoparticle manipulation [14], but these two procedures cannot be performed at the same time. This process is time consuming, because it requires a lot of line scanning.


R. Szewczyk et al. (Eds.): AUTOMATION 2020, AISC 1140, pp. 248–256, 2020.
https://doi.org/10.1007/978-3-030-40971-5_23

## 2 Manipulation in Microscale

In this part of the article, different techniques of manipulation in microscale are presented. Possibilities of micromanipulation are beneficial in many areas and one of them is a single-cell manipulation and analysis for research purposes. Some types of living cell manipulation are based on the use of microelectromechanical systems (MEMS) and other mechanical and robotic techniques. MEMS and mechanical micromanipulation are widely used in DNA injection, drugs delivery, gene therapy and in other areas, such as in biomedicine and biomedical engineering [15–18] (Fig. 1).

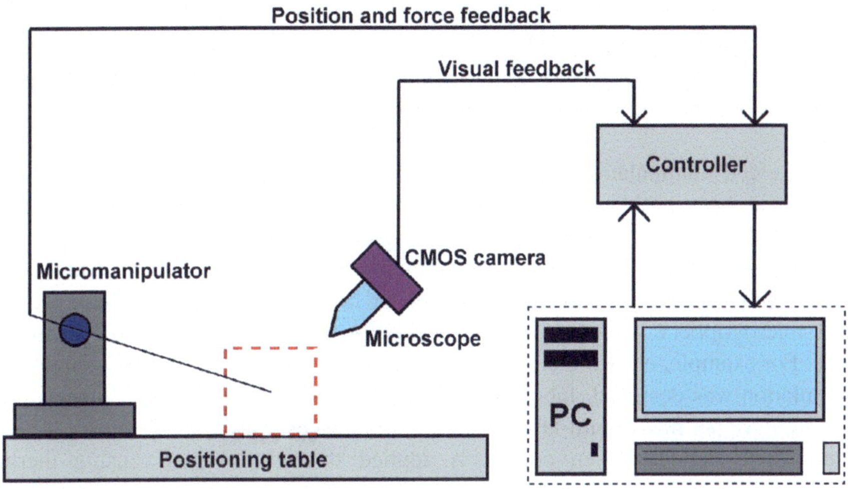


**Fig. 1.** Schematic representation of micromanipulation device.

Biological cell dimensions range from a few micrometers to hundreds, therefore their manipulation belonging to the field of micro-robotics. One of the recently invented types of micromanipulation is wireless micromanipulation system [19]. This system uses micro-pusher made of ferromagnetic material, which tracks the movement of a cylindrically shaped permanent magnet. This magnet is located under the glass slide. Furthermore, two types of micro-pusher movements were used to push cells in aqueous medium on a glass slide which are pictured in Fig. 2.

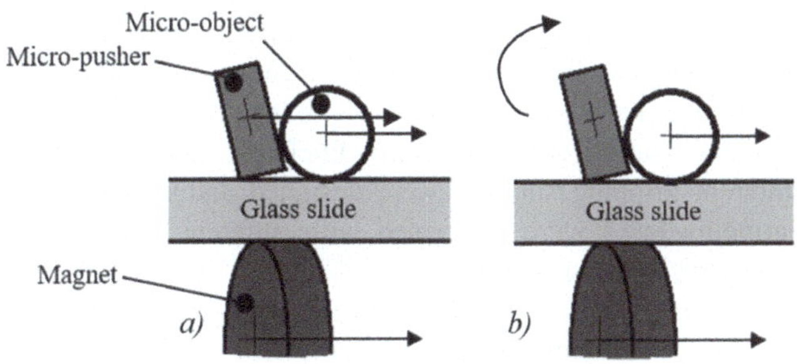


**Fig. 2.** Micromanipulation using a micro-pusher (a) translational motion (b) rotational motion.

These two kinds of micro-pusher movements can push an object and thus induce a movement: translational motion or turning the micro-pusher around the contact point (rotational motion). Rotational motion induces fine movements without friction disturbances and allows a micro-object to be placed with a precision less than 1 μm, but it operates in a narrow range. On the other hand, translational motion allows placement of a micro-object within a wide range with lower accuracy, because it is disturbed by friction effects. Using these motion techniques, experiments on polystyrene balls of 50 μm in diameter and immature human oocytes of 150 μm in diameter were performed.

An automated manipulation microsystem was created using a cluster of microrobots [20]. This cluster of microrobots was equipped with on-board electronics, sensors and wireless power supply. One robot consists of 3 DoF locomotion module, tool module, electronics module, battery, and two targets based on Moiré positioning system. For locomotion module piezoelectric actuators were used to make a bending action.

Living cell manipulation requires performing complicated tasks, which currently represents an open problem in biomedicine and bioengineering [21]. Nowadays, there are many different manipulation types for living cells, but some are still expensive, not accurate enough or inefficient. It is because the microsystem has not enough degrees of freedom, limited capabilities of manipulation or lack agility. There are designed and tested microgrippers based on MEMS technology that solve problems mentioned before. For example, electro-thermally actuated polymeric microgripper for single cells manipulation was designed, fabricated and tested with gold coated SU8 objects [22]. This microgripper has a multi-layer structure. In between two layers there is a metallic heater which deflects when current is applied through it, thus causing thermal expansion. Constant current source was used to actuate the microgripper. It was calculated that the maximum consumed power by the system is 77 mW producing displacements up to 262 μm. In micromanipulation experiments, grasping, holding and positioning techniques were successfully tested using gold coated micro-cylinders with an approximate diameter of 100 μm in a clean room environment at 20 °C. Nevertheless, living cell manipulation using this type of microgripper has not been tested.

In the other studies a new microgripping system was proposed [23]. This system differs from others because the microgripper consists of separately fabricated end-effectors (making a contact with manipulated micro-object) and actuators (induce a movement of the end-effectors). These two main parts were chosen and made separately, because every different object has its own different properties (size, roughness) and the same end-effector do not fit for every object. Furthermore, fabrication of actuators is more complex, than end-effectors, because of their complex structure. Silicon was chosen because it has better mechanical features than other materials in the micro-scales. Silicon end-effectors of 12 μm thickness, 90 μm width and 1.5 mm length with different gripping surfaces were designed and manufactured. In experiments micro-gripping tasks were performed with different shapes and materials of micro-objects in the air and in a liquid medium. The sizes of micro-object from 5 μm to 600 μm were used for manipulation. Nevertheless, the pick-and-place operations and the manipulation with living cells have not been tested.

In Table 1 the summary of all previously reviewed types of micromanipulation are summarized.

**Table 1.** Micromanipulation types

| Manipulation type | DoF | Displacement | Size of micro-object in diameter | Ref. |
|---|---|---|---|---|
| Using magnets | 2 | Less than 1 μm | 50 μm and 150 μm | [19] |
| Using microrobots | 3 | 7 nm | – | [20] |
| Using polymeric microgripper | – | Up to 262 μm | 100 μm | [22] |
| Using microgripper with separately fabricated parts | 2 | – | From 5 μm to 600 μm | [23] |

From the table, it can be seen, that the more DoF system has, the more precise is microrobot. For example, microrobot with 3 DoF can reach the precision of 7 nm, but it is not described what is the size of micro-objects that can be manipulated.

## 3 Piezo Nano Manipulation

Nano manipulation is becoming increasingly relevant today for the study of cells and for the local measurement of their electrical properties such as conductivity or mechanical properties [5]. Based on the piezoelectric motor units (SPM), a platform for the investigation and manipulation of individual particles (approximately 50 nm) can be assembled and the interaction with the particles can be carried out under direct microscopic observations [9]. Flexure-based mechanisms and piezoelectric actuators have been recognized as the most appropriate platforms and actuation devices, respectively, for accomplishing high-precision motion tasks in the field of micro-/nanomanipulation [24]. These special mechanisms and actuators offer very high (nanometric) motion resolution. They also possess several advantages over the conventional mechanical systems, which are generally based on sliding or revolute lower pairs. These advantages include negligible friction, zero backlash, noiselessness, and easy maintenance. In recent years, there have been advances in piezoelectric actuator driven micro/nanomanipulators. For optimal mechanical performance, related studies of the micro/nanomanipulator have been conducted. Increasingly, piezoelectric actuator driven manipulation systems are employed in many advanced areas, particularly in the biomedical engineering and micro-surgery applications. At the same time, attention has also been directed towards the development of piezoelectric actuators to enhance their important role in the micro/nanomanipulation mechanisms and systems. Piezoelectric actuators are increasingly used in various nanoscale platforms. This is due to their unique characteristics, including their very high precision and short response time compared to electromagnetic data motors and drives with higher output power than electrostatic actuators. Piezoelectric motors use actuators that use the reverse piezoelectric effect feature [25]. The disadvantage of a piezo motor is the complexity of the design: the more complex the piezo motor, the more complex becomes the control system, therefore, there is no ability to move quickly. This problem can be solved by combining a stepper motor system for fast moving and a piezo motor for precise positioning. The use of new technologies enables the creation of a platform for nanomanipulation using high-precision stepping motors (400 steps by 360°) and optical encoders for feedback. In combination with high-precision electronic control systems that will divide the step of the engine by 256 steps, the theoretical accuracy of positioning accuracy will be: 1 mm/(400 $\times$ 256) = 9.7 $\times$ $10^{-6}$ mm = 9.7 nm.

The performance of the nanoscale system is closely combined with the quality of the mechanical design. A good mechanical design will minimize most position errors and improve overall accuracy and performance. Poor mechanical construction, on the other hand, can lead to more errors than problems related to electronics, control systems and other components. High-precision ball screws are used to transfer torque from the motor to the work tool (Table 2).

**Table 2.** Nanomanipulation properties.

| Manipulation type | DoF | Displacement | Size of micro-object in diameter | Ref. |
|---|---|---|---|---|
| Using AFM | 1 | 0.1 | 8 nm | [26] |
| Using STM | 1 | 0.1 | 8 nm | [24] |
| Using SEM | 1 | >5 nm | 20 nm | [27] |
| Using piezoelectric actuators | 2 | >5 nm | 60 nm | [24] |
| Using SECM with manipulator | 2 | >12 nm | 100 nm | [28] |

STM – Scanning tunneling microscopy; SEM – Scanning electron microscopy; SECM – Scanning electrochemical microscopy

## 4 Manipulation by AFM

The manipulation also can be accomplished under ambient environment using a commercial atomic force microscope (AFM). There are particular difficulties associated with manipulating of small particles (nanoparticles, NPs). In addition to spatial drift, the shape of an asymmetric AFM tip has to be taken into account in order to understand the intended and actual manipulation results. Furthermore, small NPs often attach to the tip via electrostatic interaction and then can be moved to desired direction (Fig. 3) [26].

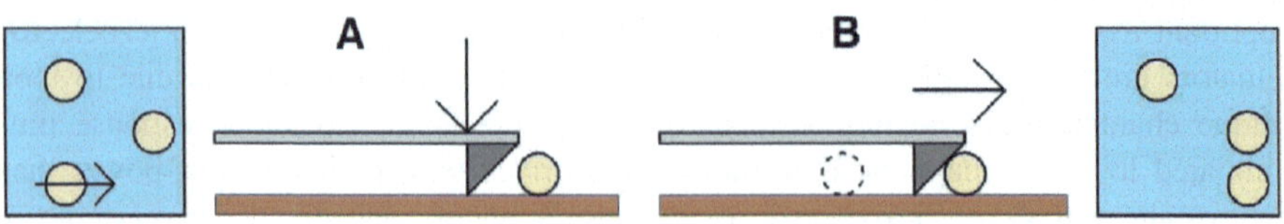


**Fig. 3.** Schematic of the method of manipulation. A – Before manipulation. B – After manipulation.

Controlled positioning nanoparticles, molecules or clusters of molecules is an important step towards researching a new class of chemicals. Working with nanometer-scale objects requires ability to recognize an object by moving it and tracking it. AFM is one of the universal methods for approaching this goal, but it is limited by low operating speed. These problems can be overcome by including real-time feedback on manipulation techniques and AFM like dynamic line scanning or contact loss detection, which provides active feedback, thereby tracking movement [29].

AFM is a conceptually simple device. Micrometer console with sharp tip (diameter – 10–50 nm) is scanned according to the sample at distances of the order of several nm. Interatomic forces between the tip and the sample are perceived by cantilever, the deviation of which is measured (usually) by a laser and a photodetector. Force tested the tip varies non-linearly depending on the separation of the tip and the sample [30].

The nanomanipulator system mainly consists of three parts that are interconnected by software: AFM (plus controller), graphics engine/interface and phantom power stylus (plus controller). When a sample is scanned in contact mode, the tip constantly touches the sample and, therefore, it is normal that lateral forces can be applied to the sample. In contrast, when using the tapping mode, the cantilever oscillates perpendicular to the surface when it scans the surface. Therefore, the tip touches the sample only periodically, and the lateral forces are almost eliminated. Nanomanipulation mode uses this important distinction between two modes. Samples can be manipulated in contact mode. With controlled normal force and they can be displayed before and after tapping manipulations without further modification [31].

AFM methodology allowed the formation of nanostructures of 30 nanoparticles with controlled manipulation of individual adsorbed particles with nanometer accuracy. In addition, it is possible to determine the true transverse particle sizes despite convolution of AFM-tip. AFM in contact mode could be used for easy manipulation of nanoparticles with the tip when it scans the surface. Often this is an undesirable effect that makes the image unstable, but this effect could be used, for moving of selected particles to any desired position on the surface [26].

The AFM manipulation method can also be used to manipulate individual carbon nanotubes to manipulate them and place them in predetermined locations. The interaction between nanotubes and the substrate on which they are located suggests that their position can be controlled by applying lateral forces of the appropriate magnitude with the AFM tip [32].

## 5 Conclusions

In this review, we provided micro and nanomanipulation methods and devices. For the micromanipulation, several tools and devices could be used, since the drives applied for such task is stepper motors, which provide micrometric resolution. For the nanomanipulation, piezo motors should be used, or low scale electromagnets. AFM is the device, which can be easy applied for nanomanipulation of nano-objects. The scanning speed is limited by a few hertz, so only relatively slow processes can be visualized, fast processes cannot be seen.

**Acknowledgements.** This Research was funded by European Social Fund (Project No 09.3.3.-LMT-K-712-16-0210).

## References

1. Sitti, M., Ceylan, H., Hu, W., Giltinan, J., Turan, M., Yim, S., Diller, E.: Biomedical applications of untethered mobile milli/microrobots. Proc. IEEE **103**, 205–224 (2015). https://doi.org/10.1109/JPROC.2014.2385105
2. Zhang, Z., Wang, X., Liu, J., Dai, C., Sun, Y.: Robotic micromanipulation: fundamentals and applications. Annu. Rev. Control Robot. Auton. Syst. **2**, 181–203 (2019). https://doi.org/10.1146/annurev-control-053018-023755
3. Li, J., Esteban Fernández de Ávila, B., Gao, W., Zhang, L., Wang, J.: Micro/nanorobots for biomedicine: delivery, surgery, sensing, and detoxification. Sci. Robot. **2** (2017). https://doi.org/10.1126/scirobotics.aam6431
4. Diller, E., Sitti, M.: Micro-scale mobile robotics. Found. Trends Robot. **2**, 143–259 (2013). https://doi.org/10.1561/2300000023
5. Ishijima, A., Harada, Y., Kojima, H., Funatsu, T., Higuchi, H., Yanagida, T.: Single-molecule analysis of the actomyosin motor using nano-manipulation. Biochem. Biophys. Res. Commun. **199**, 1057–1063 (1994). https://doi.org/10.1006/bbrc.1994.1336
6. Wang, H., Pumera, M.: Fabrication of micro/nanoscale motors. Chem. Rev. **115**, 8704–8735 (2015). https://doi.org/10.1021/acs.chemrev.5b00047
7. Martel, S., Felfoul, O., Mathieu, J.B., Chanu, A., Tamaz, S., Mohammadi, M., Mankiewicz, M., Tabatabaei, N.: MRI-based medical nanorobotic platform for the control of magnetic nanoparticles and flagellated bacteria for target interventions in human capillaries. Int. J. Robot. Res. **28**, 1169–1182 (2009). https://doi.org/10.1177/0278364908104855
8. Mastrangeli, M., Abbasi, S., Varel, C., Van Hoof, C., Celis, J.-P., Böhringer, K.F.: Self-assembly from milli- to nanoscales: methods and applications **27**, 1–19 (2014). https://doi.org/10.1037/a0032811.Child
9. Tian, Y., Shirinzadeh, B., Zhang, D.: Design and dynamics of a 3-DOF flexure-based parallel mechanism for micro/nano manipulation. Microelectron. Eng. (2010). https://doi.org/10.1016/j.mee.2009.08.001
10. Morkvenaite-Vilkonciene, I., Ramanavicius, A., Ramanaviciene, A., Morkvenaite-Vilkončiene, I., Ramanavičiene, A., Ramanavičius, A.: Atomic force microscopy as a tool for the investigation of living cells. Medicina (B Aires) **49**, 155–164 (2013)
11. Morkvenaite-Vilkonciene, I., Virzonis, D., Dzedzickis, A., Bucinskas, V., Rozene, J., Vilkoncius, R., Vaiciulis, D., Ramanaviciene, A., Ramanavicius, A.: The improvement of the accuracy of electromagnetic actuator based atomic force microscope operating in contact mode and the development of a new methodology for the estimation of control parameters and the achievement of superior image quality. Sens. Actuators A Phys. **287**, 168–176 (2019). https://doi.org/10.1016/J.SNA.2019.01.015
12. Dzedzickis, A., Bucinskas, V., Viržonis, D., Sesok, N., Ulcinas, A., Iljin, I., Sutinys, E., Petkevicius, S., Gargasas, J., Morkvenaite-Vilkonciene, I., Dzedzickis, A., Bucinskas, V., Viržonis, D., Sesok, N., Ulcinas, A., Iljin, I., Sutinys, E., Petkevicius, S., Gargasas, J., Morkvenaite-Vilkonciene, I.: Modification of the AFM sensor by a precisely regulated air stream to increase imaging speed and accuracy in the contact mode. Sensors **18**, 2694 (2018). https://doi.org/10.3390/s18082694

13. Guillaume-Gentil, O., Potthoff, E., Ossola, D., Franz, C.M., Zambelli, T., Vorholt, J.A.: Force-controlled manipulation of single cells: from AFM to FluidFM. Trends Biotechnol. **32**, 381–388 (2014). https://doi.org/10.1016/j.tibtech.2014.04.008
14. Simpson, D.A., Ampem-Lassen, E., Gibson, B.C., Trpkovski, S., Hossain, F.M., Huntington, S.T., Greentree, A.D., Hollenberg, L.C.L., Prawer, S.: A highly efficient two level diamond based single photon source. Appl. Phys. Lett. **94** (2009). https://doi.org/10.1063/1.3141450
15. Pelenis, D., Barauskas, D., Sapeliauskas, E., Vanagas, G., Mikolajunas, M., Virzonis, D.: Acoustical streaming in microfluidic CMUT integrated chip controls the biochemical interaction rate. J. Microelectromech. Syst. **26**, 1012–1017 (2017). https://doi.org/10.1109/JMEMS.2017.2701144
16. Morkvenaite-Vilkonciene, I., Virzonis, D., Vanagas, G., Krikscikas, V.: Operating point of capacitive micromachined ultrasonic transducers with sub-structural elements. Elektron Ir Elektrotechnika **18**, 43–46 (2012). https://doi.org/10.5755/j01.eee.18.9.2804
17. Mcallister, D.V., Allen, M.G., Prausnitz, R.M.: Microfabricated microneedles for gene and drug delivery. Molecules **2**, 289–313 (2000)
18. Magdanz, V., Sanchez, S., Schmidt, O.G.: Development of a sperm-flagella driven micro-bio-robot. Adv. Mater. **25**, 6581–6588 (2013). https://doi.org/10.1002/adma.201302544
19. Gauthier, M., Piat, E.: Control of a particular micro-macro positioning system applied to cell micromanipulation. IEEE Trans. Autom. Sci. Eng. **3**, 264–271 (2006). https://doi.org/10.1109/TASE.2005.861398
20. Brufau, J., Puig-Vidal, M., López-Sánchez, J., Samitier, J., Snis, N., Simu, U., Johansson, S., Driesen, W., Breguet, J.M., Gao, J., Velten, T., Seyfried, J., Estaña, R., Woern, H.: MICRON: small autonomous robot for cell manipulation applications. In: Proceedings of the IEEE International Conference on Robotics and Automation, pp. 844–849 (2005). https://doi.org/10.1109/ROBOT.2005.1570222
21. Cecchi, R., Verotti, M., Capata, R., Dochshanov, A.M., Broggiato, G.B., Crescenzi, R., Balucani, M., Natali, S., Razzano, G., Lucchese, F., Bagolini, A., Bellutti, P., Sciubba, E., Belfiore, N.P.: Development of micro-grippers for tissue and cell manipulation with direct morphological comparison. Micromachines **6**, 1710–1728 (2015). https://doi.org/10.3390/mi6111451
22. Solano, B., Wood, D.: Design and testing of a polymeric microgripper for cell manipulation. Microelectron. Eng. **84**, 1219–1222 (2007). https://doi.org/10.1016/j.mee.2007.01.153
23. Agnus, J., Hériban, D., Gauthier, M., Pétrini, V.: Silicon end-effectors for microgripping tasks. Precis. Eng. **33**, 542–548 (2009). https://doi.org/10.1016/j.precisioneng.2009.02.005
24. Liaw, H.C., Shirinzadeh, B., Smith, J., Alici, G.: Motion tracking control of piezo-driven flexure-based mechanism based on sliding mode strategy. In: 2007 IEEE/ASME International Conference on Advanced Intelligent Mechatronics, vol. 1–3, p. 144+ (2007)
25. Dong, L., Arai, F., Fukuda, T.: 3D nanorobotic manipulation of nano-order objects inside SEM. In: Proceedings of the International Symposium on Micromechatronics and Human Science, pp. 151–156 (2000). https://doi.org/10.1109/mhs.2000.903305
26. Kim, S., Shafiei, F., Ratchford, D., Li, X.: Controlled AFM manipulation of small nanoparticles and assembly of hybrid nanostructures. Nanotechnology **22**, 115301 (2011). https://doi.org/10.1088/0957-4484/22/11/115301
27. Sitti, M.: Survey of nanomanipulation systems. In: Proceedings of the IEEE Conference on Nanotechnology, pp. 75–80. IEEE Computer Society (2001)
28. Balitskyi, O., Dzedzickis, A., Makulavičius, M., Šumanas, M., Petronis, A., Vasiulis, A., Bučinskas, V., Morkvėnaitė-Vilkončienė, I.: New type microrobot for the manipulation of biological objects. In: Mechatronics 2019, 16–18 September, Warsaw, Poland (2019)

29. Chacko, J.V., Canale, C., Harke, B., Diaspro, A.: Sub-diffraction nano manipulation using STED AFM. PLoS One **8** (2013). https://doi.org/10.1371/journal.pone.0066608
30. Requicha, A.A.G., Meltzer, S., Arce, F.P.T., Makaliwe, J.H., Sikén, H., Hsieh, S., Lewis, D., Koel, B.E., Thompson, M.E.: Manipulation of nanoscale components with the AFM: principles and applications. In: Proceedings of the IEEE Conference on Nanotechnology, pp. 81–86 (2001). https://doi.org/10.1109/NANO.2001.966398
31. Guthold, M.: Controlled manipulation of molecular samples with the nano manipulator. IEEE/ASME Trans. Mechatron. **5**, 189–198 (2000). https://doi.org/10.1109/3516.847092
32. Avouris, P., Hertel, T., Martel, R., Schmidt, T., Shea, H.R., Walkup, R.E.: Carbon nanotubes: nanomechanics, manipulation, and electronic devices. Appl. Surf. Sci. **141**, 201–209 (1999). https://doi.org/10.1016/S0169-4332(98)00506-6

# Improving Positioning Accuracy of an Articulated Robot Using Deep Q-Learning Algorithms

Algirdas Petronis, Vytautas Bučinskas, Marius Šumanas, Andrius Dzedzickis(✉), Liudas Petrauskas, Nikita Edgar Sitiajev, and Inga Morkvenaite-Vilkonciene

Vilnius Gediminas Technical University, J. Basanaviciaus 28, 03224 Vilnius, Lithuania
{marius.sumanas, andrius.dzedzickis}@vgtu.lt

**Abstract.** Positioning accuracy of articulated robots decreases significantly when they are fully loaded and operated at maximum speeds due to increased inertia. Hard-coding correction algorithms using traditional methods is extremely difficult. A system, which could automatically detect patterns in deviations and offer possible corrections at every motion cycle would be preferable. This work explores the possibility to use deep q-learning algorithms to achieve this. Around forty experiments of various lengths were conducted. They were divided into three experimental groups, each of which had various parameters values and elements of algorithms. While algorithms in two experimental groups were unsuccessful in achieving improved accuracy, one offered comparable accuracy, while resulting in more stable and predictable deviations compared to uncorrected positioning.

**Keywords:** Machine learning · Deep Q-learning · Articulated robot · Positioning accuracy

## 1 Introduction

Machine learning (ML) of robotic systems belongs to a quickly developing area of applied mechatronics and robotics. In this context, ML is an artificial intelligence system, which directly controls robot's parameters to perform tasks that are difficult to program [1]. Currently, robot control is achieved by performing task analysis [2], programming or hard-coding by humans, which not always achieves best possible operating performance speed-wise and accuracy-wise [3]. This issue has various possible solutions, in most cases, there are used positioning correction algorithms based on various models [4–6] or correction algorithms based on ML technique [7–9]. Q-learning is a specific form of reinforcement learning method or algorithm [10], which in turn is a form of machine learning [11]. It's working principle can be described as follows. An agent is in an environment, which is divided into a finite number of discrete states. The agent can "sense" the environment and perform a single step per iteration so that it ends up in a different state. The agent is given a reward when


R. Szewczyk et al. (Eds.): AUTOMATION 2020, AISC 1140, pp. 257–266, 2020.
https://doi.org/10.1007/978-3-030-40971-5_24

a desirable state is achieved and given a penalty (negative reward) when an undesirable state is achieved. The rewards are defined by reward functions. In q-learning algorithm, the agent tends to perform such actions, which maximize the achieved reward [12]. In this way, it can find the most optimal sequence of actions to achieve a desired goal.

The downside of q-learning is that the agent needs to take every possible action at every possible state, which limits applications of the algorithm to a finite environment, often containing a relatively small number of states [12]. For this reason, it can be combined with an artificial neural network to form what is called a deep q-learning algorithm [13]. The neural network enables the combined algorithm to operate in much larger environment, because it's ability to recognize patterns [11] diminishes the need to try all possible actions in all states. Instead, the algorithm exploits current knowledge of what action to take in a specific state, but occasionally will perform a different action in attempt to find a better action (i.e. exploration). Such exploitation-exploration ratio can be set by a parameter called temperature [14].

Due to these reasons, deep q-learning algorithm appears to be appropriate to solve a typical problem found in robotics – decrease of positioning accuracy at maximum operational speeds when robot is fully loaded. The hypothesis is that this type of algorithm could detect patterns in positioning deviations and output such correction values for angular joint positions, that the corrected positioning would be more accurate than the uncorrected one.

# 2 Experimental Research

## 2.1 Experimental Setup

Machine learning algorithms were tested using original experimental stand (Fig. 1), which consists of 5-axis articulated robot KUKA youBot (Augsburg, Germany) and equipment for measurements of positioning accuracy.

Main applications fields of KUKA youBot are research and education, due to this access to the robot's internal hardware is available through its open interface. In this case, KUKA youBot was connected to a PC via Ethernet cable and programmed using C++ language.

Positioning accuracy in 2D was measured using USB digital microscope with zooming capability up to 1600 times. Digital microscope was attached to the robot via specially designed gripper-microscope adapter. Positioning accuracy measurements were performed by taking photos of a target, which is stationary and is mounted on a stand in parallel to the microscope. The target itself is sticker on which a black rhombus, 0.1 mm in size, is marked and placed in white background (Fig. 3).

A mass simulator, attached to the robot's gripper, was used to produce a load, which is equal to the largest load the robot is capable of carrying (0.5 kg). The purpose of applying additional load was to bring experimental conditions closer to the operating conditions of real robot.

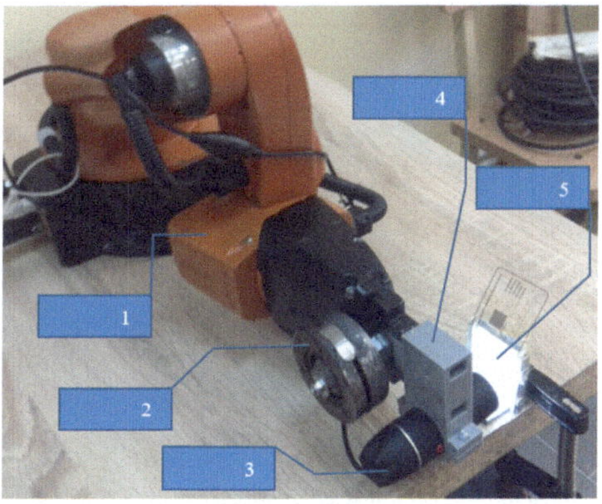


**Fig. 1.** Experimental stand: (1) KUKA youBot robot; (2) mass simulator; (3) digital microscope; (4) gripper-microscope adapter; (5) target with its stand.

## 2.2 Programming Software

The software used for the experiments was running on a PC with Debian 9 OS. Communication with the robot were carried out using appropriate EtherCAT and KUKA youBot API drivers [15] provided by the manufacturer.

The system for the experiments (Fig. 2) was created with the help of ROS (Robot Operating System) Melodic. In total, three ROS nodes were created: Botmover, AI, and Vision. Botmover was written in C++ programming language and was based on "Hello World" example [15]. The other two nodes were written in Python 3.7.1, because they required to use PyTorch 1.1 and OpenCV 4.1.1 libraries.

The Botmover node is able to: control all main parameters of the robot (angular position, maximum angular velocity, torque, and current at every joint/motor); read its internal sensor data (same parameters as for control); communicate with AI node.

The AI node coordinates the other two nodes to perform the experiments and contains the machine learning algorithm. The latter is of a deep q-learning type, which mainly consists of two parts: a q-learning algorithm, and an artificial neural network. The inputs and outputs to the algorithm vary based on what type of experiment is being carried out, but typically inputs include positioning data from Vision and Botmover, while the outputs relate to angular position corrections at joints. The reward system is also based on positioning data collected by Vision and Botmover nodes.

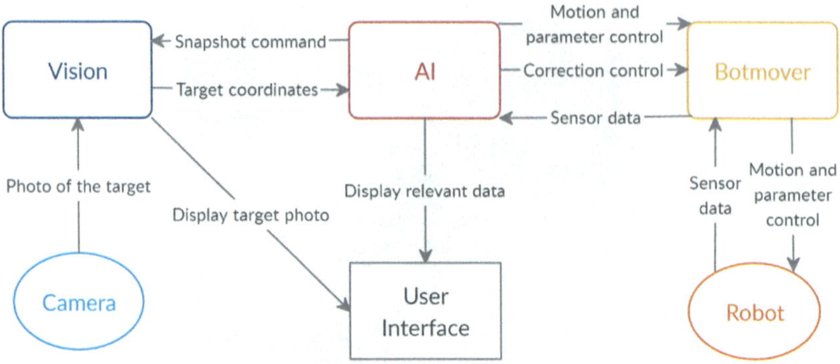


**Fig. 2.** General scheme of the system.

The purpose of the Vision node is to take a photo of the target using digital microscope, detect the target in that photo, calculate coordinates of the centroid of the target and send those coordinates to the AI node upon request. In addition, Vision displays the photo to the user, while marking the theoretical target by a blue dot and the center of actual target position by a blue dot (where the red dot should actually be after compensation) (Fig. 3). Deviation is the distance between these two points. Its value defined by evaluating deviations in vertical and horizontal directions in pixels (1 pixel = 4.5 μm).

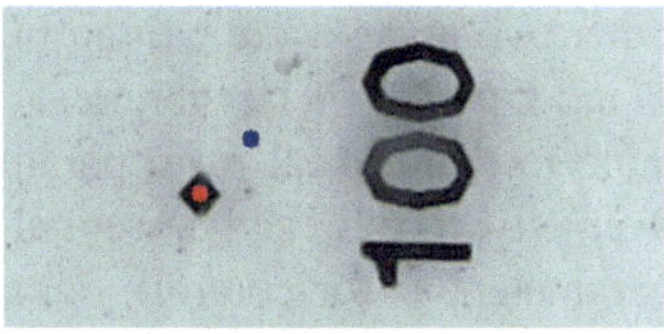


**Fig. 3.** A photo of the target made using a digital microscope: black rhombus – real target; red dot – calculated centroid of real target; blue dot is theoretical target – where the red dot should actually be after compensation.

## 2.3 Methodology

The main goal of the experiments was to determine if a deep q-learning algorithm can be used to compensate positioning deviations of an articulated robot KUKA youBot. Provided that is achieved, the next goal is to determine the most optimal configuration of the algorithm and the values of parameters.

All experiments were carried out at maximum speed that the robot can achieve (90°/s at each axis). Each experiment was carried out using identical methodology which can be divided into two main parts: positioning accuracy evaluation without correction; and training of machine learning algorithm. Evaluation of robot's positioning accuracy without correction was performed in the following order:

1. The robot is moved to “home” position (Fig. 4a).
2. The robot is moved to “target” position (Fig. 4b).
3. The Vision node takes a photo, detects target in it, calculates its centroid [16] and sends its coordinates (measured in pixels) to the AI node.
4. Steps 1–3 are repeated several times (typically 100 times).
5. The AI node calculates average coordinates of the target. The result is set to be the theoretical target coordinates.
6. Using the recorded data, the AI node calculates deviations from theoretical target at every previous cycle. This data is also written into memory.

Obtained results allows to evaluate average robot’s positioning accuracy without correction, gives information about stability of deviations and allows to create set of data which in further research can be threaten as average reference level of positioning accuracy.

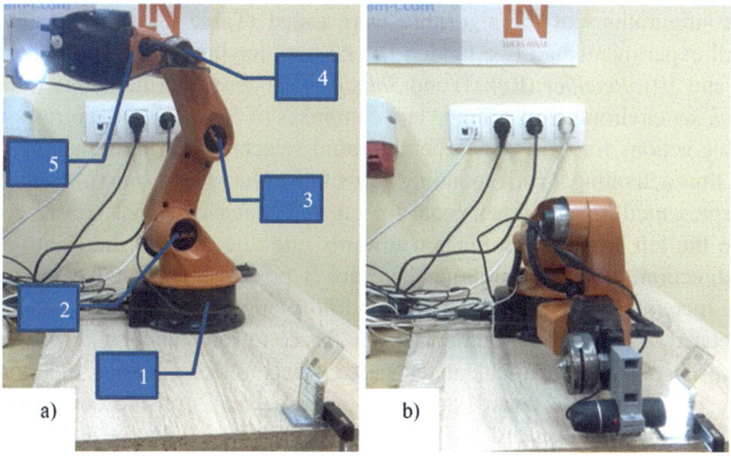


**Fig. 4.** Robot at “home” position (a) and at “target” position (b). During the experiments, the robot moves alternating between these positions. Figure at (a) also shows numbering of robot’s axes.

Experiments related to the training of the machine learning algorithm was performed in following order:

1. The robot is moved to “home” position (Fig. 4a).
2. The robot is moved to “target” position (Fig. 4b) with correction of angle.
3. The Vision node takes a photo, detects target in it, calculates its centroid (Fig. 3) and sends its coordinates to the AI node. The latter then processes the data and writes it to memory.
4. The AI node calculates positioning deviation (horizontal and vertical) and writes it into memory.
5. Based on absolute length of deviation, reward value for the deep q-learning algorithm is selected.

6. The final angular positions of the axes are measured using internal robot's sensors.
7. Horizontal and vertical deviation values, the final angular position of the axes are used in the deep q-learning algorithm.
8. The algorithm outputs angular position correction values for the axes for corrections of the following cycle.
9. Steps 1–8 are repeated until it becomes clear that either successful correction is being performed, or that there are no desirable tendencies in correction after a long (typically 1000 iterations) training.

The success of each experiment can be evaluated by comparing deviations from desired position without compensation and with compensation. Moreover, proposed methodic allows not only evaluate final results but it also provides information about nature of deviations and main tendencies.

## 2.4 Configurations of Correction Algorithm

Several configurations of the algorithm were tested (Table 1). *Artificial neural networks.* All experiments had two hidden layers, both having the same number of neurons (30 and 50). *Rectifier* (ReLU) and *Softsign* [17] activation functions were tested. *Correction set* environment is limited and comprises of 50 states and, correspondingly, 50 possible actions for each axis. The algorithm selects one of those actions at every iteration, thus activating a corresponding correction value. *Correction line* environment can be represented as one-dimensional line, at the center of which is zero correction value. To the left from it – correction towards one direction, to the right – towards another direction. At each iteration, there are 3 possible actions: increment to one direction, increment to another direction, and no motion (stay at current state). Every state is related to a corresponding correction. Three reward functions [18] were used: (i) *inverse only* – reward is calculated based solely on deviation – the smaller the deviation, the larger the reward value; (ii) *inverse plus* – reward-deviation proportionality plus additional reward for improved accuracy; (iii) *inverse plus additional* – reward is additionally increased when specific results are achieved. For instance, when a smaller deviation is achieved due to corrective action compared to uncorrected standard deviation, additional +0.5 point to reward is given; when the deviation is smaller than half of the uncorrected standard deviation, additional +1 point is given. Different temperature parameter values [15] were tested: (i) various *static* (unchanging) values; (ii) dynamic temperature setting. In latter case, a function is used, which depends on iteration. The training begins with a large bias towards exploration. Then, when the number of iterations increases, the bias moves towards exploitation. When a specific iteration is reached, the value of the temperature is fixed at large bias towards exploitation. Correction applied to the angular position of joints 2, 3, 4 (see Fig. 4a).

Testing different configurations allowed to pinpoint those parameter values and algorithm elements that improve algorithm's performance and determine how they impact general behavior of the whole experimental system. The parameter values and algorithm elements that were tested, were selected because they were deemed to be most likely achieve (at least partially) positive results, while still being practical to test time wise.

**Table 1.** Configuration of experiments.

| Exp. no | No of neurons | Activation function | Environment | Reward function | Temperature parameter | Corrected axes/axis |
|---|---|---|---|---|---|---|
| 1 | 50 | Rectifier | Correction set | Inverse only | Static values | 2, 3, 4 |
| 2 | 50 | Rectifier | Correction set | Inverse plus | Static values | 4 |
| 3 | 30 | Softsign | Correction line | Inverse plus additional | Dynamic values | 4 |

## 3 Results and Discussion

Forty experiments of various lengths were carried out. They were divided into three experimental groups based on differing sets of configurations of the algorithm used.

### 3.1 Experimental Group 1

In the first group of experiments (Table 1) angular position corrections were attempted to be carried out in three axes (joints 2, 3, 4). These were all the axes that are responsible for motion in the plane, on which 2D position can be measured using a digital microscope and a target. The experiments were carried out using various temperature values. Regardless, the results were always similar to that, which is shown in Fig. 5.

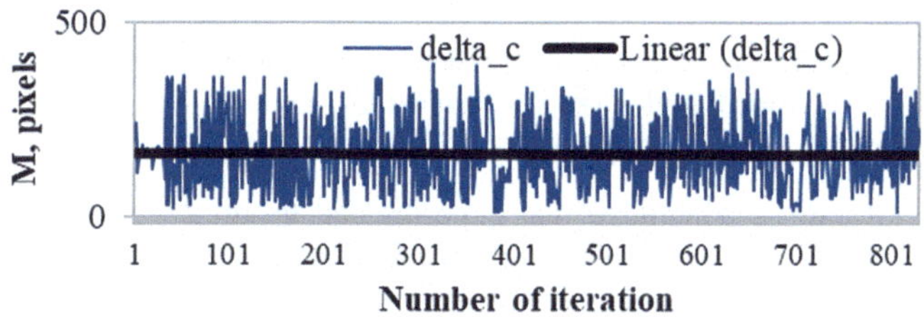


**Fig. 5.** Magnitude of deviation dependence on number of iterations: blue – variation of absolute deviation with respect to number of iterations; black – trendline.

Even though the initial experiments (which were up to 300 iterations long) also had large variations of deviations, their trendlines were sloping downwards. However, the results of longer experiments (up to 1000 iterations) have shown that in the long term, there were no changes in both variations of deviations and magnitudes of deviations, while the trendline remained mostly horizontal (Fig. 5). The results have shown that this configuration of the system and the algorithm was unable to noticeably improve positioning accuracy of the robot in up to 1000 iterations.

### 3.2 Experimental Group 2

The second group of experiments (Table 1) was carried out in order to test the performance of the algorithm performing single-axis (joint 4) corrections. In addition, the reward function was updated to additionally increase reward when corrected deviation is measured to be smaller than the uncorrected. The intention of this update is to better motivate the algorithm to correct so that the accuracy would be improved.

This set of experiments have shown that this configuration allows to achieve greater accuracy than that, which was achieved in the first experimental group. However, greater accuracy than that of uncorrected motion was not achieved.

The best results from this group are given in Fig. 6. There, it can be noted that the average corrected accuracy is similar to uncorrected. Median of deviations of uncorrected positioning was 38 pixels, while for corrected 39 pixels. Two instances of spiking deviations represent the deviations that appear as a result of exploration by the algorithm. Attempts to recreate these results in experiments with identical configuration were unsuccessful. Most of the deviations that resulted from this group of experiments were larger.

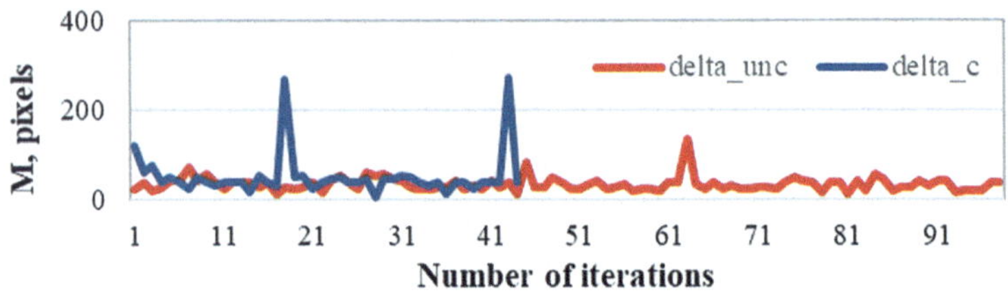


**Fig. 6.** Magnitude of deviation dependence on number of iterations: absolute deviations (red – uncorrected, blue – corrected) with respect to number of iterations.

Therefore, it can be concluded that experimental group 2 provided better results than its predecessor (group 1), even though corrections that improve accuracy were not achieved.

### 3.3 Experimental Group 3

In the third group of experiments (Table 1), different type of activation function, type of environment, and temperature setting were tested. While previous experimental groups employed Rectifier activation function in their neural networks, group 3 used a symmetrically and non-linearly behaving Softsign function [17].

In addition, to achieve a more uniform training, the environment type was changed to correction line. This change also resulted in the decreased number of output neurons, which in turn afforded a smaller number of neurons in hidden layers, thus simplifying the neural network. After a series of experiments, it was found that dynamic temperature setting significantly improved the speed of the training. Therefore, it was employed in further experiments. The results from the experiments that incorporated all previously mention changes are given in Fig. 7.

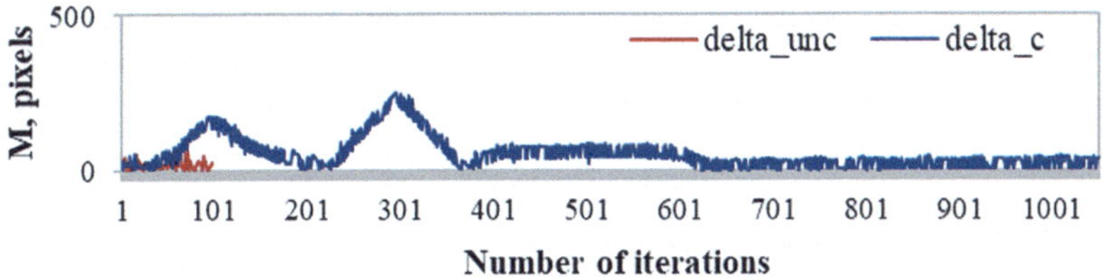


**Fig. 7.** Magnitude of deviation dependence on number of iterations: absolute deviations (red – uncorrected, blue – corrected) with respect to number of iterations.

The graph in Fig. 7 visualizes the main steps of the algorithm training. Firstly, algorithm performs exploration in one direction (iterations 30–230) and then – in another direction (iterations 230–380). A significant increase in positioning (with correction) accuracy can be seen at around 600 iterations. At that point, the algorithm starts to perform exploitation and thus, correct the positioning of the robot in one axis.

Similar results were obtained when the experiment was repeated several times, which means that the behavior of the algorithm is not random, but predictable and repeatable. Even though average deviations measured in this experimental group were similar when comparing uncorrected and corrected positioning, the ones with correction exhibited far more stable and predictable behavior.

## 4 Conclusions

Experimentally obtained results of implementing deep q-learning algorithms into control system of articulated robot demonstrated possibilities to implement ML algorithms into applied robotics in order to increase robots positioning accuracy. By carrying out numerous experiments with various algorithm and parameter configurations, it was found that not all configurations give significant improvements. Only algorithm in experimental group 3 yielded comparable positioning accuracy compared to uncorrected robot‘s positioning, where median of uncorrected deviations was 32 pixels, while for corrected was 18 pixels, which respectively corresponds to 144 μm and 81 μm. Also, results obtained in group 3, showed that around 600 iterations are required to obtain average deviations comparable to uncorrected deviations.

Moreover, obtained results shows that uncorrected deviations appeared to change randomly with time (ranging from 0 to 40 pixels, with a dip in frequency of occurrence in the middle). Meanwhile, corrected deviations were much more stable and predictable, mostly changing between two values near 5 and 40 pixels, with infrequent occurrences in between. Since the obtained results show positive tendency, further research in this field will be performed. Firstly, it's planned to perform tests of more configurations of the algorithm, performing more reliable 3-axis corrections than in experimental group 1. Secondly, tests will be performed trying to achieve better average positioning accuracy in 3D with correction, than without.

**Acknowledgement.** This work is supported by ECSEL Joint Undertaking through the Project AI4DI under grant agreement No. 826060-2. This Joint Undertaking receives support from the European Union's Horizon 2020 research and innovation program and National Authorities.

## References

1. Kober, J., Bagnell, J.A., Peters, J.: Reinforcement learning in robotics: a survey. Int. J. Robot. Res. **32**, 1238–1274 (2013)
2. de Wit, C.C., Siciliano, B., Bastin, G.: Theory of Robot Control. Springer, London (1996)
3. Zhu, W., Li, G., Dong, H., Ke, Y.: Positioning error compensation on two-dimensional manifold for robotic machining. Robot. Comput. Integr. Manuf. **59**, 394–405 (2019)
4. Shang, D., Li, Y., Liu, Y., Cui, S.: Research on the motion error analysis and compensation strategy of the delta robot. Mathematics **7**, 411 (2019)
5. Muñoz-Vázquez, A.J., Parra-Vega, V., Sánchez-Orta, A.: Control of constrained robot manipulators based on fractional order error manifolds. IFAC-PapersOnLine **48**, 118–123 (2015)
6. Dwiputra, R., Zakharov, A., Chakirov, R., Prassler, E.: Modelica model for the youBot manipulator. In: Proceedings of the 10th International Modelica Conference, Lund, Sweden, 10–12 March 2014, no. 096, pp. 1205–1212 (2014)
7. Urrea, C., Pascal, J.: Design, simulation, comparison and evaluation of parameter identification methods for an industrial robot. Comput. Electr. Eng. **67**, 791–806 (2018)
8. Marchal, P.C., Sörnmo, O., Olofsson, B., Robertsson, A., Ortega, J.G., Johansson, R.: Iterative learning control for machining with industrial robots. IFAC Proc. **47**, 9327–9333 (2014)
9. Pane, Y.P., Nageshrao, S.P., Kober, J., Babuška, R.: Reinforcement learning based compensation methods for robot manipulators. Eng. Appl. Artif. Intell. **78**, 236–247 (2019)
10. Part 2: Kinds of RL Algorithms — Spinning Up documentation. https://spinningup.openai.com/en/latest/spinningup/rl_intro2.html. Accessed 31 Oct 2019
11. Francois-Lavet, V., Henderson, P., Islam, R., Bellemare, M.G., Pineau, J.: An Introduction to Deep Reinforcement Learning (2018)
12. Watkins, C.J.C.H., Dayan, P.: Q-learning. Mach. Learn. **8**, 279–292 (1992)
13. Yang, Z., Xie, Y., Wang, Z.: A theoretical analysis of deep Q-learning. arXiv preprint arXiv:1901.00137 (2019)
14. Francois-Lavet, V., Henderson, P., Islam, R., Bellemare, M.G., Pineau, J.: An introduction to deep reinforcement learning. Found. Trends® Mach. Learn. **11**, 219–354 (2018)
15. Ishii, S., Yoshida, W., Yoshimoto, J.: Control of exploitation-exploration meta-parameter in reinforcement learning. Neural Netw. **15**, 665–687 (2002)
16. Locomotec: KUKA youBot User Manual (2011)
17. Bapat, K.: Find Center of a Blob (Centroid) Using OpenCV (C++/Python) | Learn OpenCV (2018)
18. torch.nn — PyTorch master documentation. https://pytorch.org/docs/stable/index.html. Accessed 31 Oct 2019

# Hip Twist Device for Wearable Robot

Paweł Rękas(✉)

Institute of Metrology and Biomedical Engineering,
Warsaw University of Technology,
sw. Andrzeja Boboli 8, 02-525 Warsaw, Poland
pawel.rekass@gmail.com

**Abstract.** Paper contains considerations on the implementation of the design of the hip twist module. It presents the genesis of the topic and a detailed description of the module operation. Attention is paid to user safety. The device requirements have been formulated. The loads that the module should be carrying, the kinematics and the dynamics of motion were determined – the angular range of rotation, the moment at which the limb can be rotated and the positioning time. With knowledge of the requirements that the device must meet, the initial schematic diagram of the device was made. Curvilinear rolling guides were constructed. A drive transmission system consisting of 3 different types of mechanical gears was constructed. The kinematic schematic diagram of the device was made. In the last stage, the pusher was analyzed in case of bending stresses using the FEM method. After analyzing the results, it was found that the dimensions of the pusher were selected correctly and the appropriate material was selected. Three-dimensional models of the entire module were made.

**Keywords:** Hip Twist Device · Orthotic robot · External powered exoskeleton · Exoskeleton · Rehabilitation · Paresis of lower limbs · Disability · Walking support

## 1 Introduction

The ubiquitous technological progress and the development of automotive and its dissemination caused an increase in damage to the human motor organs. Musculoskeletal disorders can be congenital or acquired. Among the disabled, 46% of them have this type of disorder. This is the second type of disorder in this group of people, immediately after cardiovascular disorders. There are about 1.3 million disabled people in Poland. A large part of them have problems with performing basic life functions. Every fifteenth disabled person will not walk alone 500 m and will not enter or go down the stairs from the first floor alone.

Restriction of mobility causes additional disorders of the skeletal system, cardiovascular system, pressure sores and deterioration of the efficiency of the whole body. Nowadays, technology meets the needs of disabled people and allows them to move and verticalise using wheelchairs and parapodies. The most developed support and upright systems are orthotic robots. They are advanced mechatronic systems performing the function of external supported skeletons. They are called Wearable Robots.


R. Szewczyk et al. (Eds.): AUTOMATION 2020, AISC 1140, pp. 267–275, 2020.
https://doi.org/10.1007/978-3-030-40971-5_25

Examples of these devices are Exoskeletons and Orthosis [3, 4, 14] and [15], but also Rehabilitation Robot for lower extremities [1]. Most of the devices in [3, 4, 14] and [15] cannot twist the hip joint. The module increases the functionality of the device described e.g. [13, 16] by the possibility of twisting hip joint. In some cases, twist of the hip may be useful when walking forward [12].

The scope of work included the creation of a hip joint twist module allowing rotation around the axis (Fig. 1):

- Parallel to the axis formed by the intersection of the frontal and sagittal planes,
- The axis contained in the frontal plane,
- Passing through the hip joint.

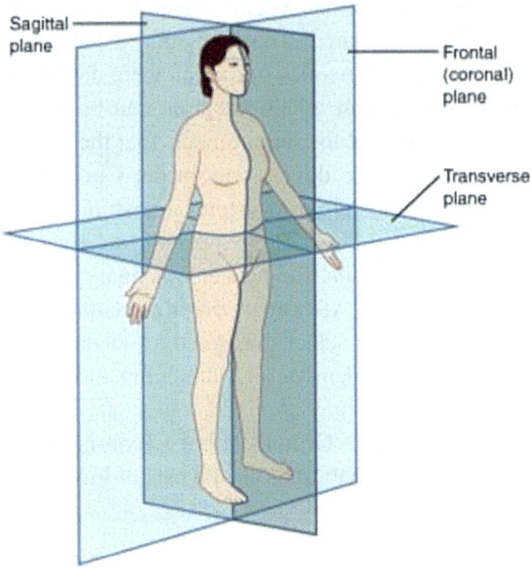


**Fig. 1.** Location of the sagittal, frontal and transverse planes [5].

Main function of device is adding a capability of realizing a turn in SPIWR orthotic robot [6] when walking forward. The project was made in several stages. The first one examines the current state of technics [19–21] and found the need for steering module. In the next stage it defines the requirements which should be suited for the module. Then, the available solutions for individual components were analyzed. At this stage solutions that should meet the established requirements were initially selected. Another important issue was to obtain the smallest size module. The resultant of these requirements allowed to choose the best solutions. After selecting specific solutions, they were analyzed in detail in terms of design and the correctness of their selection was checked. At the same time a three-dimensional model of the finished module was created. The elements with increased risk of damage due to high forces were also analyzed in terms of strength.

## 2 Structure of the System

The master robot control system will communicate with the hip control module's slave control system. Mechanism loads will be carried through the guides. The movement of the limb will be carried out through the drive, via the drive train. The idea of using this types of drive transmission is based on the actuator described in [9]. The system will be connected to the hip joint drive. The positioning will be supervised by a sub-control system via position sensors. An initial stage of choosing the motors was realized according to those procedures [10] (Fig. 2).

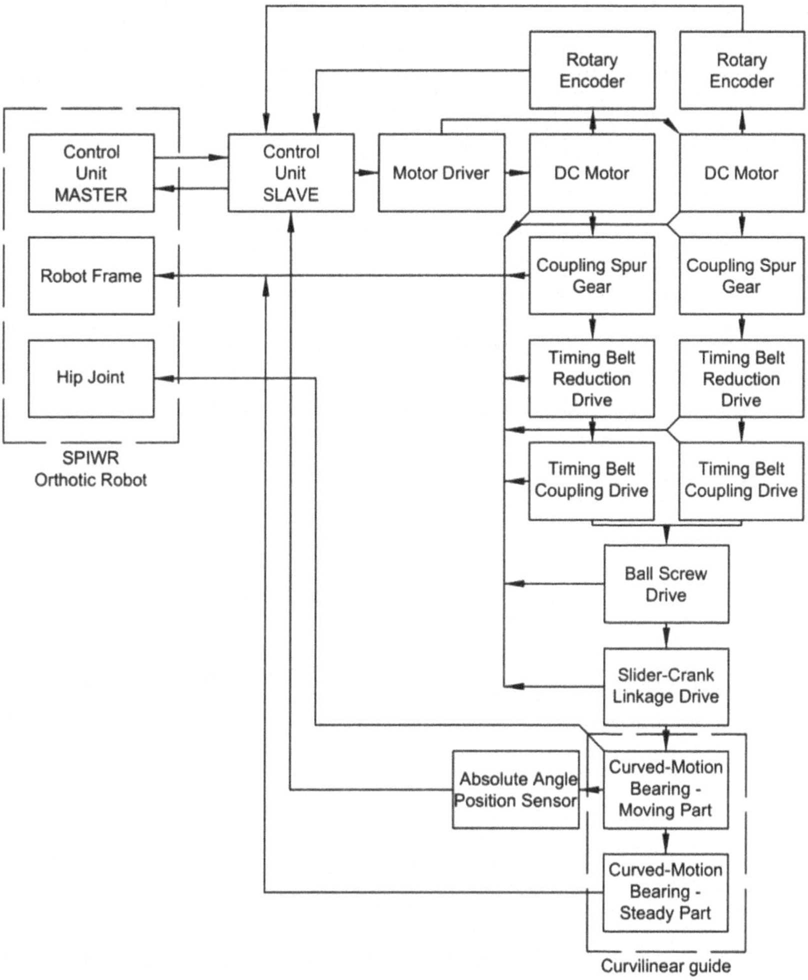


**Fig. 2.** Diagram of the hip twist device.

Designed device requires a connection to a SPIWR robot in order to work. Curved-Motion Bearing – Steady Part will be connected with robot frame. Moving part of Curved-Motion Bearing will be connected with robotic hip joint. Control units of SPIWR and Hip Twist Device will be connected via Serial Bus e.g. $I^2C$.

To obtain the optimal structure of the device, scheme should be realized according to those procedures [17]. Visualization of Hip Twist Device is shown in Fig. 6.

## 3 Curvilinear Guide

The curve guide is designed to transfer the load generated by the user along with the robot part. The case is considered when the greatest forces and moments to transfer occur during the execution of the step. The moment when the user's weight is attached to the waist belt and the robot stands on one leg were taken into account. The guide's parameters are presented in the functional requirements, these are:

- Carrying a 80 kg user load and a 30 kg robot,
- Angular range ±30°.

At the beginning the location of force was determined. The extreme case was taken into account when: with the maximum load of the user and the robot, the weight will focus on the center of the lap belt,

- the robot is in the middle of the step – the weight is transferred to one leg,
- dynamic loads equal to double static loads (Fig. 3).

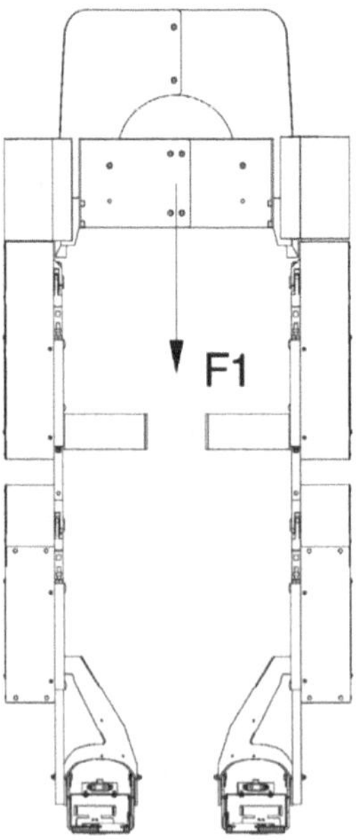


**Fig. 3.** Place of occurrence of force in a robot, based on [6].

The total robot load, as assumed in the requirements, is the sum of the mass of the user and the upper part of the SPIWR orthotic robot [6], so the value of force $F_1$ is:

$$F_1 = 2 \cdot (m_1 + m_2) \cdot g \tag{1}$$

$F_1$ means the force loading the curvilinear guides, $m_1$ means the mass of the robot user equal to 80 kg, and $m_2$ means the mass of the upper part of the robot equal to 30 kg, $g$ means standard gravity equal to 9.81 $m/s^2$. The force loading guides assumes the value of approx. 2158.2 N. Using the statics equations, the torque on the guide was determined. In an extreme case, this moment was about 639.9 Nm. The other concept of the guide is presented in [18]. The forces acting from the torque and load along with the section of the constructed guide are shown in the Fig. 4.

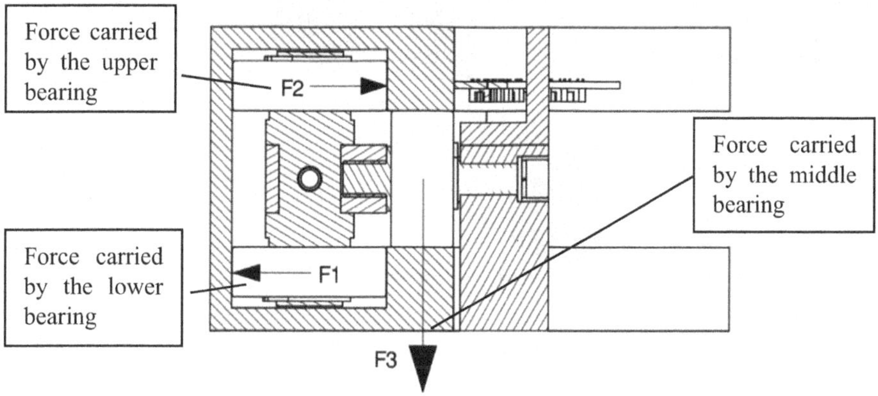


**Fig. 4.** Plan of force distribution in curvilinear guide.

# 4 Drive Transmission

Elements converting electric energy into mechanical energy are brush motors. The use of one large motor was abandoned. This was due to the free space distribution in the hip joint module and for economic reasons – cheaper motors were used. Motors speed can be controlled by PWM as described in [2]. The power from the motor shaft is transferred by a two-stage coupling gear. This is justified because no other transmission would take up so little space when the power transmission shaft is moved a short distance. The next step is a reduction belt drive with timing belt. On this gearing the incremental angular position is measured. The next step is a gear engaging with timing belt. The pulley is mounted on a ball screw. The last stage is ball screw drive. Ball screw receives power from both sides and passes on the nut. Nut is prevented from rotating for that performs linear motion along the screw. Then, using a modified slider-crank mechanism, the linear movement is converted into a rotational movement.

## 5 Pusher

The next analyzed part of the mechanism is the pusher- part of modified slider-crank mechanism. This part carries the largest loads from the Drive Transmission and was selected for analysis. It is a profiled steel rod with an approximate rectangular cross-section. It is loaded with P = 133.64 N acting at a distance of approx. 32.2 mm from the fastening. ANSYS Workbench was used for strength analysis.

The results obtained are shown in the Fig. 5.

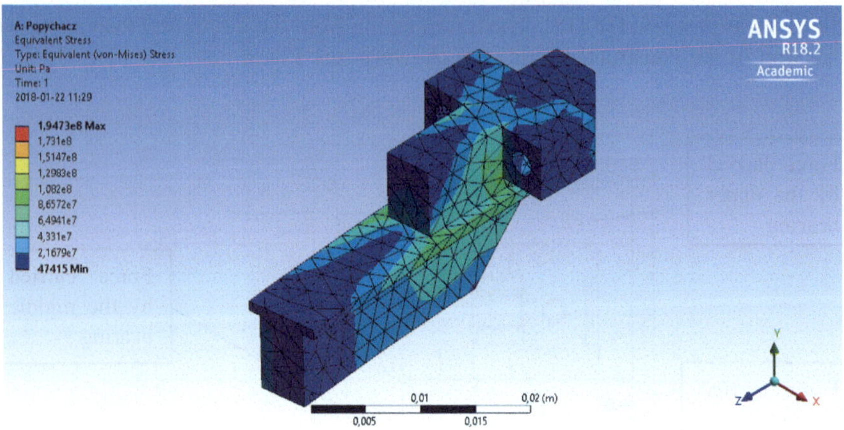


**Fig. 5.** Results of stress analysis.

The obtained results show the value of the maximum bending stress. It is $\sigma_{max}$ 195 MPa. Based on the results, C55 steel was selected as the material of the pusher, its maximum, acceptable, bending stresses are $\sigma$ = 270 MPa, which proves the correct selection of material and shape of the pusher.

## 6 Analyses

Ultimately, the assumed requirements were met. As a final result, the hip joint module was obtained, which, in addition to meeting the initial assumptions, is also small, light and compatible with the existing SPIWR robot. The obtained efficiency of the entire mechanism, calculated is:

$$\eta_{Hip\_Twist\_Device} = \eta_{wp} \cdot \eta_{s\_t} \tag{2}$$

$\eta_{wp}$ is the efficiency of all the intermediate gear between the motor and the ball screw and $\eta_{s\_t}$ is the reference efficiency ball screw drive. The efficiency of mechanical components assumes the value of approx. $\eta_{Hip\_Twist\_Device} = 0.67$. It is assumed that the mechanism is self-locking when its efficiency is less than 50%. The obtained efficiency is about 67%, which ensures the required lack of self-locking. If the low cost requirements were neglected, it would be possible to use much more expensive motors that have more power per unit of volume, which could ultimately give the possibility of further reducing the dimensions of the device. It also seems justified to reduce the scope of turnover. To determine if this would be optimal, a prototype of the hip joint module should be made and tested in a configuration with an orthotic robot. The final appearance of the Hip Twist module without casings is shown in the Fig. 6. The module security assumptions were based on the procedures described in [7].

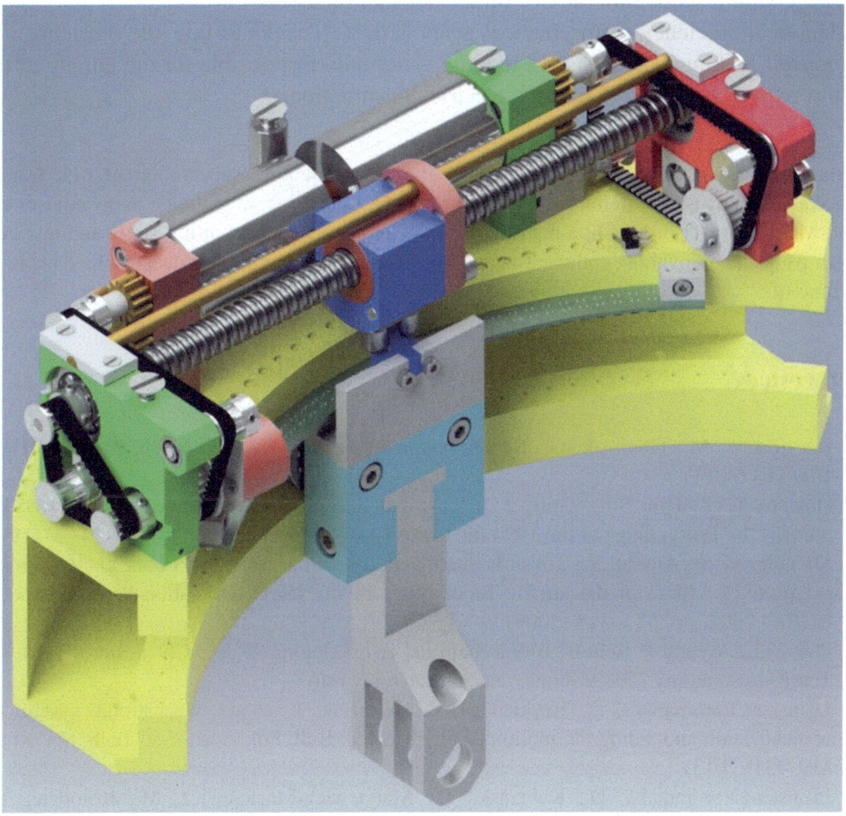

**Fig. 6.** Visualization of Hip Twist Device.

## 7 Conclusion

The work describes the design and construction process of the Hip Joint Twist Module for the SPIWR Robot. The effect of the work was to present the construction of a curvilinear guide, accepted design solutions and obtained 3D models. The following conclusions were made when constructing the device.

- Mechatronic systems significantly facilitate the functioning of a disabled people. In the future, it may be possible to integrate mechatronic systems with physically fit people to improve innate skills.
- The use of two types of sensors to measure the yaw angle of the device allowed to use a smaller and more accurate absolute sensor and a sensor with a line of opto-couplers. The electronic controller can also control the angular displacement of the module by a current sensors integrated with the motors based on the method described in [8].
- Actual angular range of device should be verified in practice. Reducing the angular range would reduce the device dimensions and its weight.
- The final efficiency of mechanical gears is 67%. Thanks to this, the mechanism is not self-locking. The designed module is not only responsible for the quality of the hip twist [11], but also reduces electric consumption.

**Acknowledgement.** The work was created under the scientific supervision of Ms. Danuta Jasińska-Choromańska. Foremost, I would like to express my sincere gratitude to my advisor and tutor, Ms. Danuta Jasińska-Choromańska, for the continuous support of my study and research, for her patience, motivation, enthusiasm, and immense knowledge. Her guidance helped me in all the time of.

## References

1. Kaczmarski, M., Granosik, G.: Rehabilitation robot RRH1. Arch. Mech. Eng. **LVIII**(1), 103–113 (2011)
2. Downes, C.G., Hill, S.L., Gray, J.O.: Distributed control of an electrically powered hip orthosis. In: Proceedings of the 1994 International Conference on Control, pp. 24–30 (1994)
3. Mohammed, S., Amirat, Y.: Towards intelligent lower limb wearable robots: challenges and perspectives - State of the art. In: Proceedings of the IEEE International Conference on ROBIO 2008, pp. 312–317 (2009)
4. Pons, J.L.: Wearable Robots: Biomechatronic Exoskeletons. Wiley, Chichester (2008)
5. TeachMe Anatomy (2019). https://teachmeanatomy.info
6. Jasińska-Choromańska, D., Szykiedans, K., Wierciak, J., et al.: Mechatronic system for verticalization and aiding the motion of the disabled. Bull. Pol. Acad. Sci. Tech. Sci. **61**(2), 419–431 (2013)
7. Jasińska-Choromdiska, D., Kabziński, B., Matyjewicz-Maciejewicz, M., Kołodziej, D.: Safety module for the system of verticalization and aiding morion of the disabled. In: Březina, T., Jabloński, R. (eds.) Mechatronics 2013, vol. 143, pp. 79–85. Springer, Cham (2014)
8. Bodnicki, M., Hawłas, H.J.: Commutation phenomena in DC micromotor as source signal of angular position transducer. In: Březina, T., Jabloński, R. (eds.) Recent Advances in Mechatronics 2008–2009, pp. 263–268. Springer, Heidelberg (2009)
9. Jasińska-Choromańska, D., Credo, W., Szykiedans, K.: Making use of anthropometric data while designing drive units of an orthotic robot. Mach. Dyn. Res. **36**(1), 90–98 (2012)
10. Wierciak, J., Bagiński, K., Credo, W.: Selection of electric driving modules for orthotic robot. In: Březina, T., Jabłoński, R. (eds.) Advanced Mechatronics Solutions, pp. 297–303. Springer, Cham (2016)

11. Bagiński, K., Wierciak, J.: Selection of drives for orthotic robot based on simulation studies Mach. Dyn. Res. **36**(1), 5–14 (2012)
12. Eng, J.J., Winter, D.A.: Kinetic analysis of the lower limbs during walking: what information can be gained from a three-dimensional model? J. Biomech. **28**, 753–758 (1995)
13. Quintero, H., Farris, R., Goldfarb, M.: Control and implementation of a powered lower limb orthosis to aid walking in paraplegic individuals. In: 2011 IEEE International Conference on Rehabilitation Robotics (ICORR), pp. 1–6, June 2011
14. Bogue, R.: Exoskeletons and robotic prosthetics: a review of recent developments. Ind. Robot Int. J. **36**(5), 421–427 (2009)
15. Dollar, A.M., Herr, H.: Lower extremity exoskeletons and active orthoses: challenges and state-of-the-art. IEEE Trans. Rob. **24**(1), 144–158 (2008)
16. Sankai, Y.: Leading edge of cybernics: robot suit HAL. In: SICE-ICASE International Joint Conference (2006). https://doi.org/10.1109/sice.2006.314982
17. Wierciak, J., Łuczak, S., Łuczak, I.: Design methodology of mechatronic systems. Mechatronic Systems: Design. Performance and Applications, pp. 111–166. Nova Science Publishers, Hauppauge (2019)
18. Osiński, D., Zaczyk, M., Jasińska-Choromańska, D.: Conception of turning module for orthotic robot. In: Jabłoński, R., Březina, T. (eds.) Advanced Mechatronics Solutions. Advances in Intelligent Systems and Computing, vol. 393, pp. 147–152. Springer, Cham (2016)
19. Zoss, A., Kazerooni, H., Chu, A.: On the mechanical design of the Berkeley Lower Extremity Exoskeleton (BLEEX). In: IEEE/RSJ International Conference on Intelligent Robots and Systems (2005). https://doi.org/10.1109/iros.2005.1545453
20. Esquenazi, A., et al.: The ReWalk powered exoskeleton to restore ambulatory function to individuals with thoracic-level motor-complete spinal cord injury. Am. J. Phys. Med. Rehabil. **91**(11), 911–921 (2012)
21. Firestone, J., Maghan, C.: Design and Implementation of the XOS2 Exoskeleton for the United States Military. University of Pittsburgh, Pittsburgh (2012)

# Measuring Techniques and Systems

# Numerical Modeling of Dynamic Disturbances Acting on the Sensitive Elements of an Instrument Navigation System

Igor Korobiichuk[1(✉)], Olena Bezvesilna[2], Yuriy Podchashinskiy[3], and Katarzyna Rzeplińska-Rykała[4]

[1] Institute of Automatic Control and Robotics, Warsaw University of Technology, Warsaw, Poland
igor.korobiichuk@pw.edu.pl
[2] National Technical University of Ukraine "Igor Sikorsky Kyiv Polytechnic Institute", Kiev, Ukraine
o.bezvesilna@gmail.com
[3] Zhytomyr State Technological University, Zhytomyr, Ukraine
ju.podchashinskiy@gmail.com
[4] ŁUKASIEWICZ Research Network – Industrial Research Institute for Automation and Measurements PIAP, Warsaw, Poland
krykala@piap.pl

**Abstract.** The numerical methods for modeling dynamic disturbances that influence a sensitive element (accelerometer) of the instrument navigation system were considered. Algorithms of reproduction of dynamic disturbances on a digital computer with a given correlation function were developed. In the paper presented the results of the studies of numerical modeling of dynamic disturbances in the instrument navigation system with acceptable reproduction accuracy of the statistical characteristics of these disturbances.

**Keywords:** Dynamic disturbances · Correlation function · Instrument navigation system · Accelerometer

## 1 Introduction

Instrument navigation systems are used to measure motion parameters, determine the navigation parameters of moving objects (airplanes, space objects, etc.). Sensitive elements (accelerometers) of such systems have measurement errors due to the fact that these sensitive elements measure the projection of a set of two signals onto the axis of sensitivity: the useful signal (acceleration of the moving base) and the interference signal, which is determined by dynamic disturbances (translational and angular accelerations).

Improvement of the accuracy of measuring the parameters of motion instrument navigation system is an important scientific and technical problem and is based on the use of updated mathematical models of objects; optimization of the parameters of this system; use of filtering methods and algorithmic compensation of errors due to dynamic


R. Szewczyk et al. (Eds.): AUTOMATION 2020, AISC 1140, pp. 279–288, 2020.
https://doi.org/10.1007/978-3-030-40971-5_26

disturbances. The study and optimization of filtering and error compensation methods are impossible without modeling dynamic disturbances influencing a sensitive element.

The purpose of the work is to analyze methods and develop algorithms for modeling dynamic disturbances influencing a sensitive element in an instrument navigation system. These algorithms can be used in the study and optimization of procedures for digital processing of the output signal of the accelerometer.

## 2 Working Conditions of a Sensitive Element in the Aviation Instrument Navigation System

In the instrument navigation system, disturbances take place that reduce the accuracy of measurements of the parameters of the motion of the base:

a. dynamic disturbances arising in the process of moving the aviation gravimetrical system (AGS) moving base due to the influence of various vibrations (from engines and other equipment) [1];
b. vertical acceleration disturbances affecting the moving base and are mainly due to atmospheric turbulence [1].

These disturbing influences were considered in theoretical and experimental studies of methods for filtering the output signal of accelerometers [1–6]. Dynamic disturbances in the instrument navigation system [1, 2] are described by Gaussian random processes having an amplitude of vibration accelerations along the transverse and normal axes of 1 $g$, along the longitudinal axis – up to 0.7 $g$. The normalized vibration range is 5–2000 Hz. In this case, the system is located on the shock absorbers and its working conditions correspond to the I or II degree of rigidity.

The influence of angular and translational vibrations on different axes on the operation of the navigation instrument system is different. In [2], the effect of dynamic disturbances on the operation of accelerometers was studied. For angular vibrations of the moving base, the systematic errors of the deflection angle of the sensitive element when measuring accelerations are $1.36 \times 10^{-8}$ rad units. For translational vibrations, these errors are $5.41 \times 10^{-6}$ rad units. The amplitudes of the mixed vibrations of the device are directly proportional to the perturbing accelerations along the sensitivity axis, and the mean square error is 0.01–0.02 mGal. When developing a device with an accuracy of 0.1 mGal, this effect of vibration accelerations must be taken into account.

The procedures for numerical modeling of dynamic disturbances should be based on the theoretical principles of statistical modeling of random processes [7–14] and the results of field recordings and statistical processing of dynamic disturbances in actual operating conditions of the instrument navigation system [1–3].

There are various methods of modeling Gaussian random processes with a given correlation function [15–19]: a filtering method, recurrent modeling algorithms, a sliding summation method, a canonical scheduling method. However, most of these algorithms either have methodological errors of reproduction of the correlation function, or require the use of complex numerical methods. Therefore, it is necessary to analyze these methods and determine those that are suitable for modeling dynamic disturbances taking into account the features of digital processing of acceleration measuring information and other motion parameters of the base on which the navigation instrument system is installed.

The need for detailed consideration and modeling of dynamic events influencing a sensitive element is due to the fact that they have a significant effect on the measurement accuracy of accelerations and navigation parameters when the base moves.

The translational and angular vibrations influencing a sensitive element are described by a normal stationary random process $\xi(t)$ with a narrow-band frequency spectrum equal to of zero, and a correlation function [2]:

$$K_{\xi i}(t_k) = D_{\xi i}e^{-\mu_i|t_k|}\left(\cos\lambda_i|t_k| + \frac{\mu_i}{\lambda_i}\sin\lambda_i|t_k|\right), \quad (1)$$

where $i = 1, \ldots, 6$ – defines one of three axes for translational vibrations and one of three angles for angular vibrations; $D_{\xi i}$ – dispersion of amplitude of vibration acceleration; $\mu_i$ – attenuation coefficient of the correlation function characterizing the degree of vibration irregularity; $\lambda_i$ – prevailing frequency of vibrations.

The work of a sensitive element and algorithms for digital processing of its output signal should be investigated [2] with vibration accelerations with amplitudes up to 1 $g$ and the prevailing frequency $\lambda_z$ = 1640 rad/s on the axis of device sensitivity. Such vibration accelerations correspond to the normal stationary random process $\xi(t)$ from $E[\xi_z(t)] = 0$, $D[\xi_z(t)] = D_{\xi z} = \sigma_\xi^2 = 0.1\,g^2$ (based on the normal amplitude distribution of vibration accelerations up to 1 $g$), $\mu_z$ = 200 rad/s and the correlation function (1).

## 3 Development of a Numerical Method for Modeling Dynamic Disturbances Based on a Shaping Filter

In order to reproduce dynamic disturbances, let's consider the method of a shaping filter and a recurrent algorithm based on auto regression. For them, numerical methods are relatively simple and do not have a methodological error in the reproduction of the correlation function [15, 16].

As you known, in the shaping filter method a random process $\eta(t)$ of the white noise type turns into a random process $\xi(t)$ with the desired statistical characteristics using the filter with the transfer function $H(p)$ and, accordingly, the frequency response $H(\omega)$:

$$|H(\omega)|^2 = \frac{S_{out\,\xi}(\omega)}{S_0},$$

where $S_{out\,\xi}(\omega)$ – power spectral density of a random process $\xi(t)$; $S_0 = const = 1$ – power spectral density of a white noise $\eta(t)$ with a correlation function $K_\eta(\tau) = S_0\delta(\tau)$, $\delta(\tau)$ – delta function.

Considering that $S_{out\,\xi}(\omega) = \sigma_\xi^2 S_\xi(j\omega)S_\xi(-j\omega)$, $|H(\omega)|^2 = H(j\omega)H(-j\omega)$, $j\omega = p$, we get:

$$H(p) = \frac{\sigma_\xi S_\xi(p)}{\sqrt{S_0}} = \frac{\sigma_\xi}{\sqrt{S_0}} \cdot \frac{S_{\xi 1}(p)}{S_{\xi 2}(p)}, \tag{2}$$

where $S_{\xi 1}(\omega), S_{\xi 2}(\omega)$ –polynomials of variable $p$.

Let's take into account that the correlation function (1) uses the modulus of the argument and therefore the physical spectrum of the random process must be considered. Therefore, the random process of dynamic disturbances $\xi(t)$ with a correlation function (1) has a spectral density of power [20, 21]:

$$S_{out\,\xi}(\omega) = \sigma_\xi^2 \frac{4\mu_i(\mu_i^2 + \lambda_i^2)}{(\mu_i^2 + \lambda_i^2)^2 + 2(\mu_i^2 - \lambda_i^2)\omega^2 + \omega^4}, \tag{3}$$

or in general

$$S_{out\,\xi}(\omega) = \sigma_\xi^2 \frac{\left|\beta_{\xi 0} j\omega + \beta_{\xi 1}\right|^2}{\left|(j\omega)^2 + \alpha_{\xi 1} j\omega + \alpha_{\xi 2}\right|^2}. \tag{4}$$

Hence

$$S_{\xi 1}(p) = \beta_{\xi 0} p + \beta_{\xi 1},\ S_{\xi 2}(p) = p^2 + \alpha_{\xi 1} p + \alpha_{\xi 2}.$$

A shaping filter (2), which reproduces dynamic disturbances (4), corresponds to the differential equation in the operator form [15, 22–24]:

$$S_{\xi 2}(p)\xi(p) = \frac{\sigma_\xi}{\sqrt{S_0}} \cdot S_{\xi 1}(p)\eta(p). \tag{5}$$

For numerical calculations, the form of the record (5) in the form of a system of differential equations of the first order is more convenient

$$\dot{X}_\xi = A_\xi X_\xi + B_\xi \eta(t), \tag{6}$$

where

$$A_\xi = \begin{bmatrix} 0 & 1 \\ -\alpha_{\xi 2} & -\alpha_{\xi 1} \end{bmatrix},\ B_\xi = \begin{bmatrix} \beta_{\xi 0}\sigma_\xi/\sqrt{S_0} \\ (\beta_{\xi 1} - \alpha_{\xi 1}\beta_{\xi 0})\sigma_\xi/\sqrt{S_0} \end{bmatrix}.$$

In this case, dynamic disturbances that are formed at the output of the filter correspond to the first component of the two-dimensional Markov process. $X_\xi(t) = (x_{\xi 1}(t), x_{\xi 2}(t))^T$, that is $\xi(t) = x_{\xi 1}(t)$.

Comparing the formulas (2), (3) and (4), we obtain the values of the coefficients of the transfer function of the shaping filter for dynamic disturbances with the correlation function (1):

$$\beta_{\xi 0} = 0,\ \beta_{\xi 1} = 2\sqrt{\mu_i(\mu_i^2 + \lambda_i^2)},$$

$$\alpha_{\xi 1} = 2\mu_i,\ \alpha_{\xi 2} = \mu_i^2 + \lambda_i^2. \tag{7}$$

Next, you need to get the equations by which you can generate measuring references of dynamic disturbances on a digital computer $\xi_k = \xi(k \cdot \tau)$, where $k = 1, \ldots, N_\xi$, $\tau$ – step of discretion, $N_\xi$ – number of measuring references of dynamic disturbances. These equations are the result of the numerical integration of differential equations (6). The method of solving a similar problem by numerical integration based on the Runge-Kutta method is given in [15, 16].

We will look for the solution of Eq. (6) in the form of a recurrent matrix equation

$$X_{\xi k} = A_{\xi k} X_{\xi(k-1)} + B_{\xi k} \eta_k.$$

The solution (6) is the system of equations, used to simulate dynamic disturbances in the instrument navigation system:

$$\begin{cases} x_{\xi 1,k} = a_{\xi 11} x_{\xi 1,(k-1)} + a_{\xi 12} x_{\xi 2,(k-1)} + b_{\xi 11}\eta_{1,k} + b_{\xi 12}\eta_{2,k} \\ x_{\xi 2,k} = a_{\xi 21} x_{\xi 1,(k-1)} + a_{\xi 22} x_{\xi 2,(k-1)} + b_{\xi 21}\eta_{1,k} + b_{\xi 22}\eta_{2,k} \end{cases} \tag{8}$$

Next, for a random process with a correlation function (1), the coefficients of equations (8) were calculated using the indicated method, taking into account the obtained coefficients (7). Thus, there were obtained such results:

$$\xi_k = \sigma_\xi x_{\xi 1,k},\ a_{\xi 11} = e^{-\mu_i \tau} \cdot \left(\cos(\lambda_i \tau) + \frac{\mu_i}{\lambda_i} \sin(\lambda_i \tau)\right),\ a_{\xi 12} = \frac{1}{\lambda_i} e^{-\mu_i \tau} \cdot \sin(\lambda_i \tau),$$

$$a_{\xi 21} = -\frac{\mu_i^2}{\lambda_i} e^{-\mu_i \tau} \cdot \left(1 + \frac{\mu_i^2}{\lambda_i^2}\right) \sin(\lambda_i \tau),\ a_{\xi 22} = e^{-\mu_i \tau} \pm \cdot \left(\cos(\lambda_i \tau) - \frac{\mu_i}{\lambda_i} \sin(\lambda_i \tau)\right), \tag{9}$$

$$b_{\xi 11} = \sqrt{c_{\xi 11}},\ b_{\xi 21} = \frac{c_{\xi 21}}{\sqrt{c_{\xi 11}}},\ b_{\xi 12} = 0,\ b_{\xi 22} = \sqrt{c_{\xi 22} - c_{\xi 12}^2 / c_{\xi 11}},$$

$$c_{\xi 11} = 1 - e^{-2\mu_i \tau} \cdot \left[\cos^2(\lambda_i \tau) + \frac{2\mu_i}{\lambda_i} \cos(\lambda_i \tau) \sin(\lambda_i \tau) + \left(1 + \frac{2\mu_i^2}{\lambda_i^2}\right) \sin^2(\lambda_i \tau)\right],$$

$$c_{\xi 12} = c_{\xi 21} = 2\mu_i \left(1 + \frac{\mu_i^2}{\lambda_i^2}\right) \cdot e^{-2\mu_i \tau} \cdot \sin^2(\lambda_i \tau),$$

$$c_{\xi 22} = (\mu_i^2 + \lambda_i^2)\left\{1 - e^{-2\mu_i \tau} \cdot \left[\cos^2(\lambda_i \tau) - \frac{2\mu_i}{\lambda_i} \cos(\lambda_i \tau) \sin(\lambda_i \tau) + \left(1 + \frac{2\mu_i^2}{\lambda_i^2}\right) \sin^2(\lambda_i \tau)\right]\right\}.$$

The modeling of dynamic disturbances in the instrument navigation system is performed in the following sequence:

1. Form the initial conditions $X_{\xi 0} = (x_{\xi 1,0}, x_{\xi 2,0})^T$. For this, 2 measuring references of the stationary random process are used with the normal distribution law, the expectation $E_{X0} = 0$ and the correlation matrix:

$$K_{X0} = \begin{bmatrix} 1 & 0 \\ 0 & (\mu_i^2 + \lambda_i^2) \end{bmatrix}.$$

2. Calculate the coefficients $a_{\xi\, ij}, b_{\xi\, ij}$ of the system (8) by the formulas (9).
3. Assign value $k = 1$.
4. Form 2 measuring references $\eta_k = (\eta_{1,k}, \eta_{2,k})^T$ of the normal stationary random process with mathematical expectation $E_\eta = 0$ and the correlation matrix $K_{\eta k} = \begin{bmatrix} 1 & 0 \\ 0 & 1 \end{bmatrix}$. These measuring references are a white noise model $\eta(t)$ at the input of the shaping filter.
5. Calculate the vector $X_{\xi k} = (x_{\xi 1,k}, x_{\xi 2,k})^T$ by the formula (8) and assign $\xi_k = x_{\xi 1,k}$.
6. Increase the value $k = k + 1$.
7. If $k \leq N_\xi$, then proceed to step 4, otherwise we will end the formation of the sequence $\xi_k$.

## 4 Development of a Numerical Method for Modeling Dynamic Disturbances Using a Recurrent Method Based on an Auto Regression Equation

The second possible method for the modeling of dynamic disturbances in the instrument navigation system is a recurrent algorithm based on auto regression. It is based on the equation

$$\xi_k = \sum_{j=0}^{l} b_{\xi j} \eta_{k-j} - \sum_{j=1}^{m} a_{\xi j} \xi_{k-j}, \tag{10}$$

where $\eta_{k-j}$ – measuring references of white noise; $a_{\xi j}, b_{\xi j}, l, m$ – coefficients depending on the statistical characteristics of a random process and are found by the factorization method [17, 18].

For a random process with a fractionally rational spectral density of type (4) in [15, 16] a solution (10) was found and a sequence of actions for the calculation of coefficients $a_{\xi j}, b_{\xi j}$ was given. Given that the dynamic disturbances in the instrument navigation system are given using the correlation function, we will perform modeling based on a recurrent equation of the type of auto regression-moving average:

$$\xi_k = a_{\xi 1}\xi_{k-1} + a_{\xi 2}\xi_{k-2} + b_{\xi 0}\eta_k + b_{\xi 1}\eta_{k-1}. \tag{11}$$

For the correlation function (1), according to this method, the coefficients of equation (11) were calculated and the following results were obtained:

$$a_{\xi 1} = 2e^{-\mu_i \tau}\cos(\lambda_i \tau), \quad a_{\xi 2} = -e^{-2\mu_i \tau}, \quad b_{\xi 0} = \sqrt{-c_{\xi 0}/c_{\xi 2}}, \quad b_{\xi 1} = -c_{\xi 2} b_{\xi 0}$$

$$c_{\xi 0} = a_{\xi 12}c_{\xi 12} - a_{\xi 22}c_{\xi 11}, \tag{12}$$

$$c_{\xi 1} = c_{\xi 11}(a_{\xi 22}^2 + 1) - 2c_{\xi 12}a_{\xi 12}a_{\xi 22} + c_{\xi 22}a_{\xi 12}^2,$$

$$c_{\xi 2} = -\frac{c_{\xi 1}}{2c_{\xi 0}} + sign\left(\frac{c_{\xi 1}}{2c_{\xi 0}}\right) \cdot \sqrt{\frac{c_{\xi 1}^2}{4c_{\xi 0}^2} - 1}$$

and values $a_{\xi ij}, c_{\xi ij}$ are calculated according to (9).

The modeling of dynamic disturbances based on the auto regression Eq. (11) is performed in the following sequence:

1. Form the initial conditions $(\xi_0, \xi_1)$. For this, 2 measuring references of the stationary random process are used with the normal distribution law, the expectation $E_{\xi 0,1} = 0$ and the correlation matrix $K_{\xi 0,1} = \begin{bmatrix} D_{\xi i} & 0 \\ 0 & D_{\xi i} \end{bmatrix}$. Form a discrete value $\eta_1$ of the input white noise with normal distribution, the expectation $E_\eta = 0$ and dispersion $D_\eta = 1$.
2. Calculate the coefficients $a_{\xi 1}, a_{\xi 2}, b_{\xi 0}, b_{\xi 1}$ of the Eq. (11) by the formulas (9) and (12).
3. Assign value $k = 2$.
4. Form a discrete value $\eta_k$ of the input white noise with normal distribution, the expectation $E_\eta = 0$ and dispersion $D_\eta = 1$.
5. Calculate the discrete value of dynamic disturbances $\xi_k$ according to the Eq. (11).
6. Increase the value $k = k + 1$.
7. If $k \leq N_\xi$, then proceed to step 4, otherwise we will end the formation of the sequence $\xi_k$.

## 5 Results of Numerical Modeling of Dynamic Disturbances

Modeling and calculations were performed on the basis of the given algorithms for modeling dynamic disturbances and the formulas (8), (9), (11), (12). The maximum relative error of reproduction of the correlation function for dynamic disturbances modeled by the recurrent method based on auto regression is 2.7%.

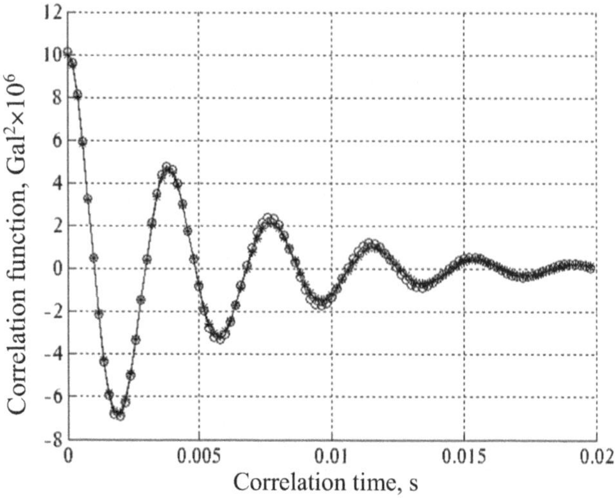


**Fig. 1.** Correlation functions of dynamic disturbances: * – calculation by the formula (1); o – calculation for disturbances modeled by the formulas (11) and (12).

The method based on the recurrent auto regression equation provides more accurate reproduction of the correlation function of dynamic disturbances than the method of the shaping filter. Figure 1 shows an example of the results of calculating estimates of the correlation functions of dynamic disturbances modeled.

## 6 Conclusions

The results of the studies confirm the possibility of numerical modeling of dynamic disturbances in the instrument navigation system with acceptable reproduction accuracy of the statistical characteristics of these disturbances. Modeling errors do not exceed 2.7%. The method based on the recurrent auto regression equation provides a more accurate reproduction of the correlation function of dynamic disturbances than the shaping filter method.

The results of numerical modeling of dynamic disturbances can be used in the development of gravimetric measuring instruments [4, 5] and in studying the effect of vibrations of the moving base of an instrument navigation system to a sensitive element that measures the motion parameters of this basis.

## References

1. Bezvesilna, O.M.: Measuring Acceleration: Tutorial. Libid: Kyiv, 264 p. (2001)
2. Bezvesilna, O.M.: Aviation gravimetric systems and gravimeters. ZhSTU: Zhytomyr, 604 p. (2007)

3. Korobiichuk, I., Bezvesilna, O., Tkachuk, A., Chilchenko, T., Nowicki, M., Szewczyk, R.: Design of piezoelectric gravimeter for automated aviation gravimetric system. J. Autom. Mob. Robot. Intell. Syst. **10**(1), 43–47 (2016). https://doi.org/10.14313/JAMRIS_1-2016/6
4. Bezvesilna, O., Korobiichuk, I., Nechai, J., Podchashynskyi, Yu.: Gravity meter, Patent UA78620 (C2) (2007)
5. Korobiichuk, I., Nowicki, M., Szewczyk, R.: Design of the novel double-ring dynamical gravimeter. J. Autom. Mob. Robot. Intell. Syst. **9**(3), 47–52 (2015). https://doi.org/10.14313/jamris_3-2015/23
6. Korobiichuk, I., Podchashinskiy, Yu., Bezvesilna, O., Nechay, S., Shavurskiy, Yu.: Three-coordinate gravimeter with exhibition of axis sensitivity based on digital videoimages. In: Proceeding ICIGP 2019 Proceedings of the 2nd International Conference on Image and Graphics Processing, Singapore, pp. 89–93 (2019). https://doi.org/10.1145/3313950.3314187
7. Korobiichuk, I.: Mathematical model of precision sensor for an automatic weapons stabilizer system. Measurement **89**, 151–158 (2016). https://doi.org/10.1016/j.measurement.2016.04.017
8. Ermakov, S.M.: Monte Carlo Method and Related Issues. Science, Moskva, 472 p. (1975)
9. Chen, D., Zhou, N., Li, C., Guo, H., Cui, T.: A dynamic power flow model considering the uncertainty of primary frequency regulation of system. Adv. Intell. Syst. Comput. **902**, 425–438 (2020). https://doi.org/10.1007/978-3-030-12082-5_39
10. Baltagi, B.H., Fingleton, B., Pirotte, A.: A time-space dynamic panel data model with spatial moving average errors. Reg. Sci. Urban Econ. **76**, 13–31 (2019). https://doi.org/10.1016/j.regsciurbeco.2018.04.013
11. Hendricks, E., Jannerup, O., Sørensen, P.H.: Linear systems control: deterministic and stochastic methods, 555 p. (2008). https://doi.org/10.1007/978-3-540-78486-9
12. Tomashevsky, V.M.: Systems Modeling: A Tutorial, p. 352. BHV, Kyiv (2005)
13. Rohani, S., Wu, Y.: Theoretical Process Dynamic Modeling. Coulson and Richardson's Chemical Engineering: Vol. 3B: Process Control, pp. 39–94 (2017). https://doi.org/10.1016/b978-0-08-101095-2.00003-5
14. Lu, C.X., Rees, N.W., Young, P.C.: Simulation model emulation in control system design. In: System Identification, Environmental Modelling, and Control System Design, 9781430237051, pp. 583–597 (2013). https://doi.org/10.1007/978-0-85729-974-1_28
15. Shalygin, A.S., Palagin, Yu.I.: Applied methods of statistical modeling. Mechanical Engineering: Lviv, 320 p. (1986)
16. Schabenberger, O., Gotway, C.A.: Statistical methods for spatial data analysis, 488 p. (2017). https://doi.org/10.1201/9781315275086
17. Bykov, V.V.: Digital Modeling in Statistical Radio Engineering, 326 p. Soviet Radio, Moscow (1971)
18. Verdugo Muñoz, E., Da Silva Mello, L., Almeida, M.P.C.: Results of medium wave HD radio mobile reception measurements in a dense urban region. In: IET Conference Publications (2018). CP741
19. Saadoune, N., Radi, B.: Probabilistic study an embedded system. In: Colloquium in Information Science and Technology, CIST, art. no. 7804988, pp. 756–761 (2017). https://doi.org/10.1109/cist.2016.7804988
20. Korobiichuk, I.: Automatic control devices. Ed. PIAP: Warsaw, p. 104 (2018). ISBN 978-83-61278-34-4
21. Korobiichuk, I., et al.: Synthesis of optimal robust regulator for food processing facilities. In: Szewczyk, R., Zieliński, C., Kaliczyńska, M. (eds.) Automation 2017. ICA 2017. Advances in Intelligent Systems and Computing, vol. 550. Springer, Cham (2017). https://doi.org/10.1007/978-3-319-54042-9_5

22. Padoan, A., Astolfi, A.: Moments of random variables: a system-theoretic interpretation. In: Proceedings of the American Control Conference, art. no. 7963089, pp. 1035–1040 (2017). https://doi.org/10.23919/acc.2017.7963089
23. Vargas, R.E.: Fuzzy logic: theory, programming and applications, pp. 1–448 (2010)
24. Volos, C., Stouboulos, I., Kyprianidis, I., Vaidyanathan, S.: Design of a chaotic random bit generator using a Duffing – van der Pol system. In: Artificial Intelligence: Concepts, Methodologies, Tools, and Applications, vol. 2, pp. 841–860 (2016). https://doi.org/10.4018/978-1-5225-1759-7.ch034

# Theoretical Estimates of the Accuracy of Determination of Geometric Parameters of Objects on Digital Images

Igor Korobiichuk[1](✉), Yuriy Podchashinskiy[2], Oksana Luhovykh[2], Vitalii Levkivskyi[2], and Katarzyna Rzeplińska-Rykała[3]

[1] Institute of Automatic Control and Robotics, Warsaw University of Technology, Warsaw, Poland
igor.korobiichuk@pw.edu.pl

[2] Zhytomyr State Technological University, Zhytomyr, Ukraine
ju.podchashinskiy@gmail.com, ksyxon@gmail.com, levkivskij.vl@gmail.com

[3] ŁUKASIEWICZ Research Network – Industrial Research Institute for Automation and Measurements PIAP, Warsaw, Poland
krykala@piap.pl

**Abstract.** The article provides theoretical estimates of the accuracy of determining the geometric parameters of objects in digital images. The geometric parameter error estimates take into account a digital video conversion with measurement information about the geometric parameters. This transformation takes place in the process of imaging by a digital video camera, which is a part of the measurement of mechanical quantities. Therefore, the article considers the measurement errors caused by the quantization of video signal level and sampling, and their statistical characteristics. The results can be used to calculate geometrical measurement errors and measurements of object motion parameters, as well as to formulate requirements for hardware and software for processing experimental data.

**Keywords:** Digital image · Digital video conversion · Geometric parameters · Motion parameters · Measurement errors

## 1 Introduction

Logging results from a scientific experiment or production process can contain a significant amount of visual information in the form of a set of digital images. The inclusion of images in the experimental data significantly increases the informativeness and clarity of the data, and the use of appropriate computer processing of these images can automate the measurement of the parameters of the process being logged. In many experimental studies, the ultimate goal is to determine the geometric and motion parameters of objects in images, therefore, it is necessary to determine the effect of digital conversion processes on the errors of these measurements [1, 2].

In the process of measuring geometric and object motion parameters, a series of operations are performed to convert visual information about these parameters, which


R. Szewczyk et al. (Eds.): AUTOMATION 2020, AISC 1140, pp. 289–299, 2020.
https://doi.org/10.1007/978-3-030-40971-5_27

results in errors that have a significant impact on the accuracy of the measurements [3–6]. The general methods and mathematical models of these transformations are given in the literature [7–9]. But these methods and mathematical models do not fully take into account the features of images that contain measurement information about the geometric parameters of objects. Therefore, it is necessary to improve these methods and mathematical models based on the problems that are solved in the development and practical use of measuring instruments of mechanical quantities.

In computer processing of experimental data (images), it is often necessary to perform the operations of calculating the geometric dimensions of the digital image elements. The results of determining the geometric dimensions contain some errors. One of the components of these errors is the error caused by the digital image discretion and the error caused by the quantization of the video signal by its amplitude level [10]. They arise in the process of forming an image in a digital camcorder. When calculating geometric errors, it is necessary to determine the probabilities of the characteristics of all components of the errors, including those errors. The results of this calculation can be used to evaluate the accuracy of the results obtained from the processing of the experimental data and to determine the hardware and software requirements for computer-aided processing of the experimental data (images).

## 2 Video Amplitude Quantization by Level

Determine the probabilities of the error (quantization noise) characteristic arising from the digital conversion of the video signal coming from the camcorder's light-signal converter and forms a digital image of the measurement objects for which geometric and motion parameters are to be determined by their images.

The amount of useful video signal and the quantization noise at the input of an analog-to-digital converter (ADC) is some random process. It is known [11] that a random process, stationary in the broad sense, must have constant mathematical expectation and variance, and its correlation function should depend only on the time interval between the first and second arguments. In the general case, the video signal at the ADC input cannot be considered a stationary random process by changing the brightness within a single image or a sequence of images.

The volatility of sum of the video signal and the quantization noise can be illustrated by an example of a measurement object with a background. In the part of the line of the image relating to the background, the mathematical expectation of the brightness $M[Y(x)] = Y_{bg}$, in the part of the line related to the object, $M[Y(x)] = Y_{obj}$, where $Y_{bg}$ i $Y_{obj}$ are mean the background and object brightness values.

You can turn a non-stationary random process into a stationary one by subtracting its mathematical expectation from the current values of a random process and moving to a centered random process. In this case, mathematical expectation is a useful signal of brightness. Subtracting the useful brightness signal results in only the quantization noise present in the digital video signal at the ADC output. Based on the data [12, 13], we can assume that this noise is a stationary normal random process.

In the following, we assume that the input video signal of the ADC $u(t)$ is obtained by subtracting from the total video signal of the useful brightness signal (mathematical expectation), that is, by bringing a random process at the input of the ADC to a stationary random process.

In [14] the quantization noise distribution law is found using the characteristic function of a random variable Δ, which is equal to the error of the analog-to-digital conversion for the current value of the input value: $\Delta = u_q(t) - u(t)$.

The sequence of readings of such an error forms the quantization noise.

A characteristic error function of analog-to-digital conversion [14]:

$$P_\Delta(\omega) = \int_\Delta p(\Delta) \cdot e^{j\omega\Delta} d\Delta = \sum_k P_u\left(\frac{2\pi k}{\delta_{\kappa\varepsilon}}\right) \cdot \frac{\sin[0{,}5\delta_{\kappa\varepsilon} \cdot (\omega + 2\pi k / \delta_{\kappa\varepsilon})]}{0{,}5\delta_{\kappa\varepsilon} \cdot (\omega + 2\pi k / \delta_{\kappa\varepsilon})}, \tag{1}$$

where $P_u(.)$ – a characteristic function of the ADC input signal at an argument value that is $2\pi k/\delta q$; $k$ – quantization interval number, $k = 0, \ldots, 2n - 1$; $n$ – number of ADC binary digits; $\delta_q$ – quantization step in ADC.

The characteristic function of a random variable, evenly distributed over the interval $\pm 0,\ 5\delta_q$, is equal to:

$$P(\omega) = \frac{\sin(0{,}5 \cdot \delta_{kb} \cdot \omega)}{0{,}5 \cdot \delta_{kb} \cdot \omega}. \tag{2}$$

Comparing formulas (1) and (2), we can conclude that the error (quantization noise) is evenly distributed if [14]:

$$P_u\left(\frac{2\pi k}{\delta_{\kappa\varepsilon}}\right) = \begin{cases} 1, & k = 0, \\ 0, & k \neq 0. \end{cases}$$

This condition is fulfilled if the mean square value of the input signal $\sigma_u^2 \gg \delta_q^2$.

In the case of video quantization, we have a normal distribution of the ADC input signal. Its characteristic function is:

$$P_u(\omega) = \exp\left\{-\frac{\sigma_u^2 \cdot \omega^2}{2}\right\}.$$

Determine $P_u(2\pi k/\delta q)$ for different values of the signal/noise ratio at the input of the ADC, using data [2, 15]. Signal-to-noise ratio at the ADC input $\Psi_{s/n}$ = 40–55 dB, which are typical values for a video signal. The number of ADC bits is $n = 8$, which is the typical value for digital images

$$P_u\left(\frac{2\pi k}{\delta_{кв}}\right)\Bigg|_{k=0} = 1;$$

$$P_u\left(\frac{2\pi k}{\delta_{кв}}\right)\Bigg|_{k=1,\psi_{c/ш}=40dB} \approx 0;$$

$$P_u\left(\frac{2\pi k}{\delta_{кв}}\right)\Bigg|_{k=1,\psi_{c/ш}=50dB} \approx 0.01;$$

$$P_u\left(\frac{2\pi k}{\delta_{кв}}\right)\Bigg|_{k=1,\psi_{c/ш}=55dB} \approx 0.14.$$

Thus, the error distribution (quantization noise) approaches uniform, the same

$$P_u\left(\frac{2\pi k}{\delta_{kb}}\right) \approx 0 \,\text{as}\, k \neq 0.$$

The sequence of errors $\Delta_q$ forms a quantization noise with a distribution approaching uniform and numerical characteristics:

- maximum value $\Delta_{q\,max} = \delta_q/2$;
- middle value $M[\Delta_q] = 0$;
- dispersion $\sigma^2[\Delta_q] = \delta_q^2/12$.

Determine the correlation function of the error counting (quantization noise).

In order for the discrete quantization noise values to be uncorrelated, it is necessary that the quantization noise correlation interval is less than the sampling interval in the ADC: $\tau_{oq} < \delta_{si}$.

We assume that the energy spectrum $W_u(\omega)$ of the noise present in the analog video signal is approximately uniform in the frequency range (0–5) MHz.

This assumption is possible, since this noise is generated by the passage of thermal and fractional noise through electronic bands with a limited frequency band [12, 15, 16].

If such an input signal inputs to the ADC input, then the error correlation interval is equal [11]:

$$\tau_{oq} = \frac{W_q(0)}{4K_q(0)} = \frac{3\delta_{kb}}{\pi^2 \cdot \Delta_\omega \cdot \sigma_u} \cdot \sqrt{\frac{3}{2\pi}} \cdot \sum_{n=1}^{\infty} \frac{1}{n^3}, \tag{3}$$

or

$$\tau_{oq} \approx \frac{1}{12} \cdot \frac{\delta_{kb}}{\sigma_u} \cdot \frac{\pi}{\Delta_\omega}, \tag{4}$$

where $W_q(0) = W_q(\omega)|_{\omega=0}$ – power spectral density (energy spectrum) of error; $K_q(0) = K_q(\tau)|_{\tau=0}$ – correlation error function; $\Delta_\omega = 2\pi f_{in}$ – the bandwidth of the input signal.

In this case, for $f_{in}$ = 5 MHz and $\Psi_{s/n}$ = 40–55 dB we have: $\tau_{oq} = (1.1-6.1)\cdot 10^{-9}$s.

Sampling interval is equal to $\delta_s = (7.22-6.50)\cdot 10^{-8}$s [15]. Thus $\tau_{oq} < \delta_s$ and discrete error counts (quantization noise) are uncorrelated. We define the mutual correlation function of the error and the input useful signal.

For an input signal with a normal distribution law, the mutual correlation function of error and signal [11, 12]:

$$K_{qu}(\tau) \approx 2K_u(\tau)\cdot\sum_{n=1}^{\infty}(-1)^n\cdot\exp\left\{-\frac{2\pi^2 n^2}{\delta_{kb}^2}\cdot\sigma_u^2\right\}, \tag{5}$$

where $K_u(\tau)$ – correlation function of the input signal.

For the signal-to-noise ratio at the ADC input 40–45 dB $\delta_q < \sigma_u$ and according to [11], the error and the input signal can be considered uncorrelated, since $K_{qu}(_\tau) \approx 10^{-8}$.

For the signal-to-noise ratio at the ADC input 50–55 dB $0,5\delta_q \le \sigma_u < \delta_q$ and you can use the formula [9]:

$$K_{qu}(\tau) \approx 2K_u(\tau)\cdot\exp\left\{-\frac{2\pi\sigma_u^2}{\delta_{kb}^2}\right\}. \tag{6}$$

As a result, we get:

- at $\Psi_{s/n}$ = 50 dB $K_{qu}(\tau) = K_u(\tau)\cdot 10^{-5}$;
- at $\Psi_{s/n}$ = 55 dB $K_{qu}(\tau) = K_u(\tau)\cdot 0{,}02$.

Thus, the error counts when converting the video signal to digital form (quantization noise) and the input useful signal of the image brightness are almost uncorrelated.

## 3 Errors of Determining the Coordinates of the Contours of Objects in a Digital Image and Distortion of the Video Signal

The errors in measuring geometry and object motion parameters in digital images are resulted from many of the factors. When calculating the total error of geometric measurements the effect of the errors that occur during such image actions should be taken into account:

- digital image conversion;
- calculation of the geometric parameters of objects based on the coordinates of the contour points.

The initial data for the calculations are:

- specifications of imaging tools (e.g. digital camcorder);
- calculation algorithm and data bit when determining geometric parameters.

When determining the geometric parameters of objects based on the threshold segmentation of the image, the coordinate point of the object contour is considered to be the horizontal coordinate value $x_c$, for which the brightness is equal to the threshold value ($Y(x_c) = Y_n$). However, in a digital image, a coordinate can only take discrete values with a sampling step $\delta_s$, equal to the distance between adjacent points. As a result, an error $\Delta_d$, occurs due to the discrete nature of the digital image.

This error can be represented by some random variable, which is characterized by a certain law of distribution, including numerical characteristics (mathematical expectation, variance, range of possible values). These characteristics are used in the calculation of the total error in determining the coordinates of points on a digital image [10].

An error due to the discrete nature of a digital image can be regarded as a magnitude similar to that of a digital video signal (quantization noise). For a discrepancy error, the input value is a continuous (exact) coordinate value of the point $x_c$, for which $Y(x_c) = Y_n$; discrete coordinate values are the coordinates of digital pixels. Therefore, all formulas and ratios given above for error (quantization noise) can be extended to the error due to the discrete nature of the digital image. It is necessary to replace the input value (brightness $Y(x)$) with the horizontal coordinate $x$, and the quantization step $\delta_q$ – with the sampling step $\delta_s$.

Based on this approach, we determine the probabilities of the sampling error characteristic for the images (Fig. 1) containing the object of measurement surrounded by the background of the image.

We will assume that some point belonging to the contour of the object is arbitrarily selected. The coordinate of this point is a random variable with a certain law of distribution over a certain range of values. The distribution law and the range of possible coordinate values are determined for a particular image or image type [3–6, 10, 17]. In this case, a possible range of values of the horizontal coordinate of the contour points ($x_{obj1}$, $x_{obj2}$).

Discrete contour points are formed when crossing the contour with horizontal lines corresponding to the image lines. Since the vertical coordinate of the contour points varies from 0 to $V$–1 and it takes each discrete value once, and the number of these values is quite large ($V$ = 300–1200 discrete points), we can assume that the vertical coordinate of the contour points has a uniform distribution in the interval (0, $V$–1).

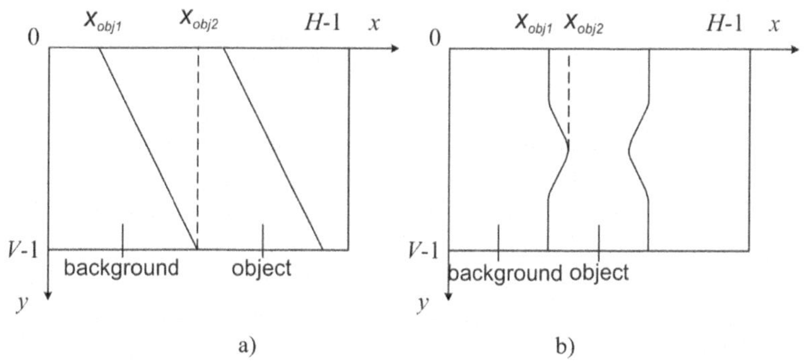


**Fig. 1.** Images used to measure geometric parameters: (a) test image; (b) the image of the sample material in the process of plastic deformation.

The contour excluding errors for the image in Fig. 1, *a* is a straight line at an angle to the vertical:

$$x = a\left(y - x_{obj1}\right).$$

Thus, the horizontal coordinate $x$ of some random contour point is a linear function of the vertical coordinate $y$. As a result, we obtain a uniform distribution of the horizontal coordinate of a random contour point in the range $(x_{obj1}, x_{obj2})$. Subtracting from the horizontal coordinate values the average $E(x) = \frac{x_{obj1} + x_{obj2}}{2}$, we get the mean square horizontal coordinate value:

The error due to the discreteness of the image has a uniform distribution in the range $\pm\delta_c/2$ for any law of distribution of the input value $x$, if $\sigma^2_{x\,obj} \gg \delta^2_s$ [9, 11, 14]. Then the average of the sampling error $\overline{\Delta_s} = 0$, and the variance $\overline{\Delta^2_s} = \delta^2_s/12$.

Based on this, you can formulate the following requirements for the digital image of measurement objects: $\sigma^2_{x\,obj} = \gamma \cdot \delta^2_s$, where $\gamma = 10, \delta_s = \pm 1$ discrete point (*dp*). As a result, we have for the image in Fig. 1(a):

$$\sigma_{x\text{об}} \geq \sqrt{10} \approx 4\ dp.;$$

$$x_{obj2} - x_{obj1} \geq 2\sqrt{3} \cdot \sigma_{xobj} = 2\sqrt{3} \approx dp \tag{7}$$

To depict the mechanical deformation process in Fig. 1(b) you can use the contour information of the measurement object given in [17]. For example, for images obtained from deformation studies of steel specimens, $x_{obj2} - x_{obj1} = 1.7$ mm, $\sigma_{\text{x obj}} = 0.57$ mm. As a result, we have the following requirements for a digital image:

$$\begin{aligned} \sigma obj &\geq \sqrt{10} \approx 4\,\text{dp}; \\ x_{obj2} - x_{obj1} &\geq 10\,\text{dp}. \end{aligned} \tag{8}$$

The sequence of errors caused by the discreteness of the digital image creates some random function for the contour points. The probabilities of the characteristics defined for this error as a random variable do not depend on the vertical coordinate $y$. Therefore, a random function formed by a sequence of these errors will have the same likely characteristics. In addition, this random function has some correlation function.

If we consider the points of the contour in the sequence of their location on the image and the corresponding sequence of errors in determining the coordinates, then we get a correlation function that characterizes the determined sequence of coordinates of this contour. In this case, this deterministic sequence is not significant, since we are interested in the error of measurement of a specific coordinate, and the probabilities of the error characteristics (mathematical expectation and variance) are obtained by summing a large number of errors of individual measurements, regardless of their order of arrival and the relative location of the points at which measurement is performed. As a result, it can be assumed that equivalent additive noise, which is replaced by a discretion error, is a random process of "white" noise type.

This defines the requirements for video images used to measure geometric and object motion parameters. These requirements ensure a uniform distribution of the discretion error across the range $\pm\delta_d/2$ and reduce the overall error of determining the coordinates of the contour points. It is found that the discreteness error can be replaced by an equivalent additive noise that is added to the brightness signal. This noise is a random process of "white" noise.

Consider the component of geometric error associated with limiting the video bandwidth in electronic circuits. The difference in brightness on the video that corresponds to the outline of the object, in its initial state, is approximately a jump from the brightness of the background $Y_{bg}$ of the brightness of the object $Y_{obj}$. When passing a video signal through electronic circuits with a limited frequency band, which form and amplify the video signal, there is a distortion of the shape of the difference in brightness. As a result, the brightness increases from $Y_{bg}$ to $Y_{obj}$ over a period of time.

When determining the coordinates of the contour points, we use a comparison of the brightness of the points in the image line with a brightness threshold $Y_n$, which is equal to the arithmetic mean brightness of the object and background. Therefore, to calculate the errors, it is necessary to determine the delay time, i.e. the time during which the brightness increases from $Y_{bg}$ to $Y_n$ (Fig. 2).

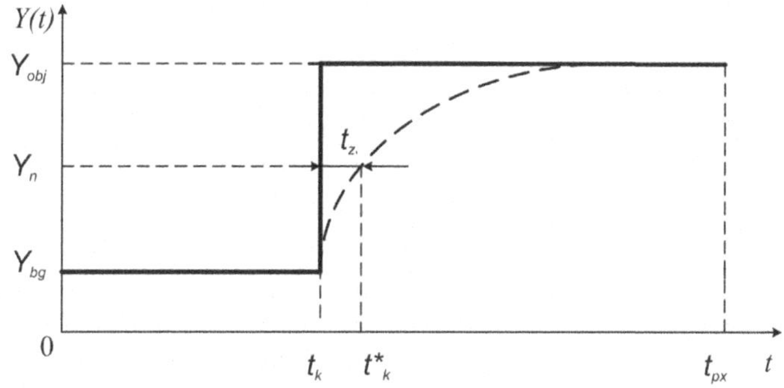


**Fig. 2.** Distortion of the video signal shape while limiting its frequency band.

For wideband and pulse amplifiers of video signal [18] the duration of the front (rise time) of the pulse at the output when applying to the input of a rectangular pulse is equal to: $t_{fr}$ = (0.35–0.5)/$f_{in}$. The amplifier in the upper frequency region can be replaced by an equivalent circuit in the form of an integrating circuit and a rise time $t_{fr} = 2.2T_i$, where $T_i$ – is the constant of the integrating circuit time.

From here $T_i$ = (0.159–0.227)/$f_{in}$.

Transient characteristic of the integrating circuit [13, 19] $h(t) = 1{-}\exp(-t/T_i)$. In this case $h(t_z) = 0{,}5$.

From here you can define the delay time:

$$t_z = \frac{-\ln 0.5 \cdot (0.159\ldots0.227)}{f_b} \approx \frac{0.134}{f_b}. \tag{9}$$

The duration of the visible part of a line of a television raster $t_{px}$ (the time of a straight line scan) is about 52 μs [20–22], and the maximum size of a digital image horizontally is $H$ = 720–800 discrete points [2].

The result is an expression for calculating the error associated with the limitation of the bandwidth of the brightness signal:

$$\Delta_{FH} = \frac{t_z \cdot H}{t_{px}} \approx \frac{0.134 \cdot H}{f_в \cdot t_{px}}$$

The error $\Delta_{FH}$ is a part of the systematic part of the error of determining the horizontal coordinate of a contour point in a video obtained from a camcorder.

In the worst case $f_{in}$ = (3–5) MHz. If $f_{in}$ = 3 MHz and $H$ = 800 discrete points, than $\Delta_{FH}$ = 0.69 discrete points. If $f_{in}$ = 3 MHz and $H$ = 640 discrete points, than $\Delta_{FH}$ = 0.55 discrete points.

## 4 Conclusions

1. The process of digitizing a video signal containing the measurement information about the geometric parameters of the objects is investigated. This video also contains noise generated during the imaging process and is a stationary normal random process. The signal to noise ratio at the ADC input is 40–55 dB, which are typical values for a signal from a digital camcorder, and the number of ADC bits is 8, which is the typical value for digital images.
2. An error (random quantization noise) occurs when converting a video signal to digital form. It has the following statistical characteristics: the distribution of values approaches uniform; the range of values is $\pm\delta_q/2$, the mathematical expectation is 0, the variance is $\delta_q^2/12$.
3. The error correlation interval is determined by formulas (3) and (4). In this case, the error correlation interval is smaller than the video sampling period in the ADC. The reciprocal correlation function of the error readings and the useful luminance signal is determined by formulas (5) and (6). The error readings and the useful brightness signal are almost uncorrelated.
4. The errors caused by the digital image discretion and arising in determining the coordinates of the points of the contour of the measurement objects are investigated. It is determined that these errors are evenly distributed in the range $\pm\delta_q/2$, if the horizontal coordinates of the contour points for the image satisfy conditions (7) and (8).
5. The results obtained can be used to calculate geometric errors and measurements of object motion on digital images, as well as to determine the requirements for hardware and software for processing experimental data.

## References

1. Kochan, R., Kochan, O., Sachenko, A., Kochan, V.: Increasing metrological reliability of measuring channels for distributed automated control systems. In: Proceedings of 2019 10th International Conference on Dependable Systems, Services and Technologies, DESSERT 2019, pp. 34–40 (2019). https://doi.org/10.1109/DESSERT.2019.8770052. Article no. 8770052
2. Shams-Nateri, A., Hasanlou, E.: Computer vision techniques for measuring and demonstrating color of textile. Appl. Comput. Vis. Fashion Textile, 189–220 (2018). https://doi.org/10.1016/b978-0-08-101217-8.00008-7
3. Korobiichuk, I., Podchashinskiy, Yu., Shapovalova, O., Shadura, V., Nowicki, M., Szewczyk, R.: Precision increase in automated digital image measurement systems of geometric values. In: Advances in Intelligent Systems and Computing Book Series (AISC), vol. 393, pp. 335–340. Springer (2016). https://doi.org/10.1007/978-3-319-23923-1_51
4. Korobiichuk, I., Podchashinskiy, Yu., Elnikova, T., Jus, A.: Geometrical parameter measurement and phytoplankton process modeling based on video images of water samples from reservoirs. Meas. J. Int. Meas. Confed. **114**, 226–232 (2018). https://doi.org/10.1016/j.measurement.2017.09.048

5. Korobiichuk, I., Podchashinskiy, Yu., Lugovyh, O., Nowicki, M., Kachniarz, M.: Algorithmic compensation of video image dynamic errors with measurement data about geometric and object motion parameters. Measurement **105**, 66–71 (2017). https://doi.org/10.1016/j.measurement.2017.04.009
6. Korobiichuk, I., Podchashinskiy, Yu., Bezvesilna, O., Nechay, S., Shavurskiy Yu.: Three-coordinate gravimeter with exhibition of axis sensitivity based on digital videoimages. In: ICIGP 2019: Proceedings of 2nd International Conference on Image and Graphics Processing, pp. 89–93 (2019). https://doi.org/10.1145/3313950.3314187
7. Measurements in LabVIEW. Application Guide. Training Center "Technology Center National Instruments". NSTU, Novosibirsk (2006). 360 p.
8. Parente, F.R., Santonico, M., Zompanti, A., Benassai, M., Ferri, G., D'Amico, A., Pennazza, G.: An electronic system for the contactless reading of ECG signals. Sensors **17**(11) (2017). https://doi.org/10.3390/s17112474. Article no. 2474
9. Korobiichuk, I., Lysenko, V., Opryshko, O., Komarchyk, D., Pasichnyk, N., Juś, A.: Crop monitoring for nitrogen nutrition level by digital camera. In: Advances in Intelligent Systems and Computing, vol. 743, pp. 595–603 (2018). https://doi.org/10.1007/978-3-319-77179-3_56
10. Podchashinsky, Yu.O., Lugovikh, O.O., Shavursky, Yu.O.: Measurement of motion of objects based on computerized video processing. ZHDTU, Zhytomyr (2018). 192 p.
11. Levin, B.R.: Theoretical Foundations of Statistical Radio Engineering, 3rd edn., Revised and add. Radio and Communications, Moscow (1989). 656 p.
12. Barclay, L.: Fading and statistics. In: Propagation of Radiowaves, 3rd edn., pp. 45–59 (2012). https://doi.org/10.1049/pbew056e_ch4
13. Frenzel Jr., L.E.: Electronics Explained: Fundamentals for Engineers, Technicians, and Makers, 2nd edn., pp. 1–378. Newnes, San Diego (2017)
14. Bryukhanov, Yu.A.: A method of analysis of periodic processes in nonautonomous discrete-time systems with quantization. J. Commun. Technol. Electron. **53**(7), 807–813 (2008). https://doi.org/10.1134/S1064226908070115
15. Lebedev, A.I.: Physics of Semiconductor Devices. Fizmatlit, Moscow (2008). 488 p.
16. Buckingham, M.: Noises in Electronic Devices and Systems. Mir, Moscow (1986). 399 p.
17. Grabar, I., Kolodnitska, R., Podchashinsky, Yu.: Hardware-software complex for research of kinetics of elastic - plastic deformations and destructions of rigid bodies. In: Proceedings of the International Scientific Conference "Mechanics 2000", pp. 103–108 (2000)
18. Keränen, P., Jansson, J.-P., Mäntyniemi, A., Kostamovaara, J.: Design principles for accurate, long-range interpolating time-to-digital converters. In: CMOS Time-Mode Circuits and Systems: Fundamentals and Applications, pp. 141–176 (2015). https://doi.org/10.1201/b19228
19. Sarwin, A.A.: Systems of non-contact measurements of geometric parameters. University Press of Leningrad, Leningrad (1983). 144 p.
20. Vinogradov, N.A., Yakovlev, V.N., Voskresensky, V.V., et al.: Handbook of Digital Information Processing Devices. Tekhnika, Kyiv (1988). 415 p.
21. Suh, Y.-W., Kim, S.-H., Kim, M.-S., Choi, J.-Y., Seo, J.-S.: A novel integrated measurement and analysis system for digital broadcasting. IEEE Trans. Consum. Electron. **55**(1), 56–62 (2009). https://doi.org/10.1109/TCE.2009.4814414
22. Shishigin, I.V., Shulman, M.G., Kolesnichenko, O.V., Zolotarev, S.A.: How to Choose a Video Camera. Lan, St. Petersburg (1996). 512 p.

# Magnetic Permeability Tensor with Saturation Flux Density Description for 2D Materials with Uniaxial Anisotropy

Roman Szewczyk(✉)

Łukasiewicz Research Network – Industrial Research Institute for Automation and Measurements PIAP, Al. Jerozolimskie 202, 02-486 Warsaw, Poland
rszewczyk@onet.pl

**Abstract.** During the development of advanced models of mechatronics devices with cores made of anisotropic soft magnetic materials, magnetic permeability has to be described as tensor. However commonly used tensor magnetic permeability descriptions doesn't consider saturation phenomenon for flux density. As a result accuracy of such simulations is significantly limited, especially for higher values of magnetizing field. Paper is proposing new method of magnetic permeability description considering saturation flux density. Proposed concept is based on the anisotropy energy analyze and was verified for magnetization curve of soft magnetic materials with uniaxial anisotropy. Results of modeling confirm good agreement with the results experimental measurements for soft magnetic materials with uniaxial anisotropy.

**Keywords:** Magnetic anisotropy · Soft magnetic materials · Tensor description of magnetic permeability

## 1 Introduction

Relative magnetic permeability $\mu$ is the most important property of soft magnetic materials determining their functional properties. Relative magnetic permeability is commonly described as scalar value. However, most of magnetic materials utilized in technical applications, such as amorphous alloys [1] or electrical steels [2], exhibit strongly anisotropic properties. This means, that relative magnetic permeability of such materials have to be described by tensor, not scalar. Moreover, during the magnetization process of anisotropic magnetic materials, the direction of flux density $B$ is not always parallel to the magnetizing field $H$ [3]. This phenomena is not intuitive, however, it is crucial for principled of operation of different mechatronic devices and sensors, such as pressductors [4] or tensductors [5].

Tensor description of relative magnetic permeability $\mu$ is commonly implemented in software for finite elements modelling of magnetic systems, such as commercial COMSOL [6] or open-source ELMER FEM [7]. On the other hand, the significant property of all magnetic materials is nonlinear characteristic of magnetization process. The most important phenomenon connected with nonlinearity is saturation flux density $B_s$. Different models considering saturation flux density $B_s$ in isotropic materials were


R. Szewczyk et al. (Eds.): AUTOMATION 2020, AISC 1140, pp. 300–308, 2020.
https://doi.org/10.1007/978-3-030-40971-5_28

presented previously [8–10]. However, the model incorporating the saturation flux density $B_s$, suitable for anisotropic materials with tensor description of relative magnetic permeability $\mu$ seems to be not presented in the literature. The lack of such model is significant barrier in modelling the inductive components based on anisotropic magnetic materials in the magnetic field in nearly-saturation region.

Paper is filling this gap in the state of the art. Proposed model utilizes the physical principles determining the magnetization vector rotation as well as the description of magnetic anisotropy based on energy analyses. As a result proposed model has connection with physical properties of magnetic materials as well as enables efficient modelling of nonlinear magnetization characteristics of anisotropic materials in nearly-saturation region.

## 2 Proposed Concept of 2D Relative Magnetic Permeability Tensor Description Considering Saturation Flux Density

Planar coordinates visualization of relative magnetic permeability tensor for materials with uniaxial anisotropy presented in Fig. 1. The major axis of the ellipse $\mu_x$ is determined by the permeability in the magnetization easy axis X, whereas permeability in hard axis Y is equal $\mu_y$.

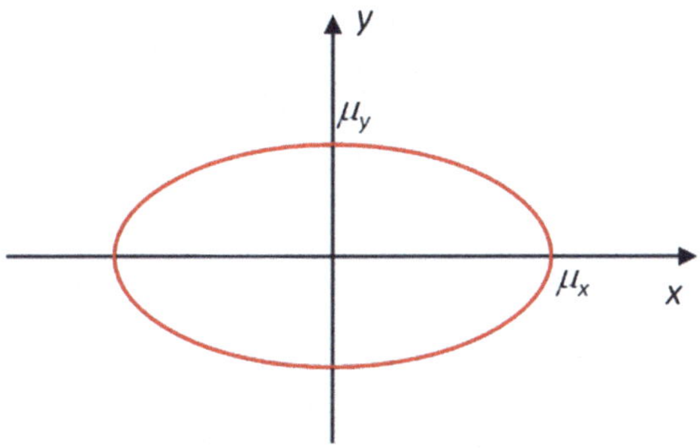


**Fig. 1.** Visualization of relative magnetic permeability $\mu$ tensor in the case of uniaxial anisotropy.

In the case of two-dimensional magnetization process, relative magnetic permeability tensor describing uniaxial anisotropy is given by the following matrix:

$$\boldsymbol{\mu} = \begin{bmatrix} \mu_x & 0 \\ 0 & \mu_y \end{bmatrix} \tag{1}$$

However, such tensor is suitable only for linear systems, whereas all ferromagnetic materials exhibit saturation phenomenon [3] connected with saturation flux density $B_s$. Figure 2 presents the simplified magnetization curves of material magnetized in the direction of magnetization easy axis (red line) and in the direction of magnetization hard axis (blue line). Relative magnetic permeability in both these direction is resented as $\mu_x$ and $\mu_y$ respectively. It should be highlighted, that the value of saturation flux density $B_s$ is intrinsic property of magnetic material and is not dependent on magnetization direction [11].

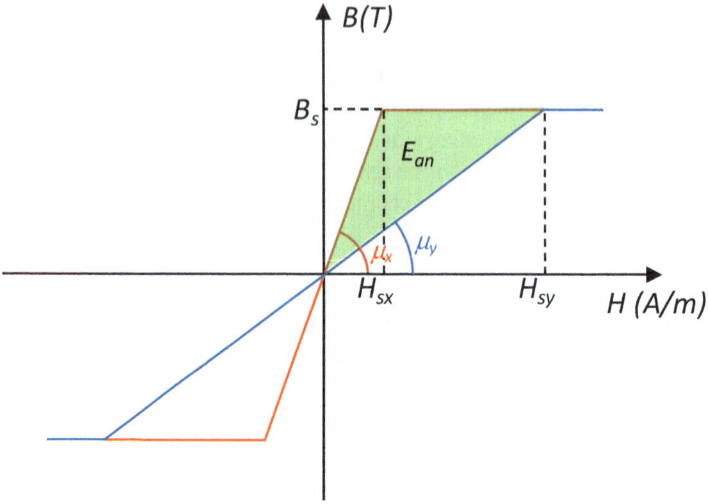


**Fig. 2.** Simplified magnetization curves of material magnetized in the direction of magnetization easy axis (red line) and in the direction of magnetization hard axis (blue line). Saturation flux density $B_s$ is the same for all directions of magnetization.

In magnetization curve model presented in Fig. 2, saturation flux density $B_s$ is given by the following equations

$$B_s = \mu_0 \mu_x H_{sx} \tag{3}$$

$$B_s = \mu_0 \mu_y H_{sy} \tag{4}$$

where $\mu_0$ is magnetic constant (equal to the permeability of the vacuum). For the magnetic materials with uniaxial anisotropy, the average anisotropy energy density $K_{an}$ is defined as the energy required for magnetizing material from easy to hard axis [12]. As a result average anisotropy energy density $K_{an}$ is represented in the Fig. 1 as green field and can be calculated from simple geometry dependences:

$$E_x = \frac{B_s H_{sx}}{2} = \frac{B_s^2}{2\mu_0 \mu_x} \tag{5}$$

$$E_y = \frac{B_s H_{sy}}{2} = \frac{B_s^2}{2\mu_0 \mu_y} \tag{6}$$

$$E_{an} = E_y - E_x = \frac{B_s^2}{2\mu_0}\left(\frac{1}{\mu_y} - \frac{1}{\mu_x}\right)\left[\frac{J}{m^3}\right] \tag{7}$$

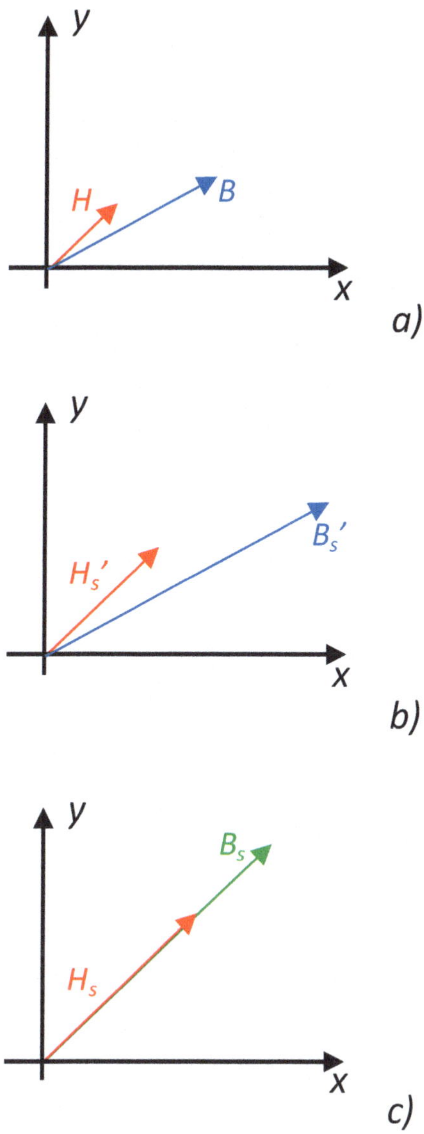


**Fig. 3.** Three phases of magnetization process for material with uniaxial anisotropy: (a) linear magnetization region, (b) material is saturated, however, flux density $\vec{B}$ vector is not parallel to magnetizing field $\vec{H}$, (c) fully saturated material.

Different stages of magnetization process [3] are presented in Fig. 3. For initial values of magnetizing field $\vec{H}$ (as it is presented in Fig. 3a), the value of flux density $\vec{B}$ can be calculated form the following equation:

$$\overrightarrow{B} = \mu_0 \begin{bmatrix} \mu_x & 0 \\ 0 & \mu_y \end{bmatrix} \overrightarrow{H} \tag{8}$$

It should be highlighted, that for anisotropic materials, flux density $\overrightarrow{B}$ can be nonparallel to the vector of magnetizing field $\overrightarrow{H}$. This phenomenon is the principle of operation of different mechatronic sensor dependent of stress-induced uniaxial anisotropy, such as tensductors [5].

Linear magnetization process can be observed until the length of flux density $\overrightarrow{B}$ reach value of saturation flux density $B_s$. For higher values of magnetizing field $\overrightarrow{H}$, process of magnetization is connected with rotation of flux density $\overrightarrow{B}$ vector towards the direction of magnetization $\overrightarrow{H}$, as it is presented in Fig. 3b.

In such a case, not fully rotated, but saturated flux density vector $\overrightarrow{B_s'}$ can be calculated as:

$$\overrightarrow{B_s'} = \mu_0 \begin{bmatrix} \mu_x & 0 \\ 0 & \mu_y \end{bmatrix} \overrightarrow{H_s} \tag{9}$$

where

$$\overrightarrow{H_s} = \overrightarrow{H} \frac{\left|\overrightarrow{B}\right|}{B_s} \tag{10}$$

To determine the rotation of flux density $\overrightarrow{B}$ in transitional region presented in Fig. 3b, the total free energy-based analyse [3] should be carried out accordingly to Fig. 4.

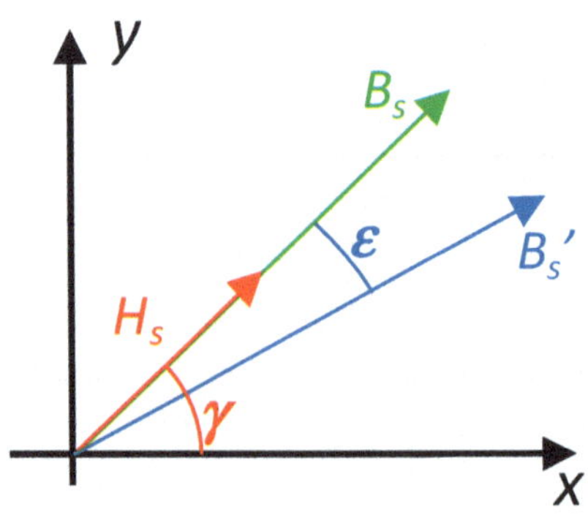


**Fig. 4.** Partially saturated material, where magnetizing field $\overrightarrow{H}$ is provided with the angle $\gamma$ to the magnetization easy axis.

Energy $E_{rot}$ passed for the rotation of flux density $\overrightarrow{B}$ vector is equal to the difference between total magnetic field energy connected with magnetizing field $\overrightarrow{H}$ and magnetic field energy required to reach the saturation flux density $B_s$:

$$E_{rot} = \frac{B_s\left(\left|\overrightarrow{H}\right| - \left|\overrightarrow{H_s}\right|\right)}{2} \tag{11}$$

Total energy $E_{an}(\varepsilon)$ required to rotate flux density $\overrightarrow{B}$ vector by the angle $\varepsilon$, to the direction of magnetizing field $\overrightarrow{H}$ is equal:

$$E_{an}(\varepsilon) = \frac{\varepsilon}{\pi/2} E_{an} \tag{12}$$

where anisotropy energy $E_{an}$ is stated by the Eq. 7. As a result, for $E_{rot} < E_{an}(\varepsilon)$, rotation angle $\varphi_{rot}$ for vector $\overrightarrow{B'_s}$ is equal:

$$\varphi_{rot} = \varepsilon \frac{E_{rot}}{E_{an}(\varepsilon)} \tag{13}$$

whereas, for $E_{rot} \geq E_{an}(\gamma)$, magnetic material can be considered as fully saturated. In such a case:

$$\varphi_{rot} = \varepsilon \tag{14}$$

and flux density $\overrightarrow{B}$ is equal $\overrightarrow{B_s}$ parallel to the magnetizing field direction $\overrightarrow{H}$ (as it is presented in Fig. 3c). Moreover, due to saturation, further increase of the length of $\overrightarrow{H}$ doesn't change the length or direction of flux density vector $\overrightarrow{B}$.

From practical point of view, then rotation of saturated vector $\overrightarrow{B'_s}$ can be carried out with the use of rotation matrix $\boldsymbol{R}$:

$$\overrightarrow{B} = R \times \overrightarrow{B'_s} = \begin{bmatrix} \sin(\varphi_{rot}) & cos(\varphi_{rot}) \\ cos(\varphi_{rot}) & -\sin(\varphi_{rot}) \end{bmatrix} \times \overrightarrow{B'_s} \tag{15}$$

In finite elements modelling system, for the material in partial saturation state, relative magnetic permeability tensor $\boldsymbol{\mu}_s$ can be calculated from the following equation:

$$\boldsymbol{\mu}_s = \begin{bmatrix} \mu_{sx} & 0 \\ 0 & \mu_{sy} \end{bmatrix} \tag{16}$$

where $B_{sx}$ and $B_{sy}$ are components of $\overrightarrow{B}$ vector after rotation by the angle $\varphi_{rot}$:

$$\mu_{sx} = \frac{B_{sx}}{\mu_0 H_x} \tag{17}$$

$$\mu_{sy} = \frac{B_{sy}}{\mu_0 H_y} \tag{18}$$

Such form of relative permeability matrix enables efficient modelling of 2D magnetic systems magnetized in saturation region.

## 3 Verification of Proposed Method of Modelling the Magnetic Permeability Tensor for Materials with Uniaxial Anisotropy

The results of modelling of $B(H)$ magnetization curves for different angle of magnetization field are presented in Fig. 5. Value of flux density $B$ on the curve is calculated as the projection of the vector $\overrightarrow{B}$ on the direction of magnetizing field of $\overrightarrow{H}$ what represents the situation when magnetizing coil is coaxial with the sensing coil on the magnetic sample.

In presented results both linear magnetization and rotation magnetization areas are clearly visible. However, for the magnetization in the direction of major and minor axes

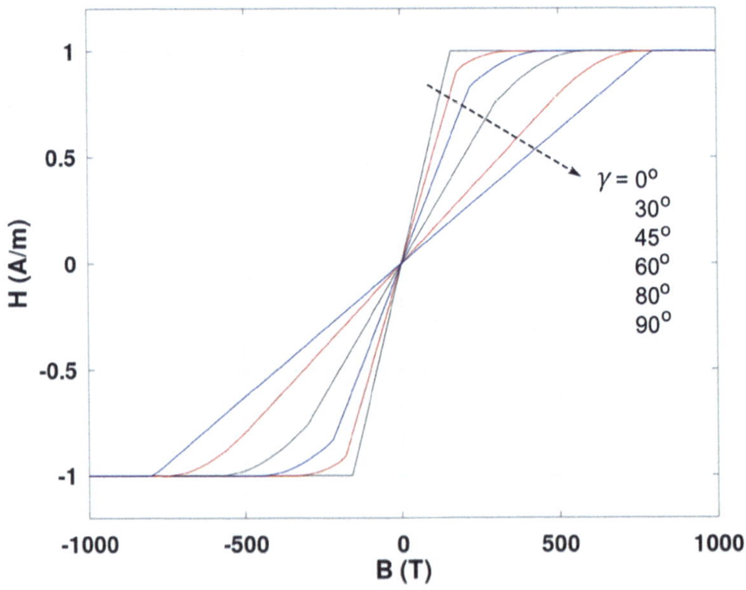


**Fig. 5.** The results of modelling $B(H)$ magnetization curves determined for different angles $\gamma$ between the magnetization easy axis and the direction of magnetizing field $\overrightarrow{H}$.

of relative magnetic permeability tensor m, rotation of the saturated flux density vector $\vec{B}$ doesn't occur. This observation is in good agreement with the results of experimental measurements presented previously for amorphous magnetic alloys with uniaxial anisotropy induced during the thermomagnetic annealing [13].

## 4 Conclusions

The method of description of 2D relative permeability tensor with uniaxial anisotropy presented in the paper utilized the physical principles of magnetization process. In proposed solution material is the subject of linear magnetization process first. When it reaches the saturation, further magnetization process is connected with the rotation of flux density vector. Proposed quantitative model is based on the magnetic energy analyse, what connects it with the physical principles of the magnetization process.

Magnetic magnetization curves determined for different angles between the magnetization easy axis and the direction of magnetizing field $\vec{H}$ are in good agreement with experimental results achieved for measurements of magnetization curve of amorphous magnetic alloys subjected to thermomagnetic treatment inducing uniaxial anisotropy in different directions. As a result, proposed concept may be used for finite elements based modelling of inductive components with cores made of anisotropic magnetic materials, such as magnetic field sensors. Moreover, due to the fact, that uniaxial anisotropy is often stress-induced, proposed method may be utilized in modelling the characteristics of magnetoelastic stress and force sensors operating in the nearly-saturation region.

## References


1. Gibbs, M.R.J.: Anisotropy and magnetostriction in amorphous alloys. J. Magn. Magn. Mater. **83**, 329–333 (1990). https://doi.org/10.1016/0304-8853(90)90536-Y
2. Chwastek, K., Baghel, A.P.S., Wodzynski, A., Kulkarni, S.: Anisotropic properties of electrical steel. In: Computational Problems of Electrical Engineering CPEE 2015, Lviv, Ukraine (2015). https://doi.org/10.1109/CPEE.2015.7333328
3. Jiles, D.C.: Introduction to Magnetism and Magnetic Materials. CRC Press, Boca Raton (2015)
4. Dahle, O.: The pressductor and the torductor — two heavy-duty transducers based on magnetic stress sensitivity. IEEE Trans. Commun. Electron. **83**, 752–758 (1964). https://doi.org/10.1109/TCOME.1964.6592601
5. Nowicki, M.: Tensductor—amorphous alloy based magnetoelastic tensile force sensor. Sensors **18**, 4420 (2018). https://doi.org/10.3390/s18124420
6. https://www.comsol.com
7. https://www.csc.fi/web/elmer
8. Ramesh, A., Jiles, D.C., Roderik, J.: A model of anisotropic anhysteretic magnetization. IEEE Trans. Magn. **32**, 4234–4236 (1996)
9. Szewczyk, R.: Validation of the anhysteretic magnetization model for soft magnetic materials with perpendicular anisotropy. Materials **7**, 5109–5116 (2014). https://doi.org/10.3390/ma7075109

10. Frydrych, P., Szewczyk, R., Nowicki, M., Charubin, T.: Application of anisotropic vector Preisach model for bulk materials. Acta Phys. Pol. A **131**, 618–620 (2017). https://doi.org/10.12693/APhysPolA.131.618
11. O'Handley, R.C.: Modern Magnetic Materials: Principles and Applications. Wiley, New York (1999)
12. Buttino, G., Poppi, M.: Dependence on the temperature of magnetic anisotropies in Fe-based alloys of Finemet. J. Magn. Magn. Mater. **170**, 211–218 (1997). https://doi.org/10.1016/S0304-8853(97)00005-X
13. Svec Sr., P., Szewczyk, R., Salach, J., Jackiewicz, D., Svec, P., Bieńkowski, A., Hosko, J.: Magnetoelastic properties of selected amorphous systems tailored by thermomagnetic treatment. J. Electr. Eng. **65**, 259–261 (2014). https://doi.org/10.2478/jee-2014-0040

# Estimation of Uncertainties in Indirect Multivariable Measurements: Part 1. Case of Correlated Quantities

Zygmunt Lech Warsza[1(✉)] and Jacek Puchalski[2]

[1] ŁUKASIEWICZ Research Network – Industrial Research Institute of Automation and Measurements (PIAP), Warsaw, Poland
zlw1936@gmail.com

[2] Central Office of Measures (GUM), Warsaw, Poland
jacek.puchalski@gum.gov.pl

**Abstract.** This two-part paper presents an improved version of the method of evaluation uncertainties in the multiparameter indirect measurements stated in the Supplement 2 to GUM guide. In this part 1 it was done for the case of correlated quantities on the example of two-parameter (2D) jointed measurements. It consists the correlation of individual components of deviations from the estimator of result with the type A and/or type B uncertainty of measurands. The general formulas for the covariance matrix, final uncertainties and correlation coefficient were determined as well as formulas for several specific cases. The graphs show the correlation coefficients of the output quantities as a function of the type B contribution in the uncertainty of the input quantities. The inclusion of correlations of uncertainty components makes the uncertainty evaluations of output quantities more reliable and accurate.

**Keywords:** Multivariate measurements · Multi-measurand · Vector propagation of uncertainties · Covariance matrix · Correlation coefficient · Correlations of deviations of type A or type B · Uncertainties of processing function

## 1 Introduction

The measurement result of any single quantity is presented in the form of two estimates, which describe their value and accuracy. Both such metrological parameters are determined based on measurement observations, which are repeated many times in the same or similar conditions, that is on the data of the measurement sample. These estimates depend on the probability distribution describing the random dispersion of the values of observed quantities. In multi-variable measurements of jointed quantities named generally measurand, beside the accuracy of each, the deterministic and statistical relationships (correlation) between them should also be considered.

To unify the estimation of measurement results and their accuracy, seven international organizations developed jointly a special document named "Evaluation of measurement data – Guide to the expression of uncertainty in measurement", firstly published in 1993. It is commonly known under the English acronym GUM. The last


R. Szewczyk et al. (Eds.): AUTOMATION 2020, AISC 1140, pp. 309–325, 2020.
https://doi.org/10.1007/978-3-030-40971-5_29

improved version appeared in 2008 [1] and it is further developed by including Supplements. Up to now four of them are published with numbers 1, 2, 4, 6 (S.1–S.6).

## 2 GUM Concept of Measurement Uncertainty

In the GUM guide, a concept called "measurement uncertainty" (MU) was introduced to estimate the accuracy of measurements. It is the width of the interval, and for multi-parameter measurements – a description of the boundaries of the multi-dimensional coverage area [1, 10], in which the estimators of values of the measurand elements, obtained after processing of the measurement results, can occur with a certain probability. The measurement uncertainty $u$ assessment is based on the determination of its type A and type B components, designated as $u_{\mathrm{A}}$ and $u_{\mathrm{B}}$, respectively [1, 3–12]. They are defined as standard deviations of the resultant distribution with a probability density function $p(x)$ constituting a convolution of two statistically independent density distributions $p(\Delta x_A)$ and $p(\Delta x_{\mathbf{B}})$ of errors (as deviations from the estimated measurand value). The first one describes a random spread of the values of experimental observations. The second one is a hypothetical distribution randomizing the assumed changes in the value of many systematic errors of different origin, unknown in value during measurements in each experiment. These unknown errors may occur randomly during the long-term in testing of the measuring object and use of instrument or measuring system due various permissible ambient conditions other than during provided calibration or control measurements. The standard deviation of the convoluted $p(x)$ distribution, i.e. its uncertainty $u$, is the geometric sum of component uncertainties $u_{\mathrm{A}}$ and $u_{\mathrm{B}}$, i.e.:

$$u = \sqrt{u_A^2 + u_B^2}. \tag{1}$$

The uncertainty of any single quantity described by formula (1) has the same form as for the uncertainty of the sum of two uncorrelated random variables, when values of one of them have only type A, and the other – type B uncertainty.

If a function describing the density distribution of $p(x)$ is known, e.g. the Gaussian function for a normal distribution, the expanded uncertainty $U$ with the required coverage probability, e.g. 95% or 99%, is determined analytically basing on standard deviation $u$ and the expansion coefficient. For distributions of observation values obtained in experiments, the expanded uncertainty $U$ can be determined also by the numerical Monte Carlo method according to Supplement 1 to the GUM guide.

Standard uncertainty $u_{\mathrm{A}}$, which depends on the distribution of the measurement observations, is determined statistically. On the other hand, the $u_B$ values can only be estimated heuristically based on knowledge about predicted ranges and distributions of values of influencing quantities and their interaction functions. In the known operating conditions with a limited range of changes, the uncertainty component $u_B$ may be significantly smaller than that assumed during the calibration of the device for the full permissible conditions of its application [4].

The measurement uncertainty (MU) is also applicated as a basic element of conformity assessment of products with requirements [1, S6, 2]. It allows to compare the results of various tests, check the exceeding of limits or meet the tolerance requirements of products. Thanks to the possibility of comparing the calibration results with their requirements, it is also the basis for metrological acceptance of the measuring equipment. Laboratories accredited according to ISO/IEC 17025 [2] must calculate the MU for each test method used in the granted accreditation scope.

There is, however, a certain discrepancy between the description of measurement accuracy through uncertainty with a certain probability and description of the accuracy of measuring devices by maximum permissible error MPE used by manufacturers.

The metrological properties of the measuring equipment and the parameters of many devices and processes that fulfill the responsible functions, besides careful calibration, also require periodic inspection during their operation. For this purpose, measurement experiments are carried out, including the study of these parameters as single and multiparameter combined measurands. They are rarely performed by one, and usually by many instruments, under the same or different environmental conditions and at the spread of the values of repeated measurement observations caused by common or different causes. This practical multi-variant situation creates different relationships between individual type A and type B component uncertainties of measured quantities. Many such situations do not include recommendations for estimating uncertainty of measurements, given in GUM [1] and all its Supplements and in other international regulations, e.g. [2–4]. Zakharow in [7] considered the non-heuristic estimates of the correlation of the quantities influencing the uncertainty of single parameter measurements, called "logical". In the literature on multiparameter measurements, we have not found an analysis of impact of the correlation of deviations of the input quantities with uncertainty type A or/and type B.

The purpose of this part of publication is to develop a method for determining the relationship between the uncertainty of the measurement results of several output values, when deviations with the individual type A or/and type B of uncertainty components of few input quantities are correlated. In this work the impact of various cases of correlation between deviations of observations of directly measured quantities on the resultant uncertainty of the quantities indirectly determined from these measurement data will be analyzed. This will be illustrated by the examples of two-parameter measurements described by a linear function or non-linear functions. Appropriate formulas will be determined. This is particularly important when the results of multiparameter measurements are used later together in various applications.

A proposal to extend the scope of the GUM guide and its Supplement 2 applications for the description of the accuracy of multivariable measuring devices by absolute and relative uncertainties is given in this Part 1 of our paper.

## 3 Uncertainty Type B Implemented by Instruments

The uncertainty type B is implemented to the procedure of measurement accuracy calculations by instruments and is estimated directly from their maximum permissible measurement errors (MPE) as descriptions of accuracy [3, 4]. The MPE value of instrument $|\Delta_x|_{max}$ is usually specified by manufacturers as the line dependence of the absolute error module on the value $x$ of measured quantity in the form of sum of components, additive and multiplicative (Fig. 1), or after relation of the MPE value to the measuring range, as it is given in formulas (2), (2a)

$$|\Delta_x| \leq |\Delta_x|_{max} = |\Delta_{x_0}|_{max} + |x - x_0| \cdot |\varepsilon_S|_{max} \tag{2}$$

$$\frac{|\Delta_x|}{|x_{max} - x_{\min}|} \leq \frac{|\Delta_{x_0}|_{max}}{|x_{max} - x_{\min}|} + \frac{|x - x_0|}{|x_{max} - x_{\min}|} \cdot |\varepsilon_S|_{max} \tag{2a}$$

Where: in (2): $|\Delta_x|, |\Delta_x|_{max}, |\Delta_{x_0}|_{max}$ - the absolute errors: real of the measured value $x$, maximum permissible for $x$ and for the beginning of range $x_0$, e.g. $x_0 = 0$; $|\varepsilon_S|_{max} \equiv \Delta_{x-x_0}/|x - x_0|$ - maximum permissible relative error of the $|x - x_0|$.

Equation (2a) gives error values related to the range $|x_{max} - x_{\min}|$ and is simplified for $x_{\min} = x_0 = 0$.

From two maximum values of MPE error $|\Delta_{x_0}|_{max}, |\Delta_x|_{max}$, which are known from instrumental technical data or from measurements under given conditions, for the $x$ value in the tested range one can determine the linear curve of maximum permissible absolute uncertainty $u_x(x)$ and normalized value (the relative uncertainty) $\delta_x|_{max}$ of $|x - x_0|$. Within the ranges of error $\pm|\Delta_x|_{max}$, the dispersion of the $x$-value with a uniform probability distribution is usually assumed. Standard deviations $\sigma_x$ of estimators of the values $x$, e.g. mean values of measurement data samples, are treated as absolute type A uncertainty $u_x$. From the formula (2) corresponds the relationship given below

$$\sigma_x \equiv u_x \leq u_x|_{max} = u_{x0}|_{max} + |x - x_0| \cdot \delta_x|_{max} \tag{3}$$

where: $\delta_x|_{max}$ is the maximum relative uncertainty of the difference $|x - x_0|$.

The maximum permissible absolute error $|\Delta_x|_{max}$ from Eq. (2) and the corresponding absolute uncertainty $u_x|_{max}$ from Eq. (3) are proportional to each other with factor $(1/\sqrt{3})$. With appropriately selected scales on the $y$-axis, they run identically as linear functions of the measurand $x$ shown in Fig. 1.

This method of estimation the uncertainty type B of instruments is very simple. For example, for many of digital devices, the absolute value of maximum permissible error increases for $x \geq x_0$, i.e. $|\Delta_x|_{max} \geq |\Delta_{x0}|_{max}$. If the maximal relative error $|\varepsilon_S|_{max}$ is constant, then the increase in $u_B$ is linear and proportional to increment of $x$ (Fig. 1).

However, when estimating the uncertainty $u_{\mathrm{B}}$ of individual values $x$, one can consider the statistical nature of these values, including the correlation between them. These components are estimated considering the maximum permissible errors of instruments, and the correlation coefficient between the obtained sets of values of their observations within the measuring range should be also find. If e.g. values $x_1$, $x_2$ are

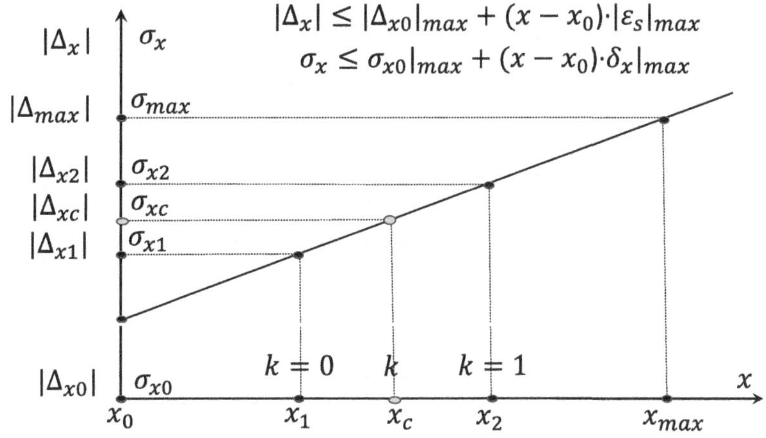


**Fig. 1.** Linear function of the maximal permissible absolute error $|\Delta_x|_{max}$ (2) and corresponding type B absolute uncertainty $u_{xB} \equiv \sigma_x = \frac{|\Delta_x|_{max}}{\sqrt{3}}$.

controlled in the same influencing circumstances and by the same instrument then the correlation coefficient between them should be assumed as equal 1. Then if the uncertainty $u_B$ of sum or difference of two values $x$ is determined with the correlation coefficient ± 1, the component uncertainties should be added algebraically. For the sum, they are larger, and for differences – smaller than for their geometric summation.

## 4 Basic Method of GUM-S2 with the Exact Processing Function

Indirect multi-parameter measurements in the general case are described by equation

$$Y = F(X) \tag{4}$$

in which: $\boldsymbol{X} = [X_1, X_2, \ldots X_n]^T$ and $\boldsymbol{Y} = [Y_1, Y_2, \ldots Y_m]^T$ are measurands in the form of vectors of the elements as random variables that respectively describe the $n$ quantities directly measured at the input and $m$ quantities determined from them at the output; $\boldsymbol{F}(\boldsymbol{X})$ – multidimensional function of processing results obtained for measurand $\boldsymbol{X}$.

The estimators $y_1, y_2, \ldots y_m$ of the values of the measurand $\boldsymbol{Y}$ elements (*observables*) are determined from the estimators $x_1, x_2, \ldots x_n$ of the directly measured measurand $\boldsymbol{X}$ according to formula (4).

Measured quantities can be correlated with each other and have uncertainty depending on the environmental conditions affecting the measurement results, the same or differently on the processing system and the tested object, depending on their location. In the general case, the $\boldsymbol{F}(\boldsymbol{X})$ function is nonlinear and its parameters may also have uncertainties. We will start from the simplest case, when the multivariate function **F**(**X**) can be treated as fully accurate. This applies to situations when the processing of **X** are carried out with a sufficiently high precision, e.g. on an external computer or used in a measuring system. This case is included in GUM Supplement S.2 [1].

The processing of measurement results use estimators $x_i$, $y_j$ of the values of elements of multi-measurands **X**, **Y** and estimators of their metrological parameters.

The small increments $\Delta x_i$ and $\Delta y_j$ constituting deviations from $x_i$, $y_i$ (formerly called experimental errors) are connected by a matrix linear relationship, which results from a total differential:

$$\boldsymbol{\Delta Y} \approx \frac{\partial \boldsymbol{F}(\boldsymbol{X})}{\partial \boldsymbol{X}} \boldsymbol{\Delta X} \equiv \boldsymbol{S}(\boldsymbol{X}) \cdot \boldsymbol{\Delta X}(\boldsymbol{X}) \tag{5}$$

$$\text{where}: \quad \boldsymbol{S} \equiv \begin{bmatrix} S_{11} & \ldots & S_{1n} \\ \ldots & \ldots & \ldots \\ S_{m1} & \ldots & S_{mn} \end{bmatrix} = \begin{bmatrix} \frac{\partial y_1}{\partial x_1} & \ldots & \frac{\partial y_1}{\partial x_n} \\ \ldots & \ldots & \ldots \\ \frac{\partial y_m}{\partial x_1} & \ldots & \frac{\partial y_m}{\partial x_n} \end{bmatrix}, \quad \text{and } S_{ji} = \frac{\partial y_j}{\partial x_i} \tag{5a}$$

Propagation of the uncertainties of these measurands is described as a matrix relation between the covariance matrices of output **Y** and input **X** variables. It occurs in the linearization of the functional **F** in the formula (4) through derivatives and has the form.

$$\boldsymbol{U_Y} = \boldsymbol{S} \cdot \boldsymbol{U_X} \cdot \boldsymbol{S}^{\mathrm{T}} \tag{6}$$

where: **S** – the sensitivity matrix ; $\boldsymbol{U_X}, \boldsymbol{U_Y}$ - covariance matrix of input **X** and output **Y** vectors described as

$$\boldsymbol{U_X} = \begin{bmatrix} u_{x1}^2 & \ldots & \rho_{x1n} u_{x1} u_{xn} \\ \ldots & \ldots & \ldots \\ \rho_{xn1} u_{xn} u_{x1} & \ldots & u_{xn}^2 \end{bmatrix}, \quad \boldsymbol{U_Y} = \begin{bmatrix} u_{y1}^2 & \ldots & \rho_{y1m} u_{y1} u_{ym} \\ \ldots & \ldots & \ldots \\ \rho_{y1m} u_{ym} u_{y1} & \ldots & u_{ym}^2 \end{bmatrix} \tag{6a, b}$$

If the **Y** vector will be further processed by any other function **G**(**Y**), i.e.:

$$\boldsymbol{Z} = \boldsymbol{G}(\boldsymbol{Y}) \tag{7a}$$

then for the indirectly determined values of the multivariable $\boldsymbol{Z} = [z_1, \ldots z_m]^{\mathrm{T}}$ with the uncertainties $u_{z1}, \ldots u_{zm}$, we get the next equation of covariance matrixes, similar as (6), describing the propagation of uncertainties

$$\boldsymbol{U_Z} = \boldsymbol{S_G} \boldsymbol{U_Y} \boldsymbol{S_G^{\mathrm{T}}} \tag{7b}$$

There are many possibilities to correlate each sets of deviations of the component type A or B uncertainties with their equivalents of other input quantities measured in a joint experiment.

The results of both directly measured values of ***X***, as well as indirectly measured values of ***Y*** and ***Z***, are determined from the collections of repeated observations obtained in randomly variable ambient conditions. If some of these conditions are constant or variable deterministically during measurements in a known manner, only then the corrections can be made to eliminate the resulting known systematic errors.

Different values of the examined quantities can be measured with different meters, or even the same quantity on different ranges of one meter. There are various permissible errors of these instruments, and therefore different values of the uncertainty component type B. In addition, the dispersion of observation values depends not only on parameters describing changes of the measurement system and the test object from ambient conditions, but also on the random and aging changes of their internal parameters. Furthermore, measurement samples from the same physical object under test (e.g. in chemical analysis) in different measurement experiments carried out under the same conditions, i.e. with the same $u_B$ uncertainty, may have different values of the uncertainty component $u_A$.

## 5 Uncertainty of Correlated Two Quantities

In multiparameter measurements, for measured values of elements of the input variable ***X*** different types of correlation of components of each uncertainty type A and type B may occur. This influences on the accuracy of the results of the output quantities ***Y***. Measurements of only the components of the vector ***X*** should be treated as a special case of multiparameter measurements related to each other only in a measurement object, i.e. when ***Y*** = ***X***. Below, some examples of determining the uncertainty of vector ***Y*** components with two values $y_1$, $y_2$ obtained from measurements of two values $y_1$, $y_2$ (or one quantity in two different experiments and under different influencing conditions) will be considered.

Both input and output measurands are two-parameter (2D). For 2D measurement of input variable $\boldsymbol{X} = [x_1, x_2]$, i.e., when $\boldsymbol{Y} = \boldsymbol{X}$, *i.e.*: $y_1 = x_1, y_2 = x_2$, two Eqs. (6a, b) and the output covariance matrix $\boldsymbol{U}_Y \equiv \boldsymbol{U}_{x1,2}$ will be obtained

$$\boldsymbol{U}_{x1,2} = \begin{bmatrix} u_{1A}^2 + u_{1B}^2 & \rho_A u_{1A} u_{2A} + \rho_B u_{1B} u_{2B} \\ \rho_A u_{1A} u_{2A} + \rho_B u_{1B} u_{2B} & u_{2A}^2 + u_{2B}^2 \end{bmatrix} \tag{8}$$

In the covariance matrix $\boldsymbol{U}_{x1,2}$ there are uncertainties according to the rule of totalizing variance, i.e. as sum of squares of both uncertainties type A and type B, i.e.:

$$u_{x1}^2 = u_{1A}^2 + u_{1B}^2, \tag{8a}$$

$$u_{x2}^2 = u_{2A}^2 + u_{2B}^2 \tag{8b}$$

and the correlation coefficient between variables $x_1$ and $x_2$ is:

$$\rho_{x1,2} = \frac{\rho_A u_{1A} u_{2A} + \rho_B u_{1B} u_{2B}}{u_{x1} u_{x2}} = \frac{\rho_A u_{1A} u_{2A} + \rho_B u_{1B} u_{2B}}{\sqrt{u_{1A}^2 + u_{1B}^2}\sqrt{u_{2A}^2 + u_{2B}^2}} \tag{8c}$$

Relations between the uncertainty components of type A and B are illustrated in Fig. 2

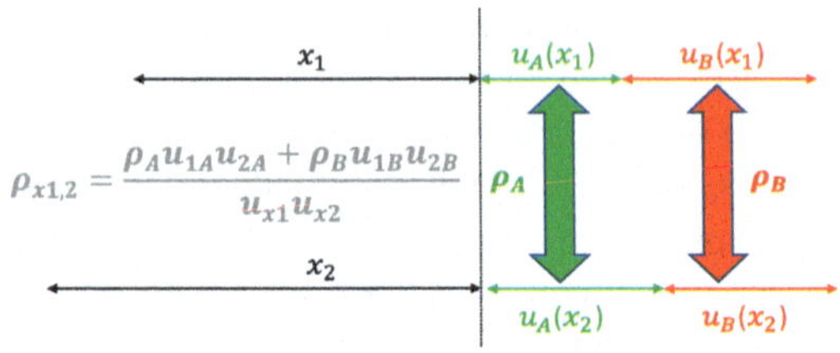


**Fig. 2.** The relations between uncertainties of Type A and/or Type B for the 2D measurand with correlated variables $x_1$, $x_2$

For ratios of uncertainty components and standard uncertainties we put designations:

$$0 \le k_{1B} \equiv \frac{u_{1B}}{u_{x1}} \le 1, \; 0 \le k_{2B} \equiv \frac{u_{2B}}{u_{x2}} \le 1.$$

From that and (8a, 8b):

$$\frac{u_{1A}}{u_{x1}} \equiv \sqrt{1 - k_{1B}^2}, \qquad \frac{u_{2A}}{u_{x2}} \equiv \sqrt{1 - k_{2B}^2}$$

and a simpler form of the pattern (8c) is obtained

$$\rho_{x1,2} = \rho_A \sqrt{1 - k_{1B}^2}\sqrt{1 - k_{2B}^2} + \rho_B k_{1B} k_{2B} \tag{9}$$

For both quantities measured in this experiment, the coefficient $\rho_{x1,2}$ in (9) depends only on the uncertainty ratios $k_{1B}, k_{2B}$ (or $k_{1A}, k_{2A}$) because for each of them there is

$$\left(\frac{u_{iB}}{u_{xi}}\right)^2 + \left(\frac{u_{iA}}{u_{xi}}\right)^2 = k_{iB}^2 + \left(1 - k_{iB}^2\right) = 1 \qquad (\text{for } i = 1, 2)$$

The form of the function (9) is invariant to the changes of values $u_{1A}$ to $u_{1B}$, $u_{2A}$ to $u_{2B}$ and $\rho_A$ to $\rho_B$ Then it enough to analyze dependencies for one type of uncertainty A or B, because for the set of variables of the second type, will be similar relationships due to the symmetry of this pattern.

# 6 Graphs of the Correlation Coefficient $\rho_{x1,2}$

The general function of the formula (9) will be used to present diagrams of the correlation coefficient. Figure 3 shows 3D graphs of the resultant values of correlation coefficient $\rho_{y1,2} = f(k^2_{1B}, k^2_{2B})$ of the two-element (2D) measurand ***Y*** = ***X*** for three pairs of correlation coefficients $\rho_A, \rho_B$ of type A or/and type B uncertainties of elements $x_1$, $x_2$ of the 2D input measurand ***X*** [12, 13].

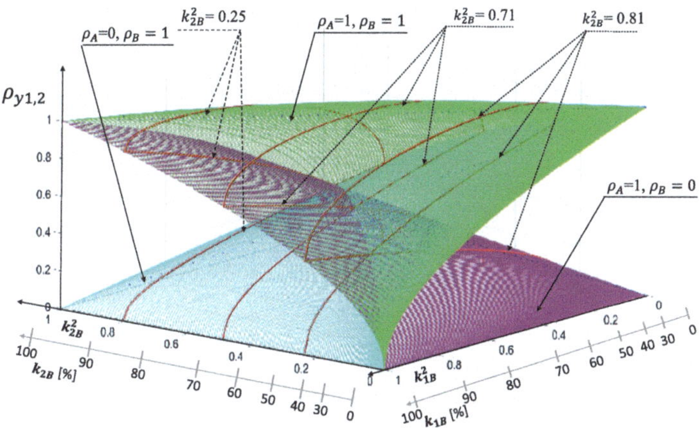


**Fig. 3.** Relations of correlation coefficient of two quantities $\rho_{y1,2} = f(k_{1B}, k_{1B})$ for three pairs of correlation coefficients $\rho_A$, $\rho_B$ of their deviation components with uncertainty type A or B

Graphs on Fig. 3 have the form of curvilinear planes. The cross-sections are marked on them for variance: $k^2_{2B} = (0.25, 0.5, 0.81)$ equal for uncertainty to $k_{2B} = (0.5, 0.71, 0.9)$ are shown on Figs. 4a, b and 5a and described by the function (10):

$$\rho_{x1,2} = f(k^2_{1B}) \tag{10}$$

or as $\rho_{x1,2} = f(k_{1B})$, if lower non-linear scales at the bottom of these drawings are used.

Conclusions resulting from the analysis of diagrams in Figs. 3, 4a, b and 5a, b are as follows:

- the largest correlation coefficient $\rho_{y1,2}$ in the entire range $k_{1B}$ has the curve of $\rho_A = 1$, $\rho_B = 1$ for $k_{1B} = k_{2B}$;
- for the value $k^2_{1B} < 1 - k^2_{2B}$ the curve for $\rho_A = 1$, $\rho_B = 0$ dominates over the curve $\rho_A = 0$, $\rho_B = 1$ (Figs. 4a, b and 5a) and approaches the curve of $\rho_A = 1$, $\rho_B = 1$ for the small values $k^2_{2B}$;

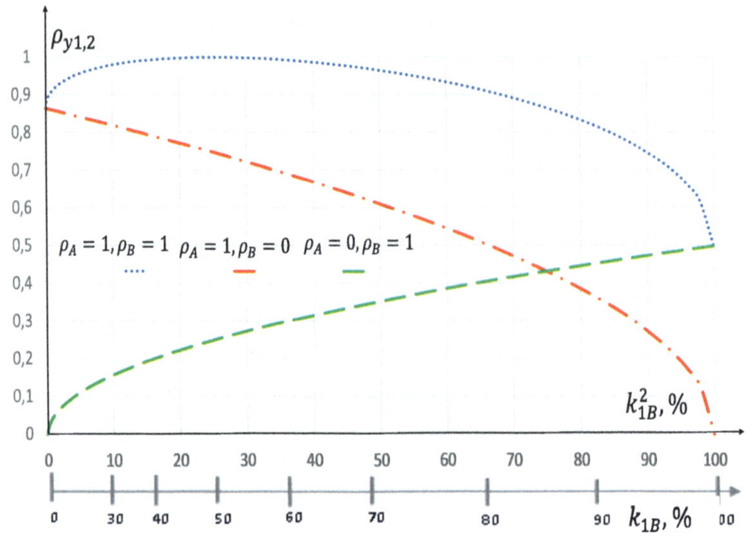


a.

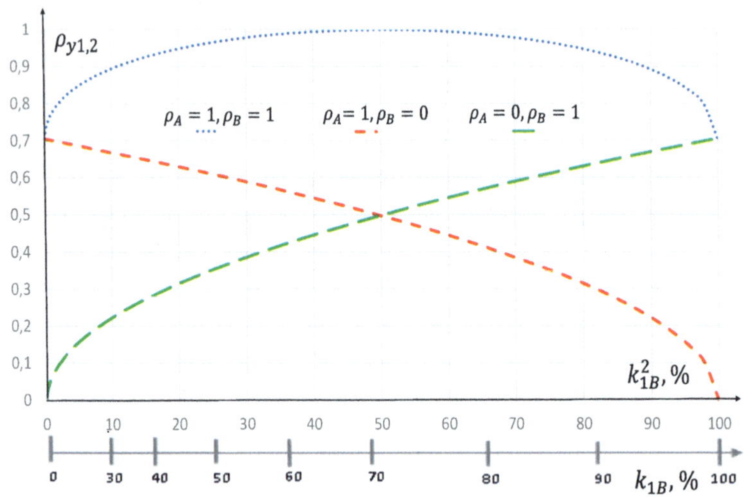


b.

**Fig. 4.** Correlation coefficient $\rho_{y1,2}$ of two quantities as function of ratio of variances $k_{1B}^2 = \left(\frac{u_{1B}}{u_{y1}}\right)^2$ (or ratio of uncertainties $k_{1B} = \frac{u_{1B}}{u_{y1}}$ – lower nonlinear scale) for three pairs of correlation coefficients $\rho_A, \rho_B$ between their deviations with uncertainty components type A or type B, and values of $k_{2B}^2$ (i.e. $k_{2B} = \frac{u_{2B}}{u_{y2}}$) are: **a.** $k_{2B}^2 = 0.25 (k_{2B} = 0.50)$; **b.** $k_{2B}^2 = 0.5$ $(k_{2B} \approx 0.71)$.

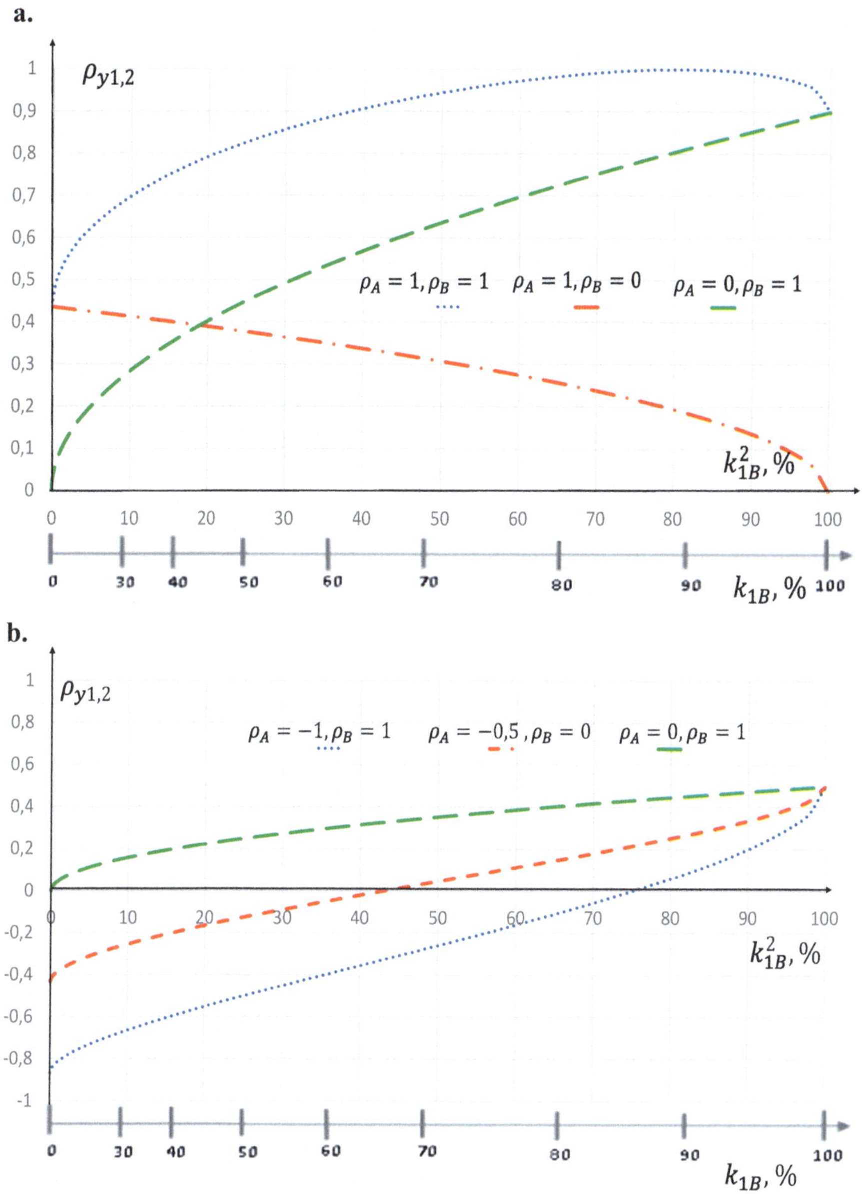


**Fig. 5.** Correlation coefficient $\rho_{y1,2}$ of two quantities as function of ratio $k_{1B}^2$ of variances (equal to ratio of uncertainties $k_{1B} = \frac{u_{1B}}{u_{y1}}$ – lower nonlinear scale) for pair of correlation coefficients $\rho_A$ and $\rho_B$ between their deviation sets with uncertainty components type A and B given on figures and $k_{2B}^2 = \left(\frac{u_{2B}}{u_{y2}}\right)^2 = 0.81$ for **a.**, and $k_{2B}^2 = 0.25$ for **b**

- for larger values of $k_{1B}^2 > 1 - k_{2B}^2$ the curve $\rho_A = 1,\ \rho_B = 0$ is below the curve $\rho_A = 0,\ \rho_B = 1$;
- at point $k_{1B}^2 = 1 - k_{2B}^2$ curves for $\rho_A = 1,\ \rho_B = 0$ and $\rho_A = 0,\ \rho_B = 1$ intersect (Figs. 4a, b and 5a). The correlation coefficient is $\rho_{y1,2} = k_{1B}\sqrt{1 - k_{1B}^2}$ and reaches the maximum value $\rho_{y1,2} = 0.5$ for $k_{1B}^2 = 0.5$;
- for the value of correlation coefficient $\rho_A < 0$, a negative coefficient $\rho_{y1,2}$ for the fragment of range of $k_{1B}$ is obtained (Fig. 5b). For example, for $\rho_A = -1,\ \rho_B = 1$ and $k_{1B}^2 < 1 - k_{2B}^2$ a negative correlation coefficient is obtained, and for $k_{1B}^2 > 1 - k_{2B}^2$ the correlation coefficient is positive. The largest range of variation $k_{1B}$ with a negative correlation occurs for small $k_{2B}$.

On this basis, some conclusions about the uncertainty of measurements can be formulated. The correlation between the values of results of two different measurement experiments associated with each other, e.g. measurements of the same value of measurand with two different meters or on two different ranges of the same meter values, depends on the ratio of the uncertainties type B and if the higher the measurement uncertainty is in relation to type A in one or both measured experiments.

The maximum correlation coefficient $\rho_{y1,2} = 1$ is achieved for the quantities of fully correlated components of type A and type B $(\rho_A = 1, \rho_B = 1)$, i.e. when $k_{1B} = k_{2B}$. This leads to the condition $\frac{u_{1B}}{u_{1A}} = \frac{u_{2B}}{u_{2A}}$. The correlation coefficient increases to 1 for $k_{1B} < k_{2B}$, but decreases for $k_{1B} > k_{2B}$. For the values of $k_{1B}^2 < 1 - k_{2B}^2$ a strong negative correlation is observed for the curves $\rho_A = (-1,\ -0.5)$, $\rho_B = (0, 1)$ – Fig. 5b.

## 7 Relative Uncertainties

The formulas for relative uncertainties are obtained by substituting $u_i = x_i\ u_{ri}$, $u_{Ai} = x_i\ u_{rAi}$, $u_{Bi} = x_i\ u_{rBi}$, for $i = 1, 2$. From (8a, 8b) relative variances of $x_1, x_2$ are:

$$u_r^2(x_1) \equiv u_{rx1}^2 = \frac{u_{A1}^2 + u_{1B}^2}{x_1^2} = u_{rA1}^2 + u_{rB1}^2 \tag{11a}$$

$$\mathrm{u}_{\mathrm{r}}^2(\mathrm{x}_2) \equiv \mathrm{u}_{\mathrm{rx2}}^2 = \frac{\mathrm{u}_{\mathrm{A2}}^2 + \mathrm{u}_{\mathrm{B2}}^2}{\mathrm{x}_2^2} \mathrm{u}_{\mathrm{rA2}}^2 + \mathrm{u}_{\mathrm{rB2}}^2 \tag{11b}$$

Relative uncertainties $u_{rx1}$, $u_{rx2}$ as functions of relative uncertainties $u_{rA1}$, $u_{rA2}$, $u_{rB1}$, $u_{rB2}$ and their correlation coefficient $\rho_{x1,2}$ are expressed in (12a, 12b) and (13) as

$$u_{rx1} = \sqrt{u_{rA1}^2 + u_{rB1}^2}, \tag{12a}$$

$$u_{rx2} = \sqrt{u_{rA1}^2 + u_{rB1}^2} \tag{12b}$$

$$\rho_{x1,2} = \frac{\rho_A u_{rA1} u_{rA1} + \rho_B u_{rB1} u_{rB1}}{u_{rx1} u_{rx2}} \quad (13)$$

In general case for types A and B of relative uncertainties $u_{rA}$ and $u_{rB}$, in a similar way as before for $u_A$ and $u_B$, a covariance matrix $\boldsymbol{U}_{rX}$ for components of relative uncertainties can be created. If the relative uncertainties of the input quantities are known, e.g. the same for the entire range, you can directly use their matrix propagation equation of similar structure as for absolute uncertainties

$$\boldsymbol{U}_{rY} = \boldsymbol{S}_r \cdot \boldsymbol{U}_{rX} \cdot \boldsymbol{S}_r^{\mathrm{T}} \quad (14)$$

where: $\boldsymbol{U}_{rX}, \boldsymbol{U}_{rY}$ and $\boldsymbol{S}_r = \left[\frac{x_i}{y_j}\frac{\partial y_j}{\partial x_i}\right]$ – covariance matrices for relative uncertainties and the sensitivity matrix with elements marked with indices $i = 1, 2, 3$ for lines and $j = 1, 2, 3$ for columns.

## 8 Uncertainties of Output Quantities

When instruments and measuring systems in various environmental conditions are used and measurement observations are different randomly distributed, it may also be necessary to find the type A and B uncertainties of elements of the output measurand $\boldsymbol{Y}$ and correlations between the pair of their values. We analyze this on the example of 2D measurand based on (8a) and (8b) equations.

The covariance matrix of measurand $\boldsymbol{X} = [x_1, x_2]^{\mathrm{T}}$ is

$$\boldsymbol{U}_X = \begin{bmatrix} u_{1A}^2 + u_{1B}^2 & \rho_A u_{1A} u_{2A} + \rho_B u_{1B} u_{2B} \\ \rho_A u_{1A} u_{2A} + \rho_B u_{1B} u_{2B} & u_{2A}^2 + u_{2B}^2 \end{bmatrix} \quad (15)$$

In calculations of $\boldsymbol{U}_Y$ the matrix $\boldsymbol{U}_X$ (15) can be used directly, as it is given on Fig. 6a, or after transformation to the sum of two matrices: $\boldsymbol{U}_{XA}$ and $\boldsymbol{U}_{XB}$ for type A and type B uncertainty components (Fig. 6b):

$$\boldsymbol{U}_X = \boldsymbol{U}_{XA} + \boldsymbol{U}_{XB} \quad (16)$$

$$\text{where : } \boldsymbol{U}_{XA} = \begin{bmatrix} u_{1A}^2 & \rho_A u_{1A} u_{2A} \\ \rho_A u_{1A} u_{2A} & u_{2A}^2 \end{bmatrix}, \quad \boldsymbol{U}_{XB} = \begin{bmatrix} u_{1B}^2 & \rho_B u_{1B} u_{2B} \\ \rho_B u_{1B} u_{2B} & u_{2B}^2 \end{bmatrix}, \quad (16\mathrm{a, b})$$

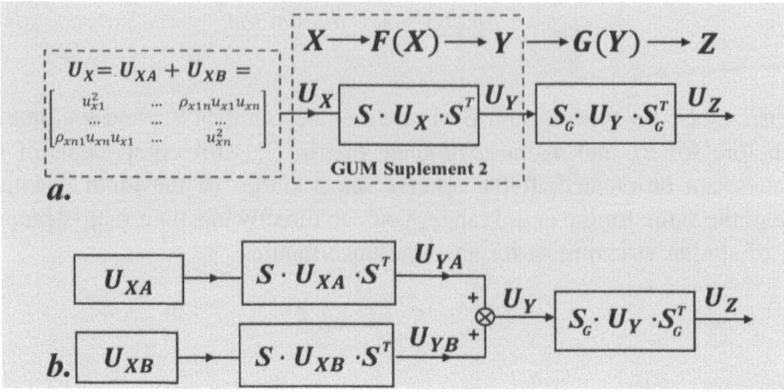


**Fig. 6.** Schemes of the uncertainty processing in multi-parameter indirect measurements and two methods: a, b for determining the output covariance matrix $\boldsymbol{U}_Y$ and its component matrixes $\boldsymbol{U}_{YA}$, $\boldsymbol{U}_{YB}$ according to the matrix law of variance propagation

The transformation of the covariance matrices $\boldsymbol{U}_{XA}$ and $\boldsymbol{U}_{XB}$ of the uncertainties of measurand $X$ is performed after the linearization of the functional $\boldsymbol{Y} = \boldsymbol{F}(\boldsymbol{X})$ according to formula (2) as follows:

$$\boldsymbol{U}_{Y_A} = \boldsymbol{S}\boldsymbol{U}_{XA}\boldsymbol{S}^{\mathrm{T}} \quad \text{and} \quad \boldsymbol{U}_{Y_B} = \boldsymbol{S}\boldsymbol{U}_{XB}\boldsymbol{S}^{\mathrm{T}} \tag{17a, b}$$

The covariance matrix $\boldsymbol{U}_Y$ can be determined in two ways shown on Fig. 6a, b:

- estimate the matrix $\boldsymbol{U}_X = \boldsymbol{U}_A + \boldsymbol{U}_B$ and determine from it directly $\boldsymbol{U}_Y$ as follows

$$\boldsymbol{U}_Y = \boldsymbol{S}\boldsymbol{U}_X\boldsymbol{S}^{\mathrm{T}} = \boldsymbol{S}(\boldsymbol{U}_{XA} + \boldsymbol{U}_{XB})\boldsymbol{S}_F^{\mathrm{T}} \tag{18}$$

- or from the matrix $\boldsymbol{U}_{XA}$ and $\boldsymbol{U}_{XB}$ find the matrices $\boldsymbol{U}_{Y_A}$ and $\boldsymbol{U}_{Y_B}$ of both components of uncertainties $u_{yi}$ of output quantities and then $\boldsymbol{U}_Y$. It is received

$$\boldsymbol{U}_Y = \boldsymbol{U}_{Y_A} + \boldsymbol{U}_{Y_B} = \boldsymbol{S}\boldsymbol{U}_{XA}\boldsymbol{S}^{\mathrm{T}} + \boldsymbol{S}\boldsymbol{U}_{XB}\boldsymbol{S}^{\mathrm{T}} \tag{19}$$

If $\boldsymbol{U}_Y$ is only designated, then both methods are equivalent. This was checked for both two-dimensional (2D) measurands $\boldsymbol{X}$ and $\boldsymbol{Y}$. On the other hand, matrices $\boldsymbol{U}_{Y_A}$ and $\boldsymbol{U}_{Y_B}$ for the uncertainties of type A and B components of the output $\boldsymbol{Y}$ measurand and the correlation coefficients of their each type can be determined from the formula (19) not found in the literature.

## 9 Uncertainty of Sum and Difference

If $Z = \boldsymbol{G}(\boldsymbol{Y}) = [y_1 + y_2, y_1 - y_2]^{\mathrm{T}}$, then absolute uncertainties are equal:

$$\begin{aligned} u_{z1}^2 &= u_{1A}^2 + u_{1B}^2 + u_{2A}^2 + u_{2B}^2 + 2(\rho_A u_{1A} u_{2A} + \rho_B u_{1B} u_{2B}) \\ u_{z2}^2 &= u_{1A}^2 + u_{1B}^2 + u_{2A}^2 + u_{2B}^2 - 2(\rho_A u_{1A} u_{2A} + \rho_B u_{1B} u_{2B}) \end{aligned} \tag{20a, b}$$

For $u_0^2 \equiv u_{1A}^2 + u_{1B}^2 + u_{2A}^2 + u_{2B}^2$, output variances in the absence of correlation are:

$$\begin{aligned} u_{y_1+y_2} &= \sqrt{u_0^2 + 2(\rho_A u_{1A} u_{2A} + \rho_B u_{1B} u_{2B})} \\ u_{y_1-y_2} &= \sqrt{u_0^2 - 2(\rho_A u_{1A} u_{2A} + \rho_B u_{1B} u_{2B})} \end{aligned} \tag{21a, b}$$

It is obtained that the uncertainty increases for the sum and decreases for the difference in relation to the model with zero input correlation. The values $y_1 + y_2$ and $y_1 - y_2$ are correlated with a coefficient equal to:

$$\rho_{y_1+y_2, y_1-y_2} = \frac{u_{y1}^2 - u_{y2}^2}{u_{y_1+y_2} u_{y_1-y_2}} \tag{22}$$

If there is no correlation between type B components, i.e. for $\rho_B = 0$, we obtain a reduction in the uncertainty for the sum of the input quantities and an increase in the uncertainty for the difference of this quantities.

If for two measured quantities the correlation coefficient $\rho_{x1,2} = 1$ (measurements with the same instrument in identical conditions, and $\rho_{\mathrm{A}} \ll \rho_{\mathrm{B}}$), then:

$$\begin{aligned} u_{(x_1+x_2)} &= \sqrt{u_0^2 + 2u_{x1} u_{x2}} = u_{x1} + u_{x2} \\ u_{(x_1-x_2)} &= \sqrt{u_0^2 - 2u_{x1} u_{x2}} = |u_{x1} - u_{x2}| \end{aligned} \tag{23a, b}$$

As an example, taken from industrial practice is the uncertainty of the sum as double average value and the difference of two temperatures measured at the inlet and outlet of the heat exchanger. Some other examples are given in [9] and [13].

## 10 Summary

In the part I of this work, the summary of current results of the work are briefly presented regarding the impact of correlations in sets of deviations from estimators of directly measured quantities with uncertainty of types A or/and B on the accuracy of indirectly determined parameters of output multi-measurand.

It was assumed that in the general case deviations of the uncertainty components of each type A or/and B for quantities measured jointly in the input of measurement system can be correlated only with each other.

The general dependences of the correlation coefficient and the different mutual relations between input uncertainties with full correlation and no correlation between data of type B uncertainties were determined on the example of 2D measurands.

In cases when both input values are measured in equal or nearly similar influencing conditions, their B-type uncertainties may be strongly correlated. This should be considered in the estimation of the uncertainty of the output quantities. For example, for the sum of the output quantities and the positive correlation coefficient, the result of uncertainty will be greater than the result from the geometric summation of both uncertainties type A and type B according to GUM, and for the difference – smaller.

Conclusions can be generalized to measurements of any multi measurands.

In order of implementing this method to the practice, including the use of a uniform description of the accuracy of measuring instruments, systems and measurements using uncertainty in addition to the maximum permissible errors.

In part 2 the new extended model, which takes also into account the uncertainties of the processing function in multiparameter measurements will be presented. It can be used both, for indirect multiparameter measurements, when high accuracy of calculations are important, and also to describe the accuracy of instruments and systems that fully perform such measurements. Example of using the extended method to describe intermediately measured parameters of a two-terminal net through a four-terminal network, considering the uncertainties of its elements is presented. Final conclusions and proposed directions of further works are there also included.

## References


1. JCGM100, 2008 Evaluation of measurement data — Guide to the Expression of Uncertainty in Measurement, JCGM 100:2008, GUM 1995 with minor corrections, and Supplements: - S1. Propagation of distributions using a Monte Carlo method. JCGM 101:2008, BIPM. -S2. Extension to any number of output quantities. JCGM 102:2011 BIPM. -S6. The role of measurement uncertainty in conformity assessment JCGM 106:2012, BIPM
2. Conformity Assessment-Requirements for Competence of Testing and Calibration Labs, ISO/IEC 17025:2017
3. JCGM 200:2012 International vocabulary of metrology – Basic and general concepts and associated terms (VIM) 3rd edition
4. EA-4/02.M: 2013 Evaluation of the Uncertainty of Measurement in Calibration
5. Dorozhovets, M., Warsza, Z.L.: Upgrading calculating methods of the uncertainty of measurement results in practice. Przegląd Elektrotechniki - Electr. Rev. **1**, 1–13 (2007). (in Polish)
6. Dorozhovets, M., Warsza, Z.L.: Methods of upgrading the uncertainty of type A evaluation, Part 2. Elimination of the influence of autocorrelation of observations and choosing the adequate distribution. In: Proceedings of 15th IMEKO TC4 Symposium, Iasi, Romania, pp. 199–204 (2007)
7. Zakharow, I.P.: Estimating measurement uncertainty on the basis of observed and logical correlation. Meas. Tech. **50**(8), 808–819 (2007)
8. Warsza, Z.L., Ezhela, V.: About evaluation of multivariate measurement results. J. Autom. Mob. Robot. Intell. Syst. **6**(4), 27–32 (2012)
9. Warsza, Z.L.: Methods of Extension of the Measurement Uncertainty Analysis (Metody rozszerzenia analizy niepewności pomiarów). Monograph. PIAP, Warszawa (2016). (in Polish)

10. Warsza, Z.L.: Estimation of uncertainty of indirect measurement in multi-parametric systems with few examples. Part 1, and with Puchalski J.: Part 2. In: Proceedings of Conference (on CD): Problems and Progress of Metrology ppm'18 Conference Series: No. 22, Metrology Commission of Katowice Branch of the Polish Academy of Science (2018). ISBN 978-83-7880-541-0
11. Warsza, Z.L., Puchalski, J.: Vector method of estimation the uncertainties of correlated multi-measurands. In: Proceedings (Conference Digest) of Quantum and Precision Metrology QPM 2019 Conference Krakow, Poland, 17–19 June 2019. ISBN 978-83-939486-5-9, SIV 4
12. Warsza, Z.L., Puchalski, J.: Extended method for the estimation of uncertainties in multiparameter measurements Part 1. Theoretical basis of correlated input measurands. Pomiary Automatyka Robotyka **1**, 47–58 (2019). https://doi.org/10.14311/PAR_230/47. (in Polish)
13. Warsza, Z.L., Puchalski, J.: Extended method for the estimation of uncertainties in multiparameter measurements Part 2 examples of 2D measurements with correlated uncertainties of type A and type B. Pomiary Automatyka Robotyka **2**, 29–35 (2019). https://doi.org/10.14311/PAR_231/29. (in Polish)

# Estimation of Uncertainties in Indirect Multivariable Measurements: Part 2. Influence of the Processing Function Accuracy

Zygmunt Lech Warsza[1(✉)] and Jacek Puchalski[2]

[1] ŁUKASIEWICZ Research Network – Industrial Research Institute of Automation and Measurements PIAP, Warsaw, Poland
zlw1936@gmail.com

[2] Central Office of Measures (GUM), Warsaw, Poland
jacek.puchalski@gum.gov.pl

**Abstract.** This two-part paper presents an improved version of the method of evaluation uncertainties in the multivariable indirect measurements. It has an extended scope of application compared to the method given in Supplement 2 of the international guide with the acronym GUM [1], which assumes a perfectly accurate parameters multidimensional function of processing input measurement data. Part 1 discussed the cases of various correlations of measured values. The law of variance propagation is described in the form of dependencies between input and output covariance matrices. General formulas for uncertainty and correlation coefficients at the output are given as well as formulas for several characteristic cases. It has been shown that for pairs of all measured quantities, the effect of correlation between sets of deviations with uncertainty for each of types A and B should be considered separately. Proposed is to calculate output covariance matrices of correlated variables separately for uncertainties type A and type B An example of associated two-parameter (2D) measurements is also discussed. In this Part 2, the effect of uncertainties of the processing function parameters on the elements of output multimeasurand covariance matrix has been determined. General formulas for full correlation and its absence are given. An example of the evaluation uncertainties of indirect current and voltage measurements on the inaccessible input of the T-type impedance twoport network based on measurements at its output is considered in detail. Final conclusions are given. The proposed upgraded method is versatile, because it allows both the estimation of the accuracy of indirect multivariable measurements as well a description of the accuracy of instruments and measuring systems performing such measurements.

**Keywords:** Multivariable measurements · Multi-measurand · Propagation of uncertainties · Covariance matrix · Correlation coefficient · Correlations of deviations of type A or type B · Uncertainties of processing function


R. Szewczyk et al. (Eds.): AUTOMATION 2020, AISC 1140, pp. 326–344, 2020.
https://doi.org/10.1007/978-3-030-40971-5_30

## 1 Introduction

The concept – "**uncertainty**" to assess the accuracy of measurements was introduced in the international guide GUM [1] in 1993. The extended uncertainty $U_P$ for any single quantity is used. It is the width of the interval around the estimator of measurand value in which results of measurements can be with the certain probability *P*. The value of *U* is based on the standard uncertainty $u = \sqrt{u_A^2 + u_B^2}$ of the probability distribution of deviations from the estimated value of measurand. This distribution is the convolution of two distributions of type A and type B noncorrelated. The uncertainty $u_\mathrm{A}$ is find statistically from values of measurement observations and $u_\mathrm{B}$ is estimated heuristically.

Next in 2011, in the Supplement 2 to Guide GUM recommended was the method of estimation uncertainties of indirect multivariable measurements. For measurements of few jointed variables, their uncertainties describe the boundaries of the multi-dimensional coverage area [1, 3], in which the estimators of values of the multi-measurand elements, can occur with a certain probability. But there are only considered cases when the processing function is computed ideally, i.e. with the accuracy of negligible uncertainties and.in practice it is recognized as ideally accurate.

However, there was a certain discrepancy in the description of the accuracy of measurement by the uncertainty according to the GUM Guide and the description of the accuracy for instruments, measuring devices and systems. Manufacturers of measuring apparatus so far use the maximum permissible error (MPE) [2, 3]. Then few years ago, the first of authors initiate works [5] for implementation of the uncertainty also for the description of the accuracy of instruments and measuring systems to be uniformly as for measurements in GUM. Additional justification for this idea is as follows.

Metrological properties of the measuring equipment and parameters of many devices and processes that fulfill the responsible functions, besides careful calibration, also require periodic inspection during their operation [3]. For this purpose, measurement experiments are carried out, including the study of these parameters as single and multiparameter combined measurands. They are rarely performed by one, and usually by many instruments, under the same or different environmental conditions and at the spread of the values of repeated measurement observations caused by common or different causes. This multi-variant situation creates different relationships between types A and type B uncertainties of individual measured quantities differently changed with time and should be recognized.

In the indirect instrumental measurements, the accuracy depends on the uncertainty of the measurements taken and additionally on the uncertainty of the parameters of the function processing the acquired data into indirect output results and depends on the parameters of signal transmission paths. The impact of correlation of elements of the multivariable measurand is discussed in detail in [3], in part I of this work [4] and in few other papers given in literature of [4]. The theoretical basis of the extended method of GUM Supplement 2 for assessing the effect of uncertainties of the processing function parameters is given below. It will be also illustrated by an example of indirect measurements of voltage and current on the output terminals of the twoport T-type from measurements on its input.

## 2 General Case of Extended Model of Multivariable Measurements Including Impact of Processing Function Uncertainties

In the multivariable indirect measurements the estimators of the parameters of the output $m$-dimensional measurand $\boldsymbol{Y} = [y_1, y_2, .., y_m]^T$ result from direct measurements in input of the $n$-dimensional measurand $\boldsymbol{X} = [x_1, x_2, .., x_n]^T$. They are related by the general formula

$$\boldsymbol{Y} = \boldsymbol{F}(\boldsymbol{X}) \tag{1}$$

where: $\boldsymbol{F}(\boldsymbol{X})$ is the nominal processing function of measurand $\boldsymbol{X}$ at normal conditions.

In the reality, estimators of values and their uncertainties of elements of the multimenzurand $\boldsymbol{Y}$ depend not only on the quantities of multimeasurand $\boldsymbol{X}$, on their systematic errors, uncertainties and correlation coefficients, but depend also on errors and uncertainties of elements of the $\boldsymbol{Y}$ implemented by the processing function $\boldsymbol{F}(\boldsymbol{X})$. Parameters of this function are described by $k$-dimensional vector $\boldsymbol{P} = [\boldsymbol{p}_1, \boldsymbol{p}_2, \ldots \boldsymbol{p}_k]^{\mathrm{T}}$. Their values are also subject to small deterministic and random scattering depending on the measurement conditions of instruments and measured object, and on the procedure used in the measurement experiment. In each of such experiment, the actual $\boldsymbol{F}(\boldsymbol{X})$ function may differ from its nominal form. Taking these into account, modifications of the general Eq. (1) for values of measurand $\boldsymbol{Y}$ elements is needed as formula (1a) given below

$$\boldsymbol{Y} = \boldsymbol{F}(\boldsymbol{X}, \boldsymbol{P}) \tag{1a}$$

$$\Delta \boldsymbol{Y} = \Delta \boldsymbol{F}(\boldsymbol{X}, \boldsymbol{P}) = \boldsymbol{S}_{X,P} \cdot [\Delta \boldsymbol{X}, \Delta \boldsymbol{P}]^{\mathrm{T}} \tag{2}$$

where: $\boldsymbol{S}_{X,P}$ with the size $[(n + k) \times m]$ is a matrix linearizing the function (1).

When estimating the uncertainty of indirect measurement results, it assumes that the deviations from the estimators of the values of all measurands and changes in the parameters of the measuring system are small increments. Then the relationships are linearized, and the resultant covariance matrix is the sum of matrices describing individual random influences (variances add up). In the general case, for the randomly variable increments of parameters, all possible correlations should also be considered.

For the processing of measurement data according to function (1a), the law of propagation of variance has the general matrix form:

$$\boldsymbol{U}_Y = \boldsymbol{S}_{X,P} \cdot \boldsymbol{U}_{X,P} \cdot \boldsymbol{S}_{X,P}^T \tag{3}$$

where: $\boldsymbol{U}_{X,P}, \boldsymbol{U}_Y$ – input and output covariance matrices and $\boldsymbol{S}_{X,P}$ – sensitivity matrix.

The sensitivity matrix $\boldsymbol{S}_{X,P}$ from formula (3) linearize the function $\boldsymbol{F}(\boldsymbol{X}, \boldsymbol{P})$ and fulfills the following relationships:

$$\boldsymbol{S}_{X,P} = [\boldsymbol{S}, \boldsymbol{S}_P] \quad \text{and} \quad \boldsymbol{S}_{X,P}^T = \begin{bmatrix} \boldsymbol{S}^T \\ \boldsymbol{S}_P^T \end{bmatrix} \tag{3a, b}$$

The randomized deviations from the estimates of the values of elements of the output measurand $\boldsymbol{Y}$ arise as a result of convolutions of the deviations of measurand $\boldsymbol{X}$ and deviations of the $\boldsymbol{P}$ parameters of the measurement system performing the function $\boldsymbol{F(X, P)}$. Their covariance matrices add up. The $\boldsymbol{U_Y}$ covariance matrix results from the linearization of the function $\boldsymbol{Y} = \boldsymbol{F}(\boldsymbol{X}, \boldsymbol{P})$ and in the general case is described by the following propagation equation

$$\begin{aligned} \boldsymbol{U_Y} &= [\boldsymbol{S}, \boldsymbol{S_p}] \begin{bmatrix} \boldsymbol{U_X} & \boldsymbol{U} \\ \boldsymbol{U^T} & \boldsymbol{U_p} \end{bmatrix} \begin{bmatrix} \boldsymbol{S^T} \\ \boldsymbol{S_P^T} \end{bmatrix} = [\boldsymbol{S}, \boldsymbol{S_p}] \begin{bmatrix} \boldsymbol{U_X S^T} + \boldsymbol{U S_P^T} \\ \boldsymbol{U^T S^T} + \boldsymbol{U_p S_P^T} \end{bmatrix} \\ &= \boldsymbol{S U_X S^T} + \boldsymbol{S U S_P^T} + \boldsymbol{S_p U^T S^T} + \boldsymbol{S_p U_p S_P^T} \end{aligned} \quad (4)$$

in which:

- sensitivity matrices with the following sizes: $\boldsymbol{S}$ $[m \times n]$ and $\boldsymbol{S_p}$ $[k \times m]$ respectively

$$\boldsymbol{S} = \begin{bmatrix} \frac{\partial y_1}{\partial x_1} & \cdots & \frac{\partial y_1}{\partial x_n} \\ \cdots & \cdots & \cdots \\ \frac{\partial y_m}{\partial x_1} & \cdots & \frac{\partial y_m}{\partial x_n} \end{bmatrix} \quad \text{and} \quad \boldsymbol{S_p} = \begin{bmatrix} \frac{\partial y_1}{\partial p_1} & \cdots & \frac{\partial y_1}{\partial p_k} \\ \cdots & \cdots & \cdots \\ \frac{\partial y_m}{\partial p_1} & \cdots & \frac{\partial y_m}{\partial p_k} \end{bmatrix} \quad (4\text{a, b})$$

- covariance matrices: $\boldsymbol{U_X}$– of input measurand $\boldsymbol{X}$ with sizes $[n \times n]$ and $\boldsymbol{U_p}$ – for parameters $\boldsymbol{P}$ of the processing function $\boldsymbol{F(X, P)}$ of the measurement system with sizes [k × $k$]

$$\boldsymbol{U_X} = \begin{bmatrix} u_{x1}^2 & \cdots & \rho_{x1n} u_{x1} u_{xn} \\ \cdots & \cdots & \cdots \\ \rho_{xn1} u_{xn} u_{x1} & \cdots & u_{xn}^2 \end{bmatrix}, \quad \boldsymbol{U_p} = \begin{bmatrix} u_{p1}^2 & \cdots & \rho_{p1k} u_{p1} u_{pk} \\ \cdots & \cdots & \cdots \\ \rho_{k1} u_{pk} u_{p1} & \cdots & u_{pk}^2 \end{bmatrix} \quad (4\text{c, d})$$

- covariance matrix for correlation between input quantities $\boldsymbol{X}$ and $\boldsymbol{P}$ parameters of the processing system of size $[n \times k]$

$$\boldsymbol{U} = \begin{bmatrix} \rho_{x1p1} u_{x1} u_{p1} & \cdots & \rho_{x1pk} u_{x1} u_{pk} \\ \cdots & \cdots & \cdots \\ \rho_{xnp1} u_{xn} u_{p1} & \cdots & \rho_{xnpk} u_{xn} u_{pk} \end{bmatrix} \quad (4\text{e})$$

After entering the designation $\boldsymbol{V} \equiv \boldsymbol{S U S_P^T}$ and transformations

$$\boldsymbol{V^T} = \left(\boldsymbol{S U S_P^T}\right)^T = \boldsymbol{S_p}(\boldsymbol{S U})^T = \boldsymbol{S_p U^T S^T}$$

the general Eq. (4) takes the simpler form:

$$\boldsymbol{U_Y} = \boldsymbol{S U_X S^T} + \boldsymbol{S_p U_p S_P^T} + \boldsymbol{V} + \boldsymbol{V^T} \quad (5)$$

In part 1 of this work [4], when determining the uncertainty in multivariable measurements, it has been shown that for different quantities the correlation takes place separately for their deviations of the component uncertainties of each type A and B. There is no correlation between deviations of different types in the GUM method of evaluating uncertainties. Therefore, uncertainties for each of types A and B should be determined separately for output quantities and then combined. For a ready meter or a system for multiparameter measurements, uncertainty and type A and correlation coefficients can be determined from experimental data by running measurements synchronously several times. But B-type uncertainties and correlation coefficients should be determined heuristically and separately for different operating conditions based on MPE error analysis, predicted correlation of parameters and structure of the measuring system.

## 3 Not Correlated Variables and Parameters of Processing Function

Let us consider a simpler case, often occurring in measuring practice, e.g. in the identification of object equivalent circuits and technical diagnostics of electrical devices, when the deviations of the measured quantities of measurand $\boldsymbol{X}$ and the deviations of the parameters $\boldsymbol{P}$ of the processing function of measurement system are not correlated each other (e.g. their hardware is located in different places and the impacts of influencing quantities affecting each of them are not related). Then $\boldsymbol{V} = \boldsymbol{V}^T = \boldsymbol{0}$ and the extended covariance matrix of the input menzurand $\boldsymbol{U}_{X,P}$ with the size $[(n + k) \times (n + k)]$ from formula (4) has two blocks equal $[\boldsymbol{0}]$, i.e.

$$\boldsymbol{U}_{X,P} = \begin{bmatrix} \boldsymbol{U}_X & [\boldsymbol{0}] \\ [\boldsymbol{0}] & \boldsymbol{U}_P \end{bmatrix} \tag{6}$$

where: $\boldsymbol{U}_X$, $\boldsymbol{U}_P$ are respectively the covariance matrix of input variables $\boldsymbol{X}$ with size $[n \times n]$ and the covariance matrix of system parameters $P$ with size $[k \times k]$.

For small increments is obtained

$$\Delta \boldsymbol{Y} = \Delta \boldsymbol{F}(\boldsymbol{X}, \boldsymbol{P}) = \boldsymbol{S}_X(\boldsymbol{X}) \cdot \varDelta \boldsymbol{X} + \boldsymbol{S}_P \cdot \varDelta \boldsymbol{P} \tag{7}$$

where: matrix $\boldsymbol{S}_X = \boldsymbol{S}|_{P\text{nom}}$ with size $[n \times m]$ and matrix $\boldsymbol{S}_P$ with size $[k \times m]$ are obtained from linearization of function $\boldsymbol{F}(\boldsymbol{X}, \boldsymbol{P})$ for input variables $\boldsymbol{X}$ and system parameters $\boldsymbol{P}$.

The variance propagation equation, for $\boldsymbol{U}_{X,P}$ according to (6) takes for the form:

$$\boldsymbol{U}_Y = [\boldsymbol{S}_X, \boldsymbol{S}_P] \cdot \begin{bmatrix} \boldsymbol{U}_X & [\boldsymbol{0}] \\ [\boldsymbol{0}] & \boldsymbol{U}_P \end{bmatrix} \cdot \begin{bmatrix} \boldsymbol{S}_X^T \\ \boldsymbol{S}_P^T \end{bmatrix} = \boldsymbol{S}_X \cdot \boldsymbol{U}_X \cdot \boldsymbol{S}_X^T + \boldsymbol{S}_P \cdot \boldsymbol{U}_P \cdot \boldsymbol{S}_P^T \tag{8}$$

Both covariance matrices $\boldsymbol{U_X}$ and $\boldsymbol{U_P}$ are symmetrical, and the deviations of the type A regardless of the deviations type B can correlate separately between all elements.

The covariance matrix $\boldsymbol{U_{\delta Y}}$ for the relative uncertainties has a similar form, i.e.

$$\boldsymbol{U_{\delta Y}} = \boldsymbol{S_{\delta X}} \cdot \boldsymbol{U_{\delta}} \cdot \boldsymbol{S_{\delta X}^T} + \boldsymbol{S_{\delta P}} \cdot \boldsymbol{U_{\delta P}} \cdot \boldsymbol{S_{\delta P}^T} \tag{9}$$

From formula (6) it follows that in the case of indirect multi-variable measurements when parameters $\boldsymbol{P}$ of the measurement system and the measured values $\boldsymbol{X}$ are not correlated, this formula is simplified to the form (8). The formula (9) for relative uncertainties results from it. Output covariance matrix $\boldsymbol{U_Y}$ then becomes the sum of two expressions only. The first component depends on the uncertainty of the input measurand $\boldsymbol{X}$, as in the classic approach according to GUM-S2 [2]. The second component concerns the uncertainty and correlation coefficients of $\boldsymbol{P}$ parameters of the system processing function $\boldsymbol{F(X, P)}$.

The determination of the covariance matrix $\boldsymbol{U_{Y,P}}$ of the initial menzurand $\boldsymbol{Y}$ should be carried out separately for the deviations described by the type A and type B uncertainties. For absolute uncertainties, the formula (10a and 10b) is obtained:

$$\boldsymbol{U_Y} = \boldsymbol{S_X}(\boldsymbol{U_{XA}} + \boldsymbol{U_{XB}})\boldsymbol{S_X^T} + \boldsymbol{S_P}(\boldsymbol{U_{PA}} + \boldsymbol{U_{PB}})\boldsymbol{S_P^T} \tag{10a}$$

or

$$\boldsymbol{U_Y} = \boldsymbol{U_{YA}} + \boldsymbol{U_{YB}} = \boldsymbol{S_X U_{XA} S_X^T} + \boldsymbol{S_P U_{PA} S_P^T} + \boldsymbol{S_X U_{XB} S_X^T} + \boldsymbol{S_P U_{PB} S_P^T} \tag{10b}$$

The use of formulas (9) and (10a and 10b) has the additional advantage when measured with the same system under different influencing conditions. All you need to do is adjust the $\boldsymbol{U_{XB}}$ and $\boldsymbol{U_{PB}}$ matrices describing the B-type uncertainties accordingly. It is also possible to use the formula (10a and 10b) when $u_{i0} \neq 0$ and two-element formula (3) from part 1 [4] is used for the instrumental component of uncertainty type B, i.e.:

$$u_x|_{max} = u_{x0}|_{max} + |x - x_0| \cdot \delta_x|_{max} \tag{11}$$

where: $\delta_x|_{max}$ – maximal permissible relative uncertainty of increment $X–X_0$

For the relative uncertainties $\delta_{xi}$ of increments $X_i – X_{i0}$ or for $u_{\text{reli}}$ when the initial uncertainties $u_{i0} = 0$, the following matrix formula is obtained:

$$\boldsymbol{U_{\delta Y}} = \boldsymbol{S_{\delta X}}(\boldsymbol{U_{\delta XA}} + \boldsymbol{U_{\delta XB}})\boldsymbol{S_{\delta X}^T} + \boldsymbol{S_{\delta P}}(\boldsymbol{U_{\delta PA}} + \boldsymbol{U_{\delta PB}})\boldsymbol{S_{\delta P}^T} \tag{12}$$

Therefore, the equations of the initial covariance matrix (9) and (12) of relative uncertainties have a similar structure as for absolute uncertainties. Formula for absolute uncertainties related to the ranges, i.e. $u_x|_{max}/(X_{max}–X_0)$ can be used and find easy.

# 4 Example. Determination the Uncertainty of Voltage and Current at the Output of the T-Type Two-Port from Measurements at Input

## 4.1 Graph and Equations of the T-Type Two-Port Network

From measurements of the 2D measurand $\boldsymbol{X} = [U_{out}, I_{out}]^{\mathrm{T}}$ on output of the passive two-port circuit the values of $\boldsymbol{Y} = [U_{in}, I_{in}]^{\mathrm{T}}$ and their standard uncertainties $[u(U_{out}), u(I_{out})]$ should be find. This applies, for example, to cases where parameters of high voltage net are estimated from measurements at low voltage. Transmission equations of the passive two-port network [5] are:

$$\begin{aligned} U_{in} &= A_{11}U_{out} + A_{12}I_{out} \\ I_{in} &= A_{21}U_{out} + A_{22}I_{out} \end{aligned} \tag{13a, b}$$

or in matrix notation: $\begin{bmatrix} U_{in} \\ I_{in} \end{bmatrix} = A \begin{bmatrix} U_{out} \\ I_{out} \end{bmatrix}$

where: $\boldsymbol{A} = \begin{bmatrix} A_{11} & A_{12} \\ A_{21} & A_{22} \end{bmatrix}, \quad det(\boldsymbol{A}) = A_{11}A_{22} - A_{12}A_{21} = 1$

Let us consider the passive twoport type T given on Fig. 1.

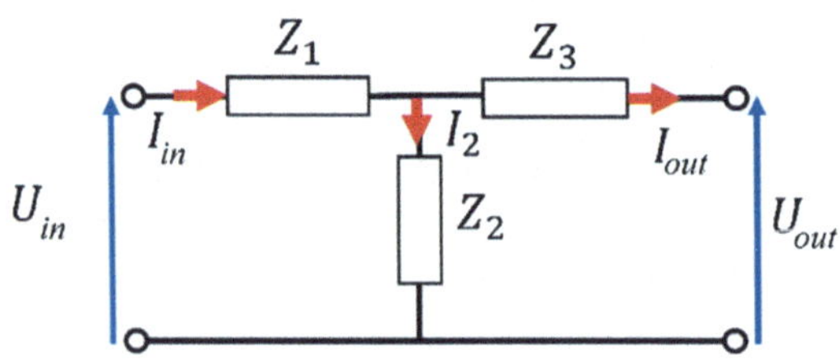


**Fig. 1.** Diagram of a passive twoport T-type circuit.

From the Kirchhoff current equation: $I_{in} = I_2 + I_{out}$ and two Kirchhoff voltage equations $U_{in} = Z_2I_2 + Z_1I_{in}$; $Z_2I_2 = U_{out} + Z_3I_{out}$ for T − type twoport [5] is obtained in input:

$$-\text{voltage}: \quad U_{in} = \left(1 + \frac{\mathbf{Z}_1}{\mathbf{Z}_2}\right)U_{out} + (Z_1 + Z_3 + \frac{\mathbf{Z}_1\mathbf{Z}_3}{\mathbf{Z}_2})I_{out} \tag{14a}$$

$$-\text{current}: \quad I_{in} = \frac{1}{\mathbf{Z}_2}U_{out} + (1 + \frac{\mathbf{Z}_3}{\mathbf{Z}_2})I_{out} \tag{14b}$$

The measurement results of $U_{out}$, $I_{out}$ and values of three impedances $Z_i$ have uncertainties (as standard deviations of the distributions of their deviations from the nominal values or from their estimators). Parameters of $Z_i$ impedances can be:

- uncorrelated, i.e.: $\rho_{zij} = 0$
- correlated.

In this example the correlation coefficient $\rho_{zij} = 1$ is only considered.

## 4.2 Input Increments Determined by the Total Differential

for any two-port: $\Delta U_{in} = \Delta(A_{11}U_{out}) + \Delta(A_{12}I_{out})$

$$\Delta U_{in} = \Delta_{A11}U_{out} + A_{11}\Delta(U_{out}) + \Delta_{A12}I_{out} + A_{12}\Delta(I_{out}) \quad (15a)$$

$$\Delta I_{in} = \Delta_{A21}U_{out} + A_{21}\Delta(U_{out}) + \Delta_{A22}I_{out} + A_{22}\Delta(I_{out}) \quad (15b)$$

where:

$\Delta_{A11} = \frac{\Delta Z_1}{Z_2} - \frac{\Delta Z_2}{Z_2^2}Z_1$; $\Delta_{A12} = \Delta Z_1\left(1 + \frac{Z_3}{Z_2}\right) - \Delta Z_2\frac{Z_1 Z_3}{Z_2^2} + \Delta Z_3\left(1 + \frac{Z_1}{Z_2}\right)$; $\Delta_{A21} = -\frac{\Delta Z_2}{Z_2^2}$;

$\Delta_{A22} = \frac{\Delta Z_3}{Z_2} - \frac{\Delta Z_2}{Z_2^2}Z_1$.

For passive twoport T

$$\Delta U_{in} = \left[1 + \frac{Z_1}{Z_2}\right]\Delta(U_{out}) + \left[Z_1 + Z_3 + \frac{Z_1 Z_3}{Z_2}\right]\Delta(I_{out}) + \left[\frac{U_{out}}{Z_2} + \left(1 + \frac{Z_3}{Z_2}\right)I_{out}\right]\Delta Z_1$$
$$- \left[\frac{Z_1}{Z_2^2}U_{out} + \frac{Z_1 Z_3}{Z_2^2}I_{out}\right]\Delta Z_2 + \left(1 + \frac{Z_1}{Z_2}\right)]I_{out}\Delta Z_3 \quad (16a)$$

$$\Delta I_{in} = \frac{1}{Z_2}\Delta(U_{out}) + \left[1 + \frac{Z_3}{Z_2}\right]\Delta(I_{out}) - \left[\frac{Z_3}{Z_2^2}I_{out} + \frac{U_{out}}{Z_2^2}\right]\Delta Z_2 + \frac{I_{out}}{Z_2}\Delta Z_3 \quad (16b)$$

## 4.3 Propagation of Uncertainties of Output Variables $I_{wy}$, $U_{wy}$ of T-Type Twoport

To determine uncertainties $u(U_{in})$, $u(I_{\text{in}})$ of input quantities as elements of the measurand $\boldsymbol{Y} = [U_{in}, I_{in}]$ for any twoport network, e.g. type T, formulas for deviations $\Delta U_{in}, \Delta I_{in}$ should be used. These deviations depend on the increments $\Delta U_{out}, \Delta I_{out}$ and $\Delta Z_1, \Delta Z_2, \Delta Z_3$. For uncertainties of input quantities with a total lack of correlation, i.e. for $\rho_{Zij} = 0$ and $\rho_{Uin,Iin} = 0$, it is enough to determine the sums of squares of these words, i.e. their variances as squares of standard uncertainties. Even in this case it is the quite tedious operation. We will use the matrix method as a general one and also as more resistant to mistakes. Indirectly determined variables $\boldsymbol{Y} = [U_{out}, I_{out}]^{\text{T}}$ at the input of the twoport T-type depend on the results of measuring variables at its output $\boldsymbol{X} = [U_{out}, I_{out}]^{\text{T}}$ and on

$\boldsymbol{P} = [Z_1, Z_2, Z_3]^{\mathrm{T}}$ parameters of the twoport circuit. This is described generally as (1a), i.e. $\boldsymbol{Y} = \boldsymbol{F}(\boldsymbol{X}, \boldsymbol{P})$. For a linear twoport network, it is a matrix relationship

$$\boldsymbol{Y} = \boldsymbol{A} \cdot \boldsymbol{X} \text{ or } \begin{bmatrix} U_{in} \\ I_{in} \end{bmatrix} = \boldsymbol{A} \cdot \begin{bmatrix} U_{out} \\ I_{out} \end{bmatrix} \tag{17}$$

where $\boldsymbol{A} \equiv \begin{bmatrix} A_{11} & A_{12} \\ A_{21} & A_{22} \end{bmatrix} = \begin{bmatrix} 1 + \frac{Z_1}{Z_2} & Z_1 + Z_3 + \frac{Z_1 Z_3}{Z_2} \\ \frac{1}{Z_2} & 1 + \frac{Z_3}{Z_2} \end{bmatrix}$

The law of propagation of variance for uncorrelated input quantities and of circuit parameters assumes the relationship (8), i.e.:

$$\begin{aligned} \boldsymbol{U}_{\boldsymbol{Y}}(\boldsymbol{X}, \boldsymbol{P}) &= [\boldsymbol{S}_{\boldsymbol{X}}, \boldsymbol{S}_{\boldsymbol{P}}] \cdot \begin{bmatrix} \boldsymbol{U}_{\boldsymbol{X}} & [\boldsymbol{0}] \\ [\boldsymbol{0}] & \boldsymbol{U}_{\boldsymbol{P}} \end{bmatrix} \cdot \begin{bmatrix} \boldsymbol{S}_{\boldsymbol{X}}^{T} \\ \boldsymbol{S}_{\boldsymbol{P}}^{T} \end{bmatrix} = \boldsymbol{S}_{\boldsymbol{X}} \cdot \boldsymbol{U}_{\boldsymbol{X}} \cdot \boldsymbol{S}_{\boldsymbol{X}}^{T} + \boldsymbol{S}_{\boldsymbol{P}} \cdot \boldsymbol{U}_{\boldsymbol{P}} \cdot \boldsymbol{S}_{\boldsymbol{P}}^{T} \\ &\equiv \boldsymbol{U}_{\boldsymbol{YX}} + \boldsymbol{U}_{\boldsymbol{YP}} \end{aligned} \tag{18}$$

where: the covariance matrices and the sensitivity matrices are defined as follows

- for uncorrelated measured variables $\boldsymbol{X}$
  Sensitivity and covariance matrices:

$$\boldsymbol{S} = \begin{bmatrix} A_{11} & A_{12} \\ A_{21} & A_{22} \end{bmatrix}, \boldsymbol{U}_{\boldsymbol{X}} = \begin{bmatrix} \sigma_{U_{out}}^2 & 0 \\ 0 & \sigma_{I_{out}}^2 \end{bmatrix}, \boldsymbol{U}_{\boldsymbol{Yx}} = \begin{bmatrix} \sigma_{U_{in}}^2 & \rho_{in}\sigma_{U_{in}}\sigma_{U_{in}} \\ \rho_{in}\sigma_{U_{in}}\sigma_{U_{in}} & \sigma_{I_{in}}^2 \end{bmatrix} \tag{18a, b, c}$$

- for uncorrelated parameters of circuit T performing processing function $\boldsymbol{F}$ covariance and sensitivity matrices:

$$\boldsymbol{U}_{\boldsymbol{P}} = \begin{bmatrix} \sigma_{Z_1}^2 & 0 & 0 \\ 0 & \sigma_{Z_2}^2 & 0 \\ 0 & 0 & \sigma_{Z_3}^2 \end{bmatrix}, \tag{19a}$$

and

$$\boldsymbol{S}_{\boldsymbol{P}} = \begin{bmatrix} \frac{U_{out}}{Z_2} + I_{out}\left(1 + \frac{Z_3}{Z_2}\right) & -\frac{Z_1}{Z_2^2}(U_{out} + I_{out}Z_3) & I_{out}\left(1 + \frac{Z_1}{Z_2}\right) \\ 0 & -\frac{1}{Z_2^2}(U_{out} + I_{out}Z_3) & \frac{I_{out}}{Z_2} \end{bmatrix}, \tag{19b}$$

After the designation for current in impedance $Z_2$ as $I_2 = (U_{out} + I_{out}Z_3)/Z_2$, from Kirchhoff's first law $I_{in} = I_{out} + I_2$ and the form of sensitivity matrix $\boldsymbol{S}_{\boldsymbol{P}}$ is simplified

$$\boldsymbol{S}_{\boldsymbol{P}} = \begin{bmatrix} I_{in} & -\frac{Z_1 I_2}{Z_2} & I_{out}\left(1 + \frac{Z_1}{Z_2}\right) \\ 0 & -\frac{I_2}{Z_2} & \frac{I_{out}}{Z_2} \end{bmatrix} \tag{20}$$

The equation of covariance matrix $\boldsymbol{U_Y}$ obtained from $\boldsymbol{U_X}$ then is

$$\boldsymbol{U_{YX}} = \boldsymbol{S_X} \cdot \boldsymbol{U_X} \cdot \boldsymbol{S_X^T} = \begin{bmatrix} A_{11}^2\sigma_{U_{out}}^2 + A_{12}^2\sigma_{I_{out}}^2 & A_{11}A_{21}\sigma_{U_{out}}^2 + A_{12}A_{22}\sigma_{I_{out}}^2 \\ A_{11}A_{21}\sigma_{U_{out}}^2 + A_{12}A_{22}\sigma_{I_{out}}^2 & A_{21}^2\sigma_{U_{out}}^2 + A_{22}^2\sigma_{I_{out}}^2 \end{bmatrix}. \quad (21)$$

and the covariance matrix of parameters $\boldsymbol{P}$ of the processing function $\boldsymbol{F}$ of twoport T is

$$\boldsymbol{U_{YP}} = \boldsymbol{S_P}\boldsymbol{U_P}\boldsymbol{S_P^T} = \begin{bmatrix} I_{in}^2\sigma_{Z_1}^2 + \frac{Z_1^2}{Z_2^2}I_2^2\sigma_{Z_2}^2 + I_{out}^2\left(1 + \frac{Z_1}{Z_2}\right)^2\sigma_{Z_3}^2 & \frac{Z_1 I_2^2}{Z_2^2}\sigma_{Z_2}^2 + I_{out}^2\frac{Z_1+Z_2}{Z_2^2}\sigma_{Z_3}^2 \\ \frac{Z_1 I_2^2}{Z_2^2}\sigma_{Z_2}^2 + I_{out}^2\frac{Z_1+Z_2}{Z_2^2}\sigma_{Z_3}^2 & \frac{I_2^2}{Z_2^2}\sigma_{Z_2}^2 + \frac{I_{out}^2}{Z_2^2}\sigma_{Z_3}^2 \end{bmatrix} \quad (22)$$

**Standard uncertainties of elements of *Y* vector:**

**for the input voltage**

- deviation:

$$\Delta(U_{in}) = \left[1 + \frac{Z_1}{Z_2}\right]\Delta(U_{out}) + \left[Z_1 + Z_3 + \frac{Z_1Z_3}{Z_2}\right]\Delta(I_{out}) + \left[\frac{U_{out}}{Z_2} + \left(1 + \frac{Z_3}{Z_2}\right)I_{out}\right]\Delta Z_1 + \\ - \left[\frac{Z_1}{Z_2^2}U_{out} + \frac{Z_1Z_3}{Z_2^2}I_{out}\right]\Delta Z_2 + I_{out}\left(1 + \frac{Z_1}{Z_2}\right)]\Delta Z_3 \quad (23)$$

variance:

$$\begin{aligned} \sigma_{U_{in}}^2 &= A_{11}^2\sigma_{U_{out}}^2 + A_{12}^2\sigma_{I_{out}}^2 + I_{in}^2\sigma_{Z_1}^2 + \frac{Z_1^2}{Z_2^2}I_2^2\sigma_{Z_2}^2 + I_{out}^2\left(1 + \frac{Z_1}{Z_2}\right)^2\sigma_{Z_3}^2 \\ &= \frac{1}{Z_2^2}[(Z_1 + Z_2)^2\sigma_{U_{out}}^2 + (Z_1Z_2 + Z_3Z_2 + Z_1Z_3)^2\sigma_{I_{out}}^2 + I_{in}^2Z_2^2\sigma_{Z_1}^2 \\ &\quad + I_2^2Z_1^2\sigma_{Z_2}^2 + I_{out}^2(Z_1 + Z_2)^2\sigma_{Z_3}^2)] \end{aligned} \quad (24)$$

**– for the input current**

- deviation

$$\Delta(I_{in}) = \frac{1}{Z_2}\Delta(U_{out}) + \left[1 + \frac{Z_3}{Z_2}\right]\Delta(I_{out}) - \left[\frac{Z_3}{Z_2^2}I_{out} + \frac{U_{out}}{Z_2^2}\right]\Delta Z_2 + \frac{I_{out}}{Z_2}\Delta Z_3 \quad (25)$$

Formulas of deviations of the matrix method and total differential are identical.

- variance:

$$\sigma_{I_{in}}^2 = A_{21}^2\sigma_{U_{out}}^2 + A_{22}^2\sigma_{I_{out}}^2 + \frac{I_2^2}{Z_2^2}\sigma_{Z_2}^2 + \frac{I_{out}^2}{Z_2^2}\sigma_{Z_3}^2 \\ = \frac{1}{Z_2^2}\left(\sigma_{U_{out}}^2 + (Z_2+Z_3)^2\sigma_{I_{out}}^2 + I_2^2\sigma_{Z_2}^2 + I_{out}^2\sigma_{Z_3}^2\right) \tag{26}$$

- correlation coefficient

$$\rho_{U_{in}I_{in}} = \frac{A_{11}A_{21}\sigma_{U_{out}}^2 + A_{12}A_{22}\sigma_{I_{out}}^2 + \frac{Z_1I_2^2}{Z_2^2}\sigma_{Z_2}^2 + I_{out}^2\frac{Z_1+Z_2}{Z_2^2}\sigma_{Z_3}^2}{\sigma_{U_{in}}\sigma_{I_{in}}} \tag{27}$$

## 4.4 The Case of Correlated Impedances $Z_1, Z_2, Z_3$

Covariance matrix of ***P***:

$$\boldsymbol{U_P} = \begin{bmatrix} \sigma_{Z_1}^2 & \rho_{Z12}\sigma_{Z_1}\sigma_{Z_2} & \rho_{Z13}\sigma_{Z_1}\sigma_{Z_3} \\ \rho_{Z12}\sigma_{Z_1}\sigma_{Z_2} & \sigma_{Z_2}^2 & \rho_{Z23}\sigma_{Z_2}\sigma_{Z_3} \\ \rho_{Z13}\sigma_{Z_1}\sigma_{Z_3} & \rho_{Z23}\sigma_{Z_2}\sigma_{Z_3} & \sigma_{Z_3}^2 \end{bmatrix} \tag{28}$$

Component $\boldsymbol{U_{YP}}$ of matrix $\boldsymbol{U_Y}$:

$$\boldsymbol{U_{YP}} = \boldsymbol{S_P}\cdot\left(\boldsymbol{U_P}\cdot\boldsymbol{S_P^T}\right) = \begin{bmatrix} I_{in} & -\frac{Z_1I_2}{Z_2} & I_{in}\left(1+\frac{Z_1}{Z_2}\right) \\ 0 & -\frac{I_2}{Z_2} & \frac{I_{in}}{Z_2} \end{bmatrix} \times \\ \begin{bmatrix} I_{in}\sigma_{Z_1}^2 - \frac{Z_1I_2}{Z_2}\rho_{Z12}\sigma_{Z_1}\sigma_{Z_2} + I_{out}\left(1+\frac{Z_1}{Z_2}\right)\rho_{Z13}\sigma_{Z_1}\sigma_{Z_3} & -\frac{I_2}{Z_2}\rho_{Z12}\sigma_{Z_1}\sigma_{Z_2} + \frac{I_{out}}{Z_2}\rho_{Z13}\sigma_{Z_1}\sigma_{Z_3} \\ I_{in}\rho_{Z12}\sigma_{Z_1}\sigma_{Z_2} - \frac{Z_1I_2}{Z_2}\sigma_{Z_2}^2 + I_{out}\left(1+\frac{Z_1}{Z_2}\right)\rho_{Z23}\sigma_{Z_2}\sigma_{Z_3} & -\frac{I_2}{Z_2}\sigma_{Z_2}^2 + \frac{I_{out}}{Z_2}\rho_{Z23}\sigma_{Z_2}\sigma_{Z_3} \\ I_{in}\rho_{Z13}\sigma_{Z_1}\sigma_{Z_3} - \frac{Z_1I_2}{Z_2}\rho_{Z23}\sigma_{Z_2}\sigma_{Z_3} + I_{out}\left(1+\frac{Z_1}{Z_2}\right)\sigma_{Z_3}^2 & -\frac{I_2}{Z_2}\rho_{Z23}\sigma_{Z_2}\sigma_{Z_3} + \frac{I_{out}}{Z_2}\sigma_{Z_3}^2 \end{bmatrix} \tag{29}$$

Then for correlation coefficients of impedances: $(\rho_{Z12}, \rho_{Z13}, \rho_{Z23}) \neq 0$ the formulas for uncertainty of output quantities have additional members, e.g. for voltage $U_{\mathrm{we}}$

$$\sigma_{U_{in}}^2 = \frac{1}{Z_2^2}\Big[(Z_1+Z_2)^2\sigma_{U_{out}}^2 + (Z_1Z_2+Z_3Z_2+Z_1Z_3)\sigma_{I_{out}}^2 + I_{in}^2Z_2^2\sigma_{Z_1}^2 + I_2^2Z_1^2\sigma_{Z_2}^2 + \\ I_{out}^2(Z_1+Z_2)^2\sigma_{Z_3}^2) - 2I_2I_{in}Z_1Z_2\rho_{Z12}\sigma_{Z_1}\sigma_{Z_2} + 2I_{in}I_{out}Z_2(Z_1+Z_2)\rho_{Z13}\sigma_{Z_1}\sigma_{Z_3} + \\ -2I_2I_{out}Z_1Z_2(Z_1+Z_2)\rho_{Z23}\sigma_{Z_2}\sigma_{Z_3}\Big] \tag{30}$$

Output correlation coefficient

$$\rho_{U_{in}I_{in}}=\frac{\begin{array}{l}A_{11}A_{21}\sigma^2_{U_{out}}+A_{12}A_{22}\sigma^2_{I_{out}}-\frac{I_2}{Z_2}\left(I_{in}\rho_{Z12}\sigma_{Z_1}\sigma_{Z_2}\right.\\ \left.-\frac{Z_1I_2}{Z_2}\sigma^2_{Z_2}+I_{out}\left(1+\frac{Z_1}{Z_2}\right)\rho_{Z23}\sigma_{Z_2}\sigma_{Z_3}\right)+\\ +\frac{I_{out}}{Z_2}\left(I_{in}\rho_{Z13}\sigma_{Z_1}\sigma_{Z_3}-\frac{Z_1I_2}{Z_2}\rho_{Z23}\sigma_{Z_2}\sigma_{Z_3}+I_{out}\left(1+\frac{Z_1}{Z_2}\right)\sigma^2_{Z_3}\right)\end{array}}{\sigma_{U_{in}}\sigma_{I_{in}}} \tag{31}$$

### 4.5 Uncertainty of the Voltage at Correlation $\rho_{Z12}=\rho_{Z13}=\rho_{Z23}=1$

Fully correlated impedances: and after taking the designations: $a=I_{in}Z_2\sigma_{Z_1}$, $b=I_2Z_1\sigma_{Z_2}, c=I_{in}(Z_1+Z_2)\sigma_{Z_3}$

$$\rho_{U_{in}I_{in}}=\frac{A_{11}A_{21}\sigma^2_{U_{out}}+A_{12}A_{22}\sigma^2_{I_{out}}+\frac{-ab+b^2-bc}{Z_1Z_2^2}+\frac{ca-bc+c^2}{(Z_1+Z_2)Z_2^2}}{\sigma_{U_{in}}\sigma_{I_{in}}} \tag{32}$$

For $a=b-c$ correlation coefficient

$$\rho_{U_{in}I_{in}}=\left(A_{11}A_{21}\sigma^2_{U_{out}}+A_{12}A_{22}\sigma^2_{I_{out}}\right)/(\sigma_{U_{in}}\sigma_{I_{in}})$$

– almost as correlation with ideal processing function.

$$\begin{aligned}\sigma^2_{U_{in}}=\frac{1}{Z_2^2}\Big[&(Z_1+Z_2)^2\sigma^2_{U_{out}}+\left(Z_1Z_2+Z_3Z_2+Z_1Z_3\right)\sigma^2_{I_{out}}+(I_{in}Z_2\sigma_{Z_1}-I_2Z_1\sigma_{Z_2})^2\\ &+(I_2Z_1\sigma_{Z_2}-I_{out}(Z_1+Z_2)\sigma_{Z_3})^2+(I_{in}Z_2\sigma_{Z_1}+I_{out}(Z_1+Z_2)\sigma_{Z_3})^2-I^2_{in}Z^2_2\sigma^2_{Z_1}+\\ &-I_2^2Z_1^2\sigma^2_{Z_2}-I^2_{out}(Z_1+Z_2)^2\sigma^2_{Z_3})\Big]\end{aligned} \tag{33}$$

Function of three variables with non-negative parameters

$$f(a,b,c)=(a-b)^2+(b-c)^2+(a+c)^2-a^2-b^2-c^2$$

reaches the minimum for $a=b-c$. This is a condition for three partial derivatives Then for $a=b-c$

$$f(b-c,b,c)=(b-c-b)^2+(b-c)^2+(b-c+c)^2-(b-c)^2-b^2-c^2=0.$$

It's easy to check that if e.g. $f(1,1,1)=1$. Therefore, the condition of minimizing the voltage measurement uncertainty of the $\boldsymbol{F}$ function, i.e. for its value zero means:

$$I_{in}Z_2\sigma_{Z_1} = I_2Z_1\sigma_{Z_2} - I_{out}(Z_1+Z_2)\sigma_{Z_3} \tag{34}$$

After substitution $I_{in} = I_{out} + I_2$ the following condition is obtained:

$$\frac{I_2}{I_{out}} = \frac{Z_2\sigma_{Z_1} + (Z_1+Z_2)\sigma_{Z_3}}{Z_1\sigma_{Z_2} - Z_2\sigma_{Z_1}} \tag{35}$$

When using the relationship $I_2 = (U_{out} + I_{out}Z_3)/Z_2$, this condition leads to the determination of the proper load impedance

$$\frac{U_{out}}{I_{out}} = Z_2\frac{Z_2\sigma_{Z_1} + (Z_1+Z_2)\sigma_{Z_3}}{Z_1\sigma_{Z_2} - Z_2\sigma_{Z_1}} - Z_3 = Z_2\frac{\delta_{Z_1} + Z_3\left(\frac{1}{Z_1} + \frac{1}{Z2}\right)\delta_{Z_3}}{|\delta_{Z_2} - \delta_{Z_1}|} - Z_3 \tag{36}$$

For relations of relative uncertainties $|\delta_{Z_2} - \delta_{Z_1}| \to 0$ the load resistance at the output in this case tends to infinity – the circuit output terminals are open and then we get the smallest value of input voltage uncertainty identical to the processing error $\boldsymbol{U_F} = \boldsymbol{0}$

$$\sigma^2_{U_{in}} = \frac{1}{Z_2^2}\left[(Z_1+Z_2)^2\sigma^2_{U_{out}} + (Z_1Z_2 + Z_3Z_2 + Z_1Z_3)\sigma^2_{I_{out}}\right] \tag{37}$$

In the event of a short circuit at the output, reaching the minimum value of the uncertainty of the input voltage measurement is possible when

$$\frac{Z_2\sigma_{Z_1} + (Z_1+Z_2)\sigma_{Z_3}}{Z_1\sigma_{Z_2} - Z_2\sigma_{Z_1}} - Z_3 = 0$$

Hence:

$$Z_2^2(\sigma_{Z_1} + \sigma_{Z_3}) + Z_2(Z_1\sigma_{Z_3} + Z_3\sigma_{Z_1}) - Z_3Z_1\sigma_{Z_2} = 0 \tag{38}$$

For relative uncertainties, this gives the relationship:

$$Z_2\left(\frac{\delta_{Z_1}}{Z_3} + \frac{\delta_{Z_3}}{Z_1}\right) + \delta_{Z_3} + \delta_{Z_1} - \delta_{Z_2} = 0$$

and finally

$$\delta_{Z_1}\left(\frac{1}{Z_3} + \frac{1}{Z_2}\right) + \delta_{Z_3}\left(\frac{1}{Z_1} + \frac{1}{Z_2}\right) = \frac{\delta_{Z_2}}{Z_2} \tag{39a}$$

or (39a) in the simplified form using the symbol of parallel connection

$$\frac{\delta_{Z_1}}{Z_2||Z_3} + \frac{\delta_{Z_3}}{Z_1||Z_2} = \frac{\delta_{Z_2}}{Z_2} \tag{39b}$$

### 4.6 Uncertainty of the Current at Correlation $\rho_{Z23} = 1$

The general variance of current from matrix $\boldsymbol{U_{YX}}$ and $\boldsymbol{U_{YP}}$ is given

$$\begin{aligned}\sigma_{I_{in}}^2 &= A_{21}^2\sigma_{U_{out}}^2 + A_{22}^2\sigma_{I_{out}}^2 + \frac{I_2^2}{Z_2^2}\sigma_{Z_2}^2 + \frac{I_{out}^2}{Z_2^2}\sigma_{Z_3}^2 - 2\rho_{Z23}\frac{I_{out}I_2}{Z_2^2}\sigma_{Z_2}\sigma_{Z_3} \\ &= \frac{1}{Z_2^2}\sigma_{U_{out}}^2 + \left(1 + \frac{Z_3}{Z_2}\right)^2\sigma_{I_{out}}^2 + \frac{I_2^2}{Z_2^2}\sigma_{Z_2}^2 + \frac{I_{out}^2}{Z_2^2}\sigma_{Z_3}^2 - 2\rho_{Z23}\frac{I_{out}I_2}{Z_2^2}\sigma_{Z_2}\sigma_{Z_3}\end{aligned} \tag{40}$$

For $\rho_{Z23} = 1$

$$\begin{aligned}\sigma_{I_{in}}^2 &= \frac{1}{Z_2^2}\left(\sigma_{U_{out}}^2 + (Z_2 + Z_3)^2\sigma_{I_{out}}^2 + I_2^2\sigma_{Z_2}^2 + I_{out}^2\sigma_{Z_3}^2 - 2I_{out}I_2\sigma_{Z_2}\sigma_{Z_3}\right) \\ &= \frac{1}{Z_2^2}\left(\sigma_{U_{out}}^2 + (Z_2 + Z_3)^2\sigma_{I_{out}}^2 + (I_2\sigma_{Z_2} - I_{out}\sigma_{Z_3})^2\right)\end{aligned} \tag{41}$$

The lowest uncertainty value is obtained when $I_2\sigma_{Z_2} - I_{out}\sigma_{Z_3} = 0$
When using relations:

$$\begin{aligned}I_2 &= (U_{out} + I_{out}Z_3)/Z_2 \\ \frac{U_{out}}{I_{out}} &= \frac{Z_2\sigma_{Z_3}}{\sigma_{Z_2}} - Z_3 = \frac{Z_2\sigma_{Z_3} - \sigma_{Z_2}Z_3}{\sigma_{Z_2}};\end{aligned}$$

for the condition $\sigma_{Z_2} \to 0$ we get infinite matching resistance at the output.

If $Z_2\sigma_{Z_3} = \sigma_{Z_2}Z_3$, the relative uncertainties are equal for $Z_2$ and $Z_3$ $\delta_{Z_2} = \delta_{Z_3}$ the matching resistance is zero - short circuit at the system output.

Measurement uncertainty of input current is identical to the case when the covariance matrix of the processing function $\boldsymbol{U_F} = \boldsymbol{0}$

$$\sigma_{I_{out}}^2 = \frac{1}{Z_2^2}\left(\sigma_{U_{out}}^2 + (Z_2 + Z_3)^2\sigma_{I_{out}}^2\right) \tag{42}$$

**Conclusion** (case of full correlation). When the two-port network at the output is open, despite the correlation of the parameters $[Z_1, Z_2, Z_3]$ for the condition $\delta_{Z_2} \to \delta_{Z_1}$, the voltage measurement uncertainties at the four-terminal input are obtained, such as in the absence of a covariance matrix $\boldsymbol{U_P}$ – i.e. ideal system parameters processing. For the uncertainty of the current measured at the two-port output and for the uncertainty $\sigma_{Z_2} \to 0$, also uncertainties as in the absence of function uncertainty are obtained, i.e. as for the ideal processing. There is correlation between current and voltage as almost for ideal processing function. For a short circuit of the system for measuring the current, for obtaining uncertainty such as in ideal processing requires the fulfillment of the condition $\delta_{Z_2} = \delta_{Z_3}$, while for measuring the voltage at the input with a short circuit at the output needed is the condition: $\frac{\delta_{Z_1}}{Z_2||Z_3} + \frac{\delta_{Z_3}}{Z_1||Z_2} = \frac{\delta_{Z_2}}{Z_2}$.

## 4.7 The Simplified Case of Twoport Type T: $Z_3 = 0$

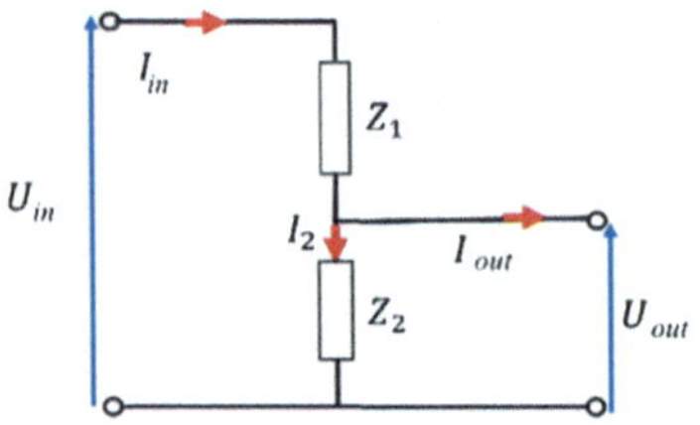


$$A = \begin{bmatrix} A_{11} & A_{12} \\ A_{21} & A_{22} \end{bmatrix} = \begin{bmatrix} 1 + \frac{Z_1}{Z_2} & Z_1 \\ \frac{1}{Z_2} & 1 \end{bmatrix}$$

$$I_2 = (U_{out} + I_{out} Z_3)/Z_2 = U_{out}/Z_2$$

**Fig. 2.** The twoport type T of $Z_3 = 0$.

**Output uncertainties**

- **for voltage**
  - in the absence of correlation $\rho_{Z12} = 0$

$$\sigma^2_{U_{in}} = \frac{1}{Z_2^2}\left[(Z_1 + Z_2)^2 \sigma^2_{U_{out}} + Z_1 Z_2 \sigma^2_{I_{out}} + I_{in}^2 Z_2^2 \sigma^2_{Z_1} + I_2^2 Z_1^2 \sigma^2_{Z_2}\right] \tag{43}$$

  - with full correlation $\rho_{Z12} = 1$

$$\sigma^2_{U_{in}} = \frac{1}{Z_2^2}[(Z_1 + Z_2)^2 \sigma^2_{U_{out}} + Z_1 Z_2 \sigma^2_{I_{out}} + (I_{in} Z_2 \sigma_{Z_1} - I_2 Z_1 \sigma_{Z_2})^2] \tag{44}$$

The minimum value of uncertainty occurs for the ideal processing function

$$I_{in} Z_2 \sigma_{Z_1} - I_2 Z_1 \sigma_{Z_2} = 0 \quad \text{and} \quad \frac{U_{out}}{I_{out}} = \frac{Z_2^2 \sigma_{Z_1}}{Z_1 \sigma_{Z_2}} = \frac{\delta_{Z_1}}{\delta_{Z_2}}. \tag{45a, b}$$

If $\delta_{Z_2} \to 0\,(\sigma_{Z_2} \to 0,\ Z_1 \sigma_{Z_2} \to 0)$ then the load resistance for the minimum voltage uncertainty is infinite - output opening, when $\delta_{Z_1} \to 0$ $(\sigma_{Z_1} \to 0,\ Z_2^2 \sigma_{Z_1} \to 0)$ then the load resistance for the minimum voltage uncertainty is zero - output short circuit (Fig. 2).

- **for current**

In the absence of correlation and full correlation, the formula is the same.

$$\sigma^2_{I_{in}} = \frac{1}{Z_2^2}\left(\sigma^2_{U_{out}} + Z_2^2 \sigma^2_{I_{out}} + \frac{U^2_{out}}{Z_2^2} \sigma^2_{Z_2}\right) \tag{46}$$

For the voltage divider, only a correlation between $Z_2$ a $Z_3$ is possible. But as $Z_3 = 0$, it is not possible to eliminate the errors in the processing of the form $\frac{U^2_{out}}{Z_2^4} \sigma^2_{Z_2}$.

## 4.8 The Simplified Case of Twoport Type T of $Z_1 = 0$

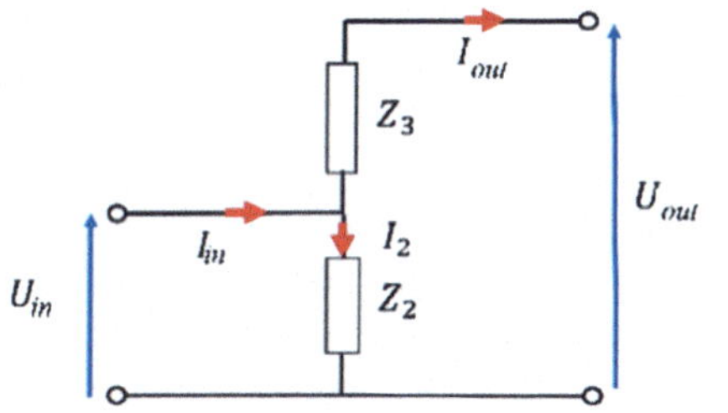


$$A = \begin{bmatrix} A_{11} & A_{12} \\ A_{21} & A_{22} \end{bmatrix} = \begin{bmatrix} 1 & Z_3 \\ \frac{1}{Z_2} & 1 + \frac{Z_3}{Z_2} \end{bmatrix}$$

**Fig. 3.** The twoport type T of $Z_1 = 0$.

**Output uncertainties**

- **for voltage**
  - in the absence of correlation $\rho_{Z12} = 0$

$$\sigma^2_{U_{in}} = \frac{1}{Z_2^2}\left[Z_2^2\sigma^2_{U_{out}} + Z_2 Z_3 \sigma^2_{I_{out}} + I^2_{out} Z_2^2 \sigma^2_{Z_3}\right] \tag{47}$$

The minimum value of uncertainty occurs for the condition of ideal processing

$$I_2\sigma_{Z_2} - I_{out}\sigma_{Z_3} = 0$$

that it is, for

$$\frac{I_2}{I_{out}} = \frac{\sigma_{Z_3}}{\sigma_{Z_2}}$$

- for $\rho_{23} = 1$ with full correlation

$$\sigma^2_{U_{in}} = \frac{1}{Z_2^2}[Z_2^2\sigma^2_{U_{out}} + Z_2 Z_3 \sigma^2_{I_{out}} + I^2_{out} Z_2^2 \sigma^2_{Z_3}] \tag{48}$$

In both cases: the full correlation and its absence, the uncertainty of input voltage measurement is the same. The uncertainty of processing given as $I^2_{out}\sigma^2_{Z_3}$ cannot get out (Fig. 3).

- **for current**

  with full correlation

$$\sigma^2_{I_{in}} = \frac{1}{Z_2^2}\left(\sigma^2_{U_{out}} + (Z_2 + Z_3)^2\sigma^2_{I_{out}} + (I_2\sigma_{Z_2} - I_{out}\sigma_{Z_3})^2\right) \tag{49}$$

If the simplified circuit of twoport T is used, which is the voltage divider, equations may contain correlation and the processing uncertainty can be eliminated, because

$$\frac{U_{out}}{I_{out}} = \frac{Z_2\sigma_{Z_3}}{\sigma_{Z_2}} - Z_3 = Z_3\left(\frac{\delta_{Z_3}}{\delta_{Z_2}} - 1\right) \tag{50}$$

If $\delta_{Z_2} \to 0(\sigma_{Z_2} \to 0)$ then the load resistance for the minimum current uncertainty is infinite – the output is open, when $\delta_{Z_3} \to \delta_{Z_2}$ then the load resistance for the minimum voltage uncertainty is zero – output short circuit. In both cases, we get the minimum current measurement uncertainty without functional processing uncertainty, i.e.

$$\sigma^2_{I_{in}} = \frac{1}{Z_2^2}\left(\sigma^2_{U_{out}} + (Z_2 + Z_3)^2\sigma^2_{I_{out}}\right) \tag{51}$$

## 4.9 Numerical Example

Now, we consider the numerical examples with input quantities and output quantities given by $\boldsymbol{X} = [U_{out}, I_{out}]^T$ and $\boldsymbol{Y} = [U_{in}, I_{in}]^T$, respectively. Based on the manufacturer's data or previous measurements of the T-type components, the following parameters of this T-adapter were obtained: $Z_1 = Z_2 = Z_3 = 100\ \Omega$ and relative uncertainties $u_{\mathrm{relA}} = u_{\mathrm{relB}}$: $\delta_{Z_1} = \delta_{Z_2} = \delta_3 = 0.2\%$ and data for direct measurements, i.e. input quantities $\boldsymbol{X}$: $U_{out} = 1$ V, $I_{out} = 1$ mA and relative uncertainties $u'_{\mathrm{relA}} = u_{\mathrm{relB}}$: $\delta_{U_{out}} = \delta_{I_{out}} = 0.2\%$. Hence the equivalent load resistance is $R = 1$ kΩ.

For $Z_1 = Z_2 = Z_3 = 100\ \Omega$ ($Y = 0.01$ S) matrix **A** takes values:

$$\boldsymbol{A} = \begin{bmatrix} A_{11} & A_{12} \\ A_{21} & A_{22} \end{bmatrix} = \begin{bmatrix} 2 & 300\,\Omega \\ 0.01\,S & 2 \end{bmatrix},$$

and the output quantities are:

$U_{in} = A_{11}U_{out} + A_{12}I_{out} = 2.3\,\mathrm{V}$; $I_{in} = A_{21}U_{out} + A_{22}I_{out} = 12\,\mathrm{mA}$, and current $I_3 = I_{in} - I_{out} = 11\,\mathrm{mA}$.

We determine the uncertainties for different cases, correlating only the input quantities or the four-terminal parameters themselves, as well as both, when they depend on the same external influences, i.e. they are also correlated with each other. We will consider two basic values of the correlation coefficient 0 and +1. Negative correlation (in the range (0, −1) rarely occurs in real systems and will not be analyzed here.

Below two tables with non-correlated (Table 1) and correlated (Table 2) variables / input voltages and current and parameters $Z_1, Z_2, Z_3$/.

**Table 1.** Summary of matrix operations without correlation of input variables and including processing errors in the absence of correlation.

| Chosen parameter | Non-correlated parameters without errors of performance functions | Non-correlated parameters with errors of performance functions |
|---|---|---|
| $\boldsymbol{U_X}$ | $\begin{bmatrix} 4\cdot10^{-6}\mathrm{V}^2 & 0 \\ 0 & 4\cdot10^{-12}\mathrm{A}^2 \end{bmatrix}$ | $\begin{bmatrix} 4\cdot10^{-6}\mathrm{V}^2 & 0 \\ 0 & 4\cdot10^{-12}\mathrm{A}^2 \end{bmatrix}$ |
| $\boldsymbol{U_P}$ | – | $\begin{bmatrix} 0.04\,\Omega^2 & 0 & 0 \\ 0 & 0.04\,\Omega^2 & 0 \\ 0 & 0 & 0.04\,\Omega^2 \end{bmatrix}$ |
| $\boldsymbol{U_{YX}}$ | $\begin{bmatrix} 1.636\cdot10^{-5}\mathrm{V}^2 & 8.24\cdot10^{-8}\mathrm{VA} \\ 8.24\cdot10^{-8}\mathrm{VA} & 4.16\cdot10^{-10}\mathrm{A}^2 \end{bmatrix}$ | $\begin{bmatrix} 1.636\cdot10^{-5}\mathrm{V}^2 & 8.24\cdot10^{-8}\mathrm{VA} \\ 8.24\cdot10^{-8}\mathrm{VA} & 4.16\cdot10^{-10}\mathrm{A}^2 \end{bmatrix}$ |
| $\boldsymbol{U_{YP}}$ | – | $\begin{bmatrix} 10.76\cdot10^{-6}\mathrm{V}^2 & 4.92\cdot10^{-8}\mathrm{VA} \\ 4.92\cdot10^{-8}\mathrm{VA} & 4.88\cdot10^{-10}\mathrm{A}^2 \end{bmatrix}$ |
| $\boldsymbol{U_Y}$ | $\begin{bmatrix} 16.36\cdot10^{-6}\mathrm{V}^2 & 8.24\cdot10^{-8}\mathrm{VA} \\ 8.24\cdot10^{-8}\mathrm{VA} & 4.16\cdot10^{-10}\mathrm{A}^2 \end{bmatrix}$ | $\begin{bmatrix} 27.12\cdot10^{-6}\mathrm{V}^2 & 13.16\cdot10^{-8}\mathrm{VA} \\ 13.16\cdot10^{-8}\mathrm{VA} & 9.04\cdot10^{-10}\mathrm{A}^2 \end{bmatrix}$ |
| $\sigma_{U_{in}}$ $(\frac{\sigma_{U_{in}}}{U_{in}}\times 100\%)$ | 4.04 mV (relative 0.18%) | 5.21 mV (relative 0.23%) |
| $\sigma_{I_{in}}$ $(\frac{\sigma_{I_{in}}}{I_{in}}\times 100\%)$ | 0.02 mA (relative 0.17%) | 0.03 mA (relative 0.25%) |
| $\rho_{U_{in}I_{in}}$ | 1 | 0.84 |

**Table 2.** Summary of matrix operations with correlation of input variables and including processing errors with full positive correlations.

| Chosen parameter | Correlated parameters without errors of performance functions | Correlated parameters with errors of performance functions |
|---|---|---|
| $\boldsymbol{U_X}$ | $\begin{bmatrix} 10^{-4}\mathrm{V}^2 & 10^{-7}\mathrm{VA} \\ 10^{-7}\mathrm{VA} & 10^{-10}\mathrm{A}^2 \end{bmatrix}$ | $\begin{bmatrix} 10^{-4}\mathrm{V}^2 & 10^{-7}\mathrm{VA} \\ 10^{-7}\mathrm{VA} & 10^{-10}\mathrm{A}^2 \end{bmatrix}$ |
| $\boldsymbol{U_P}$ | – | $\begin{bmatrix} 0.04\,\Omega^2 & 0.04\,\Omega^2 & 0.04\,\Omega^2 \\ 0.04\,\Omega^2 & 0.04\,\Omega^2 & 0.04\,\Omega^2 \\ 0.04\,\Omega^2 & 0.04\,\Omega^2 & 0.04\,\Omega^2 \end{bmatrix}$ |
| $\boldsymbol{U_{YX}}$ | $\begin{bmatrix} 2.116\cdot10^{-5}\mathrm{V}^2 & 11.04\cdot10^{-8}\mathrm{VA} \\ 11.04\cdot10^{-8}\mathrm{VA} & 5.76\cdot10^{-10}\mathrm{A}^2 \end{bmatrix}$ | $\begin{bmatrix} 2.116\cdot10^{-5}\mathrm{V}^2 & 11.04\cdot10^{-8}\mathrm{VA} \\ 11.04\cdot10^{-8}\mathrm{VA} & 5.76\cdot10^{-10}\mathrm{A}^2 \end{bmatrix}$ |
| $\boldsymbol{U_{YP}}$ | – | $\begin{bmatrix} 0.036\cdot10^{-5}\mathrm{V}^2 & -1.2\cdot10^{-8}\mathrm{VA} \\ -1.2\cdot10^{-8}\mathrm{VA} & 4\cdot10^{-10}\mathrm{A}^2 \end{bmatrix}$ |
| $\boldsymbol{U_Y}$ | $\begin{bmatrix} 2.116\cdot10^{-5}\mathrm{V}^2 & 11.04\cdot10^{-8}\mathrm{VA} \\ 11.04\cdot10^{-8}\mathrm{VA} & 5.76\cdot10^{-10}\mathrm{A}^2 \end{bmatrix}$ | $\begin{bmatrix} 2.152\cdot10^{-5}\mathrm{V}^2 & 9.84\cdot10^{-8}\mathrm{VA} \\ 9.84\cdot10^{-8}\mathrm{VA} & 9.76\cdot10^{-10}\mathrm{A}^2 \end{bmatrix}$ |
| $\sigma_{U_{in}}$ $\left(\frac{\sigma_{U_{in}}}{U_{in}}\times 100\%\right)$ | 4.60 mV (relative 0.2%) | 4.64 mV (relative 0.2%) |
| $\sigma_{I_{in}}$ $\left(\frac{\sigma_{I_{in}}}{I_{in}}\times 100\%\right)$ | 0.02 mA (relative 0.2%) | 0.03 mA (relative 0.26%) |
| $\rho_{U_{in}I_{in}}$ | 1 | 0.68 |

## 5 Summary and Conclusions

An extended method of estimating the uncertainty of indirect multi-parameter measurements has been developed. It considers the uncertainty of the parameters of the multidimensional function of processing the results of these measurements. It is presented in this part II of our work. The method recommended in Supplement 2 of the Guide to the Expression of Uncertainty in Measurement GUM, which assumes that the processing function is accurate, is includes as a special case.

In Part I, with various ratios of relative uncertainty of A and B types, the effect of correlations between directly measured quantities on elements of the covariance matrix of the multi-measurand determined indirectly was analyzed. It is shown that the correlation of deviation sets with uncertainties type A and B should be considered separately.

The extended method consider both uncertainties of measured quantities and uncertainties of the system performing the processing function as well all possible own and mutual correlations. As an example, the relationships for indirect measurements of voltage and current through T-type cross-section and its specific cases were considered. The numerical example of T-type two port terminal was also presented.

The proposed extended method is versatile, because it allows both the estimation of the accuracy of indirect multi-parameter measurements and a description of the accuracy of instruments and measuring systems performing such measurements.

## References


1. Supplement 2 to the Guide to the Expression of Uncertainty in Measurement - Extension to any number of output quantities. JCGM 102:2011 BIPM (2011)
2. Warsza, Z.L.: Estimation of uncertainty of indirect measurement in multi-parametric systems with few examples. Part 1. In: CD Proceedings of Conference: Problems and Progress of Metrology PPM 2018 Szczyrk 04–06. 09/2018. Series Conferences: No 22, Metrology Commission of Katowice Branch of Polish Academy of Science (2018)
3. Warsza, Z.L., Puchalski, J.: Upgraded method for determining the uncertainty of multi-parameter measurements. Part 1. Pomiary Automatyka Robotyka **23**(1), 47–57 (2019). https://doi.org/10.14313/PAR_231/47. (in Polish)
4. Warsza, Z.L., Puchalski, J.: Estimation of uncertainties in multivariable measurements. Part 1. Case of correlated quantities. In: Proceedings of Automation 2020, pp. 309–324 (2020)
5. Osiowski J., Szabatin J.: Basics of circuit theory. v. 3 Chapter 8: Twoport Networks (1995). (in Polish: Podstawy teorii obwodów, tom 3, rozdz. 8 Czwórniki, WNT, Warszawa)

# Utilization of Vibration Energy from Vehicle Using Smart Damper

Vytautas Bučinskas, Tadas Lenkutis(✉), Dainius Kunkis, Gediminas Vaičiūnas, Nikolaj Šešok, and Igor Iljin

Vilnius Gediminas Technical University, J. Basanaviciaus Str. 28, Vilnius, Lithuania
tadas.lenkutis@vgtu.lt

**Abstract.** Harvesting of mechanical vibration energy recently is a popular idea, but its implementation is still limited due to real size of harvesters and efficiency of the systems. Here provided an idea and realization of smart shock absorber with smart liquid, able to generate electric energy from vehicle suspension excitation and thus damping vehicle vibrations. Linear types of electric generators are not very efficient due to limited stroke, therefore their efficiency of energy transformation is far from ideal. This paper provides experimental research of energy harvesting efficiency using smart liquid core, magnetic core and ferromagnetic core of shock absorber. Comparison of these materials in the shock absorber provides clear ability to use smart liquid as ferromagnetic core in the process of power generating. Such device proves possibility to be implemented in the vehicle suspension and allows power-damping control as well as data transmission energy request. Generated power and its dependency from excitation frequency provided in the result graphs.

**Keywords:** Vibrations · Vehicle suspensions · Energy harvesting · Smart fluid

## 1 Introduction

Vehicles are most popular transport for people and loads, they takes main part of transportation resources. By Eurostat statistics more than 3 quarters of cargo in Europe was transported by land transportation – trucks (76.7% in 2017) and 17.3% by railways [1]. ESEA (Europe automobile manufacturers association) counts 308.3 million motor vehicles registered in Europe and this is about one vehicle for two persons [2]. So land transport is most popular transportation type. This statistic shows that popular transportation asks for infrastructure and vehicles need to be developed for maximum efficiency and comfort.

Every vehicle tire during its movement has contact with ground and this contact creates vibrations [3]. Standard suspension of vehicle damps vibration energy by converting it to heat and dissipating it to ambience [4]. The amount of energy in the vibrations of a vehicle depends on its dynamics, vehicle weight, driving speed and actual road profile [5]. Manufacturers of the transport means improve vibrations damping efficiency to raise comfort of driving and longer exploitation time for suspension parts [6].


R. Szewczyk et al. (Eds.): AUTOMATION 2020, AISC 1140, pp. 345–352, 2020.
https://doi.org/10.1007/978-3-030-40971-5_31

Vibration energy conversion to electricity may lower energy consumption for driving and lower transportation cost [7], environmental factor will be caused as well. Vibration energy harvesting potential by vehicle type and its partial consumption is shown in Fig. 1.

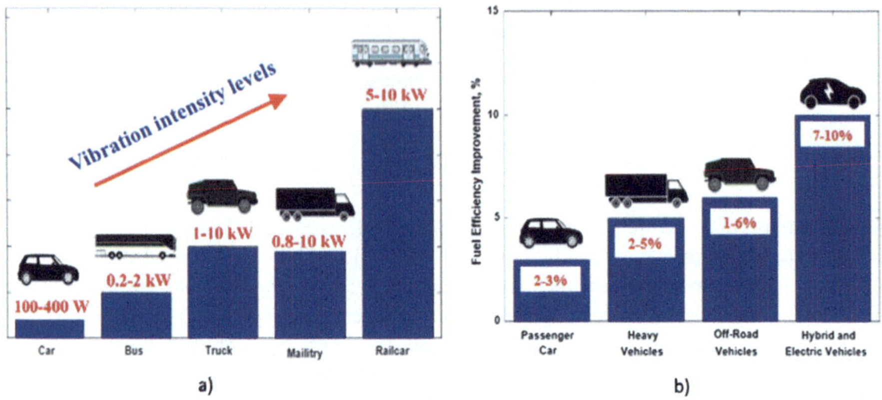


**Fig. 1.** Potential of energy harvesting: (a) use of regenerative shock absorber; (b) and fuel efficiency improvement [8].

From Fig. 1 we can see that energy harvesting potential depends on vehicle type (mass) and it can vary depending of the road quality. Even in the best road we can obtain enough energy for small electronic devices to power up and energy part in electric vehicles can obtain till 7–10% of used energy.

Regenerative shock absorbers can be classified in to two groups by energy exchange: mechanical (Fig. 2) [9] and magnetic (Fig. 3) [10].

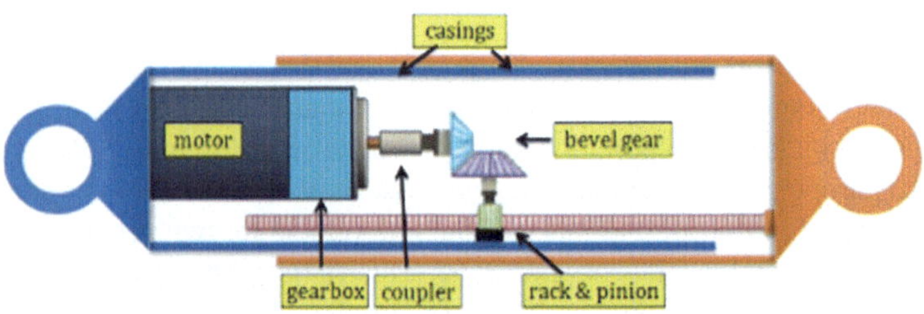


**Fig. 2.** Mechanical shock absorber [9].

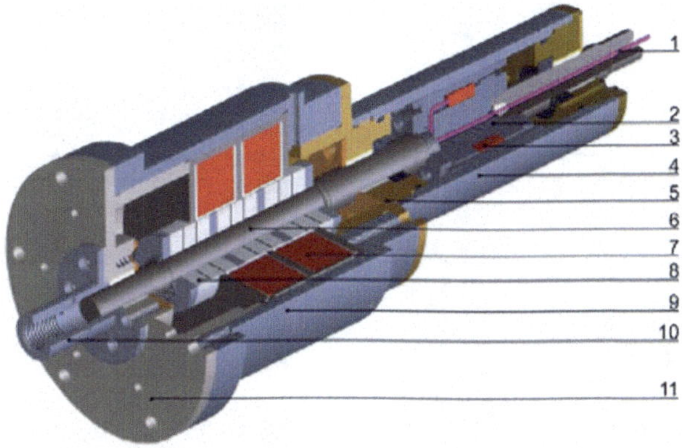


**Fig. 3.** Magnetic shock absorber [10] 1 – lead, 2 – rod, 3 – damper coil, 4 – damper housing, 5 – connector, 6 – shaft, 7 – generator coil, 8 – magnets, 9 – generator housing, 10 – generator fixture, 11 – lower cover of generator.

These two designs are not the only possible cases, there are more choices [11–14]. Nevertheless, both types of generative shock absorbers used exceptionally for power generating are inefficient in vibration damping. Ensuring of both functions – power harvesting and vibration damping – requires different design.

As possible solutions there is provided shock absorber with magneto rheological fluid [15]. A magneto-rheological (MR) fluid can change mechanical properties by using electro-magnetic field. MR fluid usage instead of hydraulic fluid let us to change damping characteristics in damper by road irregularities and gives possibility to produce electric energy instead of heat. Important function of this liquid is the possibility to lubricate sliding surfaces, thus enabling to create real an operating device. MR fluid is called smart fluid, because of its characteristics can be controlled by a magnetic field (Fig. 4) [16].

A Smart fluid is produced from oil and magnetic particles and it has the same lubricating characteristics as regular oil. In summary magneto-rheological fluid can fulfil all three requirements: Energy harvesting; Lubricating and Damping.

A Test prototype was created to perform experiments (Fig. 5). Most of the parts were made from plastics PP and PLA due to easy machining and non-magnetic characteristics. The inner volume was filled with magnetorheological fluid, where it performed three functions – lubricating sliding process, controlling damping force and energy generation.

Frame 1 is for sliding movement to direct. Inside piston 4 is sliding that is filled with magnetorheological fluid and in the top and bottom of the piston two neodymium magnets N42 are installed to create initial magnetic field. Piston stem 3 is connected to piston and second end is connected to vibrations test machine. Fixing eye 6 is fixed to the vibration testing machine too and its fixed point. Volume 5 is for controlling damping coefficient and air piston 8 compensates volume changes.

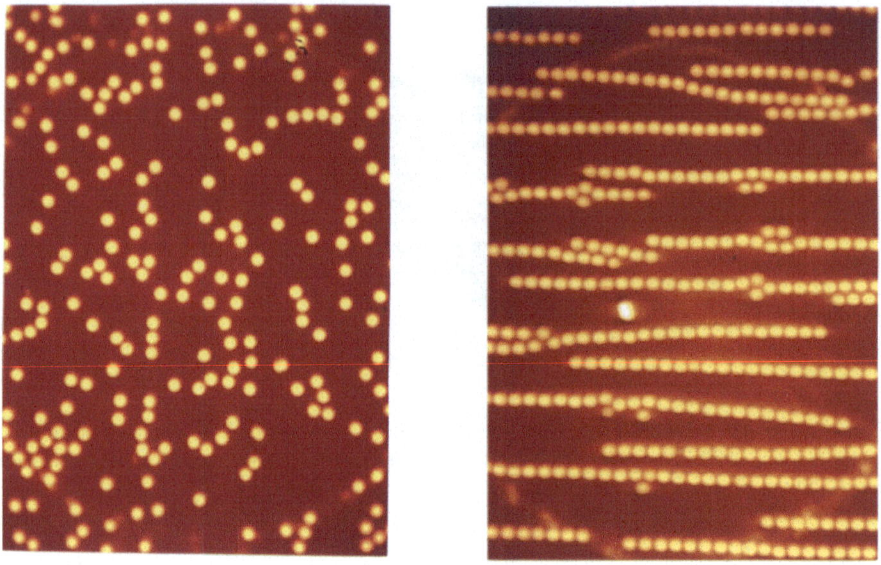

a) b)

**Fig. 4.** Activation of MR fluid. (a) no magnetic field applied, (b) magnetic field (B = 0.02 T) applied.

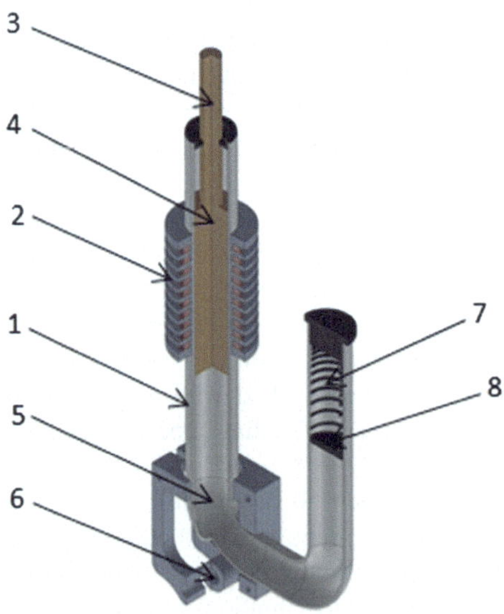


**Fig. 5.** Experiments prototype structure. 1 – frame; 2 – solenoid coil; 3 – piston stem for force application; 4 – piston; 5 – Ferro active fluid; 6 – fixing eye; 7 – compensating spring; 8 – air piston.

# 2 Experimental Research

## 2.1 Formulation of Research

The aim of the research is to find energy generation possibilities from vibrations. The Relation between vibrations amplitude, frequency and generated voltage will be examined. We are looking for voltage profile that will show general distribution of generated voltage and highest voltages dependence with oscillation frequency. Source of excitations is sine wave and it will let easy to follow response of generated energy.

## 2.2 Experimental Methodology

The prototype was examined in automobile dampers testing machine SPA PSD 2004. One part of machine was configured to insert our testing prototypes. Experiment apparatus is shown in Fig. 6.

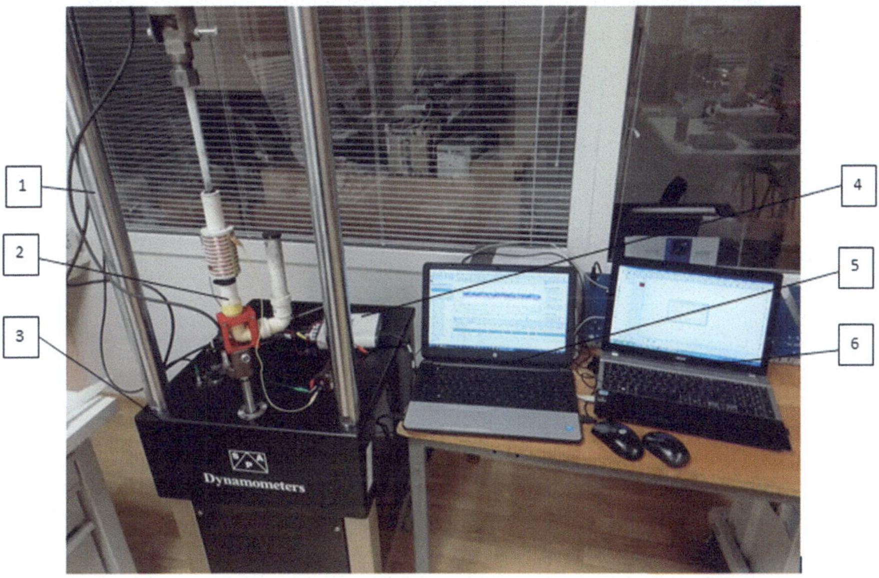


**Fig. 6.** Experiment apparatus. 1 – holding frame, 2 – damper, 3 – Vibrations source, 4 – National Instruments signal collecting device, 5 – personal computer for collect data, 6 – personal computer for control vibration source.

The experiment was performed in few configurations. Vibrations amplitude was 25 mm and 50 mm, and vibrations frequency 1 Hz, 2 Hz, 3 Hz, 4 Hz, 5 Hz, 5.5 Hz, 6 Hz, 6.5 Hz, 7 Hz and 7.5 Hz (maximum test device value). Two coils with maximum voltage values were recorded with load on the circuit and open circuit. Element 4 is National Instruments USB – 4432 data collecting system and computer 5 have National Instruments software "Signal Express" were data is collected and later was saved and processed using MS Excel software package.

## 3 Results

Experimental research provided interesting results, which shows prospective to develop such type of shock absorbers and energy harvesters. Below presented generated voltage graph in time domain. While actual stroke has 50 mm amplitude and 7.5 Hz frequency on the 100 Ω load on two coils is shown in Fig. 7. Difference of voltage appears due to actual position of magnetic chain relative to phase coils. Extra peak on 1st coil is a parasitic magnetic field influence to it and shows need to improve magnetic chain of harvester.

In Fig. 7 we can see, that the coil position with piston movement has high value on the generated energy –1.9 V. On the other hand, fast movement creates narrow, but high peaks. Maximum generated power in these parameters can reach 200 W in harmonic mode, nevertheless, real road conditions never create harmonic excitations.

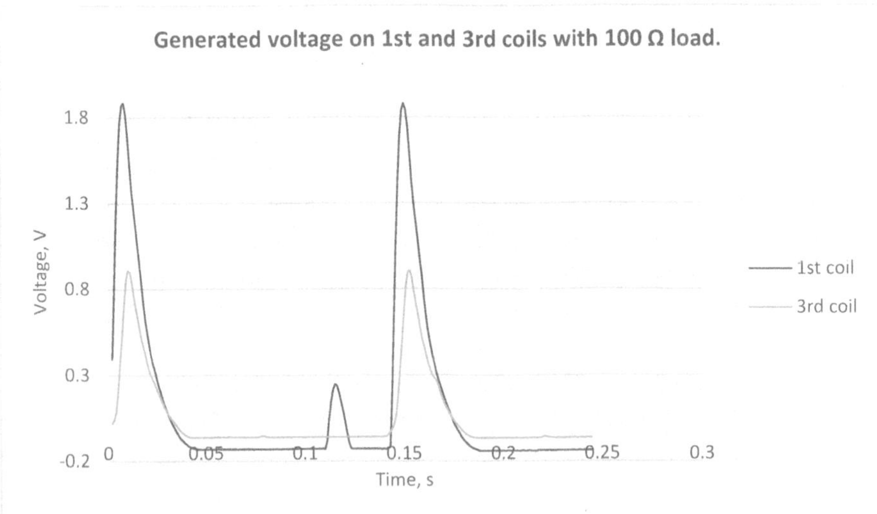


**Fig. 7.** Generated voltage on the 1st and 3rd coils with 100 Ω load, 50 mm oscillations amplitude and 7.5 Hz frequency.

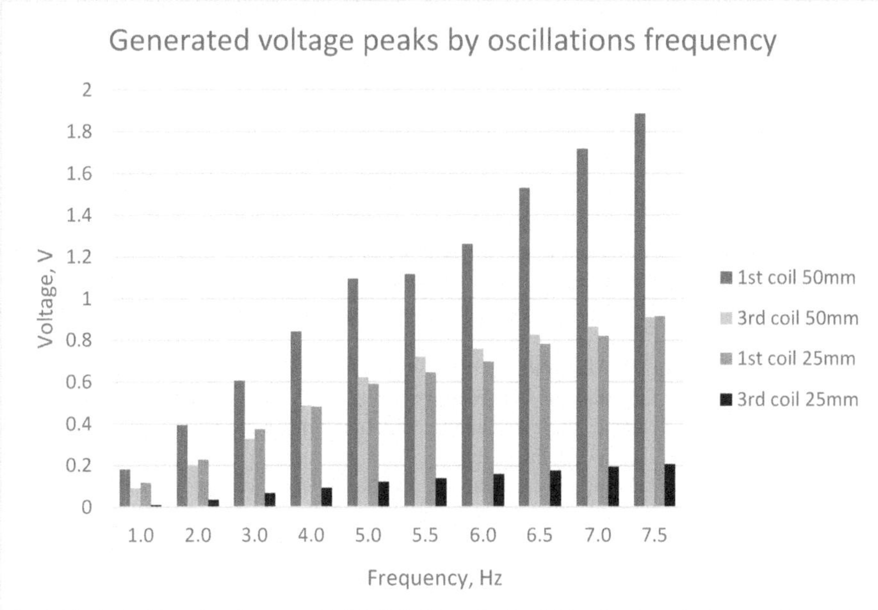


**Fig. 8.** Generated voltage peaks dependence with oscillations frequency.

In Fig. 8 we see that generated voltage dependence of frequency is linear. Circuit load is 100 Ω and in 1 Hz frequency oscillations we can obtain 20 W power and with 7.5 Hz – 190 W.

## 4 Conclusions

Performed experimental research revealed many interesting effects of implementing such design of the shock absorber. Firstly, this type of liquid core generator can operate and it is really functional. In the other hand, many features of this type of electric machines are unknown and need to be further researched. Even damping efficiency is not evaluated in this research, we note some valuables conclusions:

- Power harvesting using linear electric machine with liquid slider is possible and successfully generating power, that is up to 200 W. Efficiency of the generator can be improved, but this opens way to continue design refinement.
- Magnetic chain of the harvester should be refined for better utilization of the stroke of shock absorber piston, while extra peaks on coil appears and diminish common amount of harvested energy. This relates to position of permanent magnets as well for piston created pole distance.
- Deviation of linearity for dependency voltage/frequency is stable and not clear as effect; it can be connected with hydraulic movement of liquid and magnetization phenomena of smart liquid.

**Acknowledgment.** The research was supported by ECSEL Joint Undertaking under the project No T0709T627 Autodrive (Advancing fail-aware, fail-safe, and fail-operational electronic components, systems, and architectures for fully automated driving to make future mobility safer, affordable, and end-user acceptable).

## References

1. Eurostat: Modal split of freight transport, October 2019. http://appsso.eurostat.ec.europa.eu/nui/submitViewTableAction.do
2. ACEA: Vehicle in use, October 2019. https://www.acea.be/statistics/tag/category/vehicles-in-use
3. Kumar, A., Dwivedi, A., Jaiswal, H., Patil, P.P.: Material-based vibration characteristic analysis of heavy vehicle transmission gearbox casing using finite element analysis. In: Jain, L., Patnaik, S., Ichalkaranje, N. (eds.) Intelligent Computing, Communication and Devices, pp. 527–533. Springer, New Delhi (2015)
4. Zhang, Y., Guo, K., Wang, D., Chen, C., Li, X.: Energy conversion mechanism and regenerative potential of vehicle suspensions. Energy **119**, 961–970 (2017)
5. Katu, U.S., Desavale, R.G., Kanai, R.A.: Effect of vehicle vibration on human body–RIT experience. In: Proceedings of 11th National Conference on Machines and Mechanisms, pp. 1–9 (2003)
6. Li, G., Wang, T.: Influence of long-waved road roughness on fatigue life of dump truck frame. J. Vibroeng. **16**(8), 3862–3878 (2014)
7. Khayyam, H., Bab-Hadiashar, A.: Adaptive intelligent energy management system of plug-in hybrid electric vehicle. Energy **69**, 319–335 (2014)
8. Abdelkareem, M.A., Xu, L., Ali, M.K.A., Elagouz, A., Mi, J., Guo, S., Zuo, L.: Vibration energy harvesting in automotive suspension system: a detailed review. Appl. Energy **229**, 672–699 (2018)
9. Li, Z., Zuo, L., Luhrs, G., Lin, L., Qin, Y.X.: Electromagnetic energy-harvesting shock absorbers: design, modeling, and road tests. IEEE Trans. Veh. Technol. **62**(3), 1065–1074 (2012)
10. Sapiński, B.: Energy-harvesting linear MR damper: prototyping and testing. Smart Mater. Struct. **23**(3) (2014)
11. Tang, X., Lin, T., Zuo, L.: Design and optimization of a tubular linear electromagnetic vibration energy harvester. IEEE/ASME Trans. Mechatron. **19**(2), 615–622 (2013)
12. Guan, X., Huang, Y., Ru, Y., Li, H., Ou, J.: A novel self-powered MR damper: theoretical and experimental analysis. Smart Mater. Struct. **24**(10) (2015)
13. Satpute, N.V., Singh, S., Sawant, S.M.: Energy harvesting shock absorber with electromagnetic and fluid damping. Adv. Mech. Eng. (2014)
14. Zuo, L., Scully, B., Shestani, J., Zhou, Y.: Design and characterization of an electromagnetic energy harvester for vehicle suspensions. Smart Mater. Struct. **19**(4) (2010)
15. Bučinskas, V.: Energy harvesting shock absorber and method for controlling same. U.S. Patent 13/560,985 (2014)
16. Popplewell, J., Rosensweig, R.E.: Magnetorheological fluid composites. J. Phys. D Appl. Phys. **29**(9), 2297–2303 (1996)

# Stability of Jiles-Atherton Anhysteretic Magnetization Curve Model for Magnetic Materials with Uniaxial Anisotropy

Roman Szewczyk(✉)

ŁUKASIEWICZ Research Network – Industrial Research Institute for Automation and Measurements PIAP, Al. Jerozolimskie 202, 02-486 Warsaw, Poland
rszewczyk@onet.pl

**Abstract.** Anhysteretic magnetization curve plays the key role in modelling the characteristics of components made of soft magnetic materials. However, due to the positive feedback, for some set of parameters the most common model of magnetization curve might be unstable. Moreover, formal stability assessment of anhysteretic magnetization curve is sophisticated due to nonlinearities. Paper presents practical approach to stability assessment of Jiles-Atherton anhysteretic magnetization curve model for magnetic materials with uniaxial anisotropy. Results of this assessment enable the increase of efficiency of the process of identification of the model's parameters, and as a result enable more efficient description of functional characteristics of inductive components with cores made of anisotropic soft magnetic materials.

**Keywords:** Anhysteretic magnetization · Jiles-Atherton model · Anisotropic magnetic materials

## 1 Introduction

Anhysteretic magnetization curve [1] is one of the most informative characteristics of soft magnetic materials. This curve is defined as magnetization characteristic neglecting the magnetic hysteresis [2]. Such curve can be measured using constant magnetic field biased demagnetization [3], however, such method is difficult to implement from technical point of view. Alternative, efficient and robust methods of measurements of anhysteretic magnetization curve were presented in recent paper [4].

It should be highlighted, that anhysteretic magnetization curve is very useful for modelling the magnetization characteristics of soft magnetic materials with finite elements method, as well as with the method of moments [5]. Neglecting magnetic hysteresis phenomenon is justified especially in the case of anisotropic materials, where magnetic parameters strongly depend on the direction of magnetization [6].

Moreover, previously presented Jiles-Atherton based model of uniaxial anisotropic anhysteretic magnetization [7–9] enables modelling the inductive elements magnetized in different directions, which is especially useful for 2D magnetization process, as well as for modelling the magneto-mechanical properties of inductive cores [10]. It should


R. Szewczyk et al. (Eds.): AUTOMATION 2020, AISC 1140, pp. 353–358, 2020.
https://doi.org/10.1007/978-3-030-40971-5_32

be highlighted, that Jiles-Atherton model based approach to description of anhysteretic magnetization curve is especially valuable due to the strong connection with physical phenomena and parameters of soft magnetic material.

The most common method of identification of parameters of anhysteretic magnetization curve is stochastic optimization process [11]. Especially differential evolution algorithm seems to be useful for this purpose [12]. However, in the case of instability of the model of anhysteretic magnetization, optimization algorithm stuck on the time-consuming oscillations modelling. As a result, the method of fast detection of the possible instability may be useful for practical applications. Such method was presented for the simplified case of isotropic materials [13]. However, technically useful method of uniaxial anisotropic anhysteretic magnetization was not presented previously, which is significant barrier for efficient identification of uniaxial anisotropic anhysteretic magnetization curve parameters. For this reason, presented paper is filling the gap in the state of the art, proposing robust, technical method of detection of uniaxial anisotropic anhysteretic magnetization instabilities, which enables immediate avoiding of instable parameters during the Jiles-Atherton model based modelling of characteristics of soft magnetic materials.

## 2 Anhysteretic Magnetization Curve in Jiles-Atherton Model for Magnetic Materials with Uniaxial Anisotropy

The principles of Jiles-Atherton model are based on Maxwell's equations generalizing description of electro-magnetic phenomena [1]:

$$A = G + \mu_0 HM \tag{1}$$

$$G = U - TS + \frac{3}{2}\sigma\lambda \tag{2}$$

$$U = \frac{1}{2}\alpha\mu_0 M^2 \tag{3}$$

where: $A$ is the free Helmholtz energy, $G$ is Gibbs energy, $H$ is magnetizing field, $M$ is magnetization of material, $U$ is interparticle interaction energy, $S$ is entropy of the material, $T$ its temperature, $\sigma$ are stresses, $\lambda$ is magnetostriction, $\alpha$ is interdomain coupling as well as $\mu_0$ is magnetic constant. Considering thermodynamics of the system [2] as well as after re-arrangement od Eqs. 1–3, following dependence describing efficient magnetizing field $H_e$ in the material can be stated:

$$H_e = H + \alpha M \tag{4}$$

Efficient magnetizing field $H_e$, quantifying the influence of external magnetizing field $H$ on the magnetic material is the most important positive feedback loop connected with magnetization process. This positive feedback loop is presented in Fig. 1 for hysteretic and anhysteretic description of magnetization process.

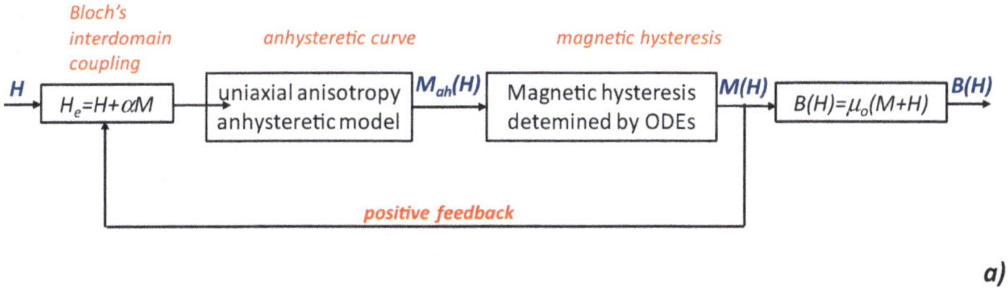


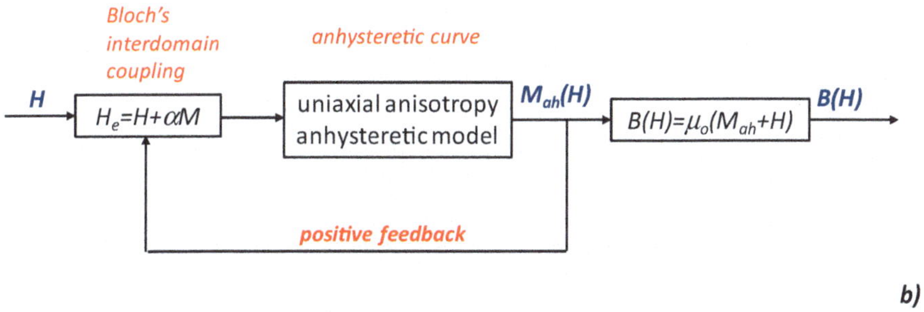


**Fig. 1.** Positive feedback loop in the model of magnetization process: (a) Jiles-Atherton model of $B(H)$ hysteresis loop, (b) the model of anhysteretic magnetization.

Obviously, in spite of positive feedback loop, the physical phenomenon of magnetization process is always stable [1] due to the small values of Bloch's interdomain coupling $\alpha$. However, the anhysteretic magnetization curve $M_{ah}(H)$ of magnetic material with uniaxial anisotropy can be calculated on the base of Maxwell-Boltzmann statistical distribution [2]. Such model of uniaxial anisotropic anhysteretic magnetization was presented previously [8, 9]:

$$M(H) = M_s \left[ \frac{\int_0^{\pi} e^{\frac{E(1)+E(2)}{2}} \sin\theta \cos\theta \cdot d\theta}{\int_0^{\pi} e^{\frac{E(1)+E(2)}{2}} \sin\theta \cdot d\theta} \right] \tag{5}$$

$$E(1) = \frac{H_e}{a} \cos\theta - \frac{K_{an}}{\mu_0 M_s a} sin^2(\psi - \theta) \tag{6}$$

$$E(2) = \frac{H_e}{a} \cos\theta - \frac{K_{an}}{\mu_0 M_s a} sin^2(\psi - \theta) \tag{7}$$

where $K_{an}$ represents average energy density of uniaxial anisotropy, which is the energy required to magnetize the material from easy to hard magnetization axis [7]. In addition $\psi$ is the angle between the easy axis of uniaxial anisotropy and direction of the external magnetizing field *H*, whereas *a* is the parameter quantifying a domain walls density and $M_s$ is saturation magnetization. In the general case saturation flux density $B_s$ is given as:

$$B_s = \mu_0(H + M_s) \tag{8}$$

However, for the high permeability materials, Eq. 8 reduces to:

$$B_s = \mu_0 M_s \tag{9}$$

It should be indicated, that for $K_{an}$ equal 0, material becomes isotropic, and Eqs. 5–7 reduce to the form of Langevin equation:

$$M(H) = M_s\left(\coth\left(\frac{H_e}{a}\right) - \frac{a}{H_e}\right) \tag{10}$$

The detailed analysis of stability of Jiles-Atherton model for simplified case of isotropic material was presented previously [13]. However, the sophisticated form of Maxwel-Boltzman distribution based anhysteretic magnetization curve for the materials with uniaxial anisotropy requires simplified approach, suitable for optimization process for identification of parameters of Jiles-Atherton model.

## 3 Method and Results of Practical Stability Tests

The practical anhysteretic model stability criterion is, that for all values of magnetizing field *H*, the magnetization *M*(*H*) should be smaller or equal saturation magnetization $M_s$. The practical observation is, that the most unstable region of magnetization curve is around the coercive field $H_c$. Moreover, the coercive field $H_c$ can be approximated by model's parameter *a* (given in A/m) [2]. It should be also pointed out, that the highest slope of anhysteretic curve is observed in the direction of the easy axis of magnetization (for $\psi$ equal 0). These assumption lead to the definition of practical stability criterion:

$$M(a)|_{\psi=0} \leq M_s \tag{11}$$

However, it should be stressed, that criterion given by Eq. 9 is a necessary condition, but insufficient for general prove of model stability. However, proposed condition can significantly reduce the number of instable cases during the identification of model's parameters.

Tests of a stability of the model stated by Eqs. 5–7 indicated, that the value of average anisotropy energy density $K_{an}$ doesn't significantly influence the model's stability. However, the stability can be significantly dependent of the value of magnetic flux density $B_s$. Figure 2 presents the stability limits of uniaxial anisotropic anhysteretic magnetization model.

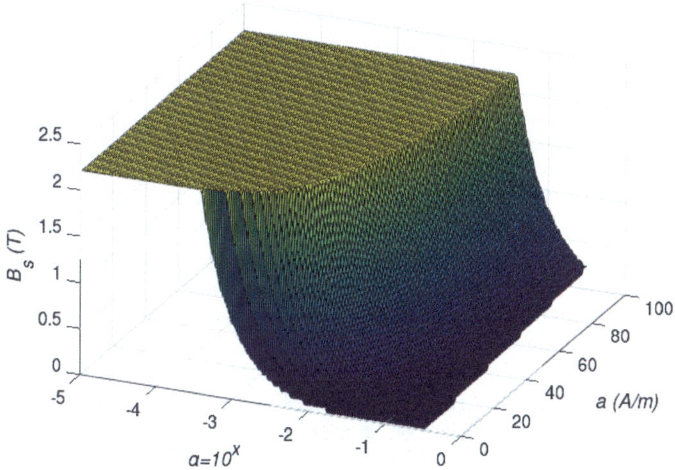


**Fig. 2.** Stability limits dependence on parameters $B_s$, $a$ and $\alpha$ for uniaxial anisotropic anhysteretic magnetization model.

It can be seen, that for values of Bloch's interdomain coupling higher than $10^{-2}$, the model is unstable for higher values of saturation flux density $B_s$. However, maximal value of analyzed saturation flux density $B_s$ is limited by technical capabilities to maximal available value achievable in special steels and equal 2.3 T. Moreover, for Bloch's interdomain coupling higher than $10^{-4}$, the model stated by Eqs. 5–7 seems to be stable for all values of practically observed saturation flux density $B_s$.

# 4 Conclusions

Presented criterion of stability of the model of uniaxial anisotropic anhysteretic magnetization is the necessary, but insufficient condition. As a result it can't be treated as general prove of model stability. However, proposed stability criterion enables radical reduction sets of instable parameters of Jiles-Atherton anhysteretic magnetization. Moreover, proposed criterion is simple and easy to implement in numerical modelling implementations.

As a result proposed criterion may significantly increase the efficiency of model's parameters identification process by reduction of the numbers of time-consuming solving the equations with unstable parameters. As a result, in spite of the fact, that proposed criterion is the necessary, but insufficient condition for stability proof, it is efficient and worth of consideration for engineering numerical systems for modelling the anhysteretic magnetic characteristics of inductive components.

## References

1. Jiles, D.C.: Introduction to Magnetism and Magnetic Materials. CRC Press, Boca Raton (2015)
2. Jiles, D.C., Atherton, D.L.: Theory of ferromagnetic hysteresis. J. Magn. Magn. Mater. **61**, 48–60 (1986). https://doi.org/10.1016/0304-8853(86)90066-1
3. Jiles, D.C., Atherton, D.: Theory of ferromagnetic hysteresis. J. Appl. Phys. **55**, 2115 (1984). https://doi.org/10.1063/1.333582
4. Nowicki, M.: Anhysteretic magnetization measurement methods for soft magnetic materials. Materials **11**, 2021 (2018). https://doi.org/10.3390/ma11102021
5. Szewczyk, R.: The method of moments in Jiles-Atherton model based magnetostatic modelling of thin layers. Arch. Electr. Eng. **67**, 27–35 (2018). https://doi.org/10.24425/118989
6. Schauerte, B., Steentjes, S., Hameyer, K.: Flexible extension of hysteresis models for magnetic anisotropy. IEEE Magn. Lett. **9**, 6606804 (2018). https://doi.org/10.1109/LMAG.2018.2874168
7. Ramesh, A., Jiles, D.C., Bi, Y.: Generalization of hysteresis modeling to anisotropic materials. J. Appl. Phys. **81**, 5585 (1997). https://doi.org/10.1063/1.364843
8. Ramesh, A., Jiles, D.C., Roderik, J.: A model of anisotropic anhysteretic magnetization. IEEE Trans. Magn. **32**, 4234–4236 (1999). https://doi.org/10.1109/20.539344
9. Szewczyk, R.: Validation of the anhysteretic magnetization model for soft magnetic materials with perpendicular anisotropy. Materials **7**, 5109–5116 (2014). https://doi.org/10.3390/ma7075109
10. Gyselinck, J., Dular, P., Sadowski, N., Leite, J.V., Bastos, J.P.A.: Incorporation of a Jiles-Atherton vector hysteresis model in 2D FE magnetic field computations: application of the Newton-Raphson method. COMPEL Int. J. Comput. Math. Electr. Electron. Eng. **23**, 685–693 (2004). https://doi.org/10.1108/03321640410540601
11. Chwastek, K., Szczyglowski, J.: Identification of a hysteresis model parameters with genetic algorithms. Math. Comput. Simul. **71**, 206–211 (2006). https://doi.org/10.1016/j.matcom.2006.01.002
12. Biedrzycki, R., Jackiewicz, D., Szewczyk, R.: Reliability and efficiency of differential evolution based method of determination of Jiles-Atherton model parameters for X30Cr13 corrosion resisting martensitic steel. J. Autom. Mob. Robot. Intell. Syst. **8**, 63 (2014). https://doi.org/10.14313/JAMRIS_4-2014/39
13. Iyer, R.V., Krishnaprasad, P.S.: On a low-dimensional model for ferromagnetism. Nonlinear Anal. Theory Methods Appl. **61**, 1447–1482 (2005). https://doi.org/10.1016/j.na.2005.01.109

# Estimation of Azimuth and Elevation Angles of Ultrasonic Signal Arrival by Indirect Phase Determination

Bogdan Kreczmer(✉)

Department of Cybernetics and Robotics, Electronics Faculty, Wrocław University of Science and Technology, Wrocław, Poland
bogdan.kreczmer@pwr.edu.pl

**Abstract.** The paper presents the concept of the method of determining the direction of ultrasonic signal arrival, i.e. the azimuth and elevation angles. This method is an extension of the previous approach which was proposed to determine only the azimuth angle. The approach is based on indirect phase determination. This makes it possible to tolerate spacing of receivers greater than half the wavelength of the received signal. At the same time, it provides increased measurement accuracy and reduced hardware requirements. For the proposed method, the preliminary implementation was performed and tested in simulations. This made it possible to estimate the value of the tolerated measurement error.

**Keywords:** Ultrasonic range finder · Sonar · Direction of arrival

## 1 Introduction

The main concept of the approach presented in this paper is based on the method described in [2] and in more detail together with the theoretical analysis in [3]. This method makes it possible to construct a 2-D sonar which can determine distance to an object and the azimuth angle of signal arrival. The approach presented in this paper extends this method to the 3-D case when the elevation angle of signal arrival is also estimated (see Fig. 1).

## 2 Related Work

There are several methods of determining direction of signal arrival (DOA). The well-known method is multiple signal classification (MUSIC) [6]. This method has several variants e.g. root-MUSIC [12], the total spectral search MUSIC method [13], or the partial spectral search one [9]. Another well-known approach to DOA determination is the estimation of signal parameters via rotational invariance techniques (ESPRIT) [4,5]. This approach also has a lot of variants and adaptations e.g. Unitary ESPRIT [1], Conjugate ESPRIT [10].


R. Szewczyk et al. (Eds.): AUTOMATION 2020, AISC 1140, pp. 359–367, 2020.
https://doi.org/10.1007/978-3-030-40971-5_33

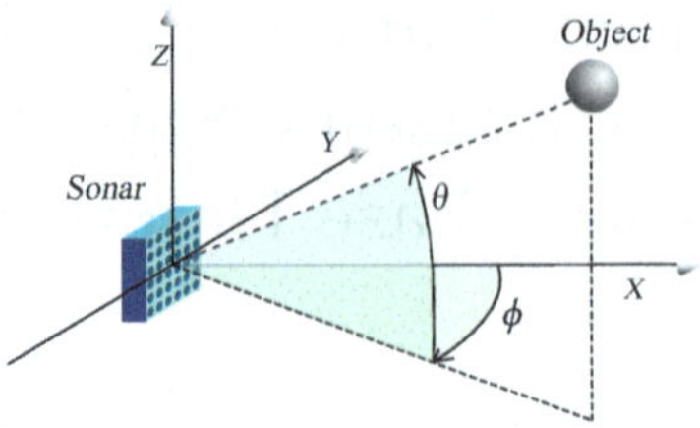


**Fig. 1.** Object localization by using a sonar which determines its distance, the azimuth angle $\phi$ and elevation angle $\theta$

Another well-known approach is beamforming which exploits array processing technique. The implementation of such a method is presented in [8]. It appeared to be very effective and made it possible to implement the system BatSLAM which was able to solve the simultaneous localization and mapping problem (SLAM) [7].

The discussed approaches involve methods that are relatively computationally expensive. They important advantage is that they are able to determine the DOA of signals that come from several sources. The approach presented in this paper is restricted to the problem of DOA determination for an echo coming from a single direction. This is the obvious drawback. However, when an emitted signal is short enough, it can be acceptable for many applications of mobile robot navigation. The important advantages are that the computational burden is reduced and the hardware requirements are minimized. Despite this, the implementation of the previous version of the method for the 2-D case has achieved very good accuracy [3].

## 3 Determining Direction of Signal Arrival

In this paper, it is assumed that the signal emitted, and then received, is narrowband. It is easy to meet this assumption when piezoelectric transducers are used. In this case, to estimate the angle of signal arrival the approach based on measurement of phase shift could be used. This can be done by using receivers arranged on a plain. When the distance between the receivers is less than half the wavelength of the signal, it is sufficient to use three receivers which are placed non-collinear (see Fig. 2). When the distances between receivers are small in relation to the distance of a signal source or object from which it was reflected, the received acoustic wave can be treated as a planewave in the vicinity of the receivers. Assuming $t_0, t_1, t_2$ are times at which the same planewave was detected by receivers $R_0$, $R_1$ and $R_2$ respectively and $t_0 \leq t_1$ and $t_0 \leq t_2$, distances from receivers $R_1$ and $R_2$ to the planewave at the moment when it reached the receiver $R_0$ are as follows

$$s_1 = v_a \tau_{01}, \qquad s_2 = v_a \tau_{02},$$

where $v_a$ is the speed of acoustic wave, $\tau_{01} = t_1 - t_0$ and $\tau_{02} = t_2 - t_0$. Because the waveplane can be modeled as a plane, it has to meet the equation

$$ax + by + cz + d = 0.$$

where $a$, $b$ and $c$ are coordinates of the vector perpendicular to the plane. It can be assumed that this vector is normalized. Hereinafter, it will be designated as $\boldsymbol{n} = (n_x, n_y, n_z)$ and

$$n_x^2 + n_y^2 + n_z^2 = 1. \tag{1}$$

Due to the convention of determining azimuth and elevation angle values, it is convenient to assume that the vector $\boldsymbol{n}$ is directed opposite to the direction of wave propagation. In this case, it means that $n_x > 0$ (see Fig. 2). Because the vector $\boldsymbol{n}$ is normalized, the absolute value of $d$ is the distance of the plane to the origin of the coordinate system. The sign of $d$ depends on which side of the plane the origin of the coordinate system is on. But it can be assumed that the origin is located at the receiver $R_0$ (see Fig. 2). In this case, when the coordinates of a point outside the plane are substituted for the plane equation, $d$ refers to its distance from the plane. Taking it into account the previous discussed assumptions, the coordinates of the receivers $R_1$ and $R_2$ are $(0, y_1, z_1)$ and $(0, y_2, z_2)$ respectively, it gives the following set of equations

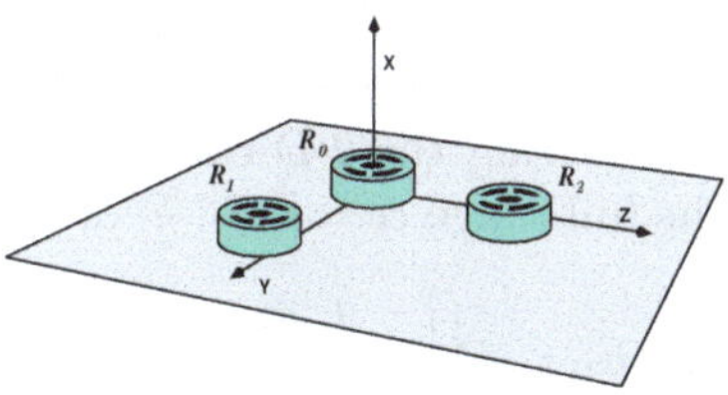


**Fig. 2.** The coordinate system for the receivers and measurement values interpretation

$$\begin{cases} n_y y_1 + n_z z_1 = s_1, \\ n_y y_2 + n_z z_2 = s_2, \\ n_y^2 + n_z^2 + n_z^2 = 1. \end{cases} \tag{2}$$

Its solution is

$$n_y = \frac{z_2 s_1 - z_1 s_2}{y_1 z_2 - y_2 z_1}, \qquad n_z = \frac{y_1 s_2 - y_2 s_1}{y_1 z_2 - y_2 z_1}, \qquad n_x = \sqrt{1 - n_y^2 - n_z^2}. \tag{3}$$

Having the coordinates of the normal vector, the azimuth $\phi$ and elevation $\theta$ can be computed using the following formulae

$$\begin{aligned} \phi &= \arcsin \frac{n_y}{\sqrt{n_y^2 + n_x^2}} = \arcsin \frac{n_y}{\sqrt{1 - n_z^2}}, \\ \theta &= \arcsin \frac{n_z}{\sqrt{n_z^2 + n_x^2}} = \arcsin \frac{n_z}{\sqrt{1 - n_y^2}}. \end{aligned} \tag{4}$$

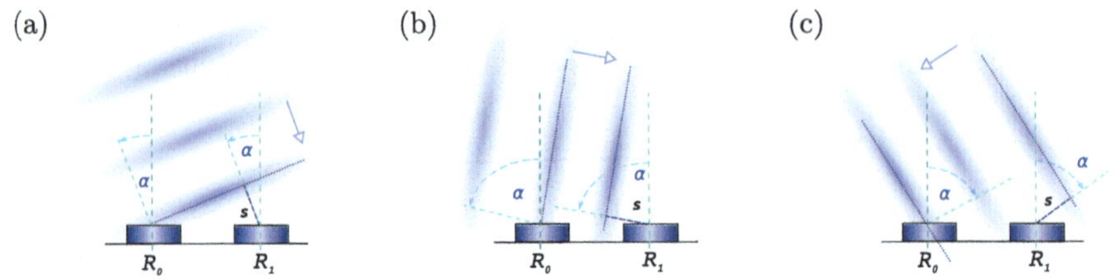


**Fig. 3.** The same measured distance $s$ can refer to different directions of signal arrival

It is worth noting that due to Eq. (1), it is no need to compute $n_x$. The azimuth and elevation angles can be determined in this way when the inter-distances between the receivers are less than half the signal wavelength. When they are larger, ambiguity occurs. To explain it, the 2-D case can be considered for simplicity (see Fig. 3). For some angles, after the detection of the wavefront by the $R_0$ receiver, the receiver $R_1$ detects a different wavefront than the one that detected by $R_0$ (compare Fig. 3a and b). When the case presented e.g. in Fig. 3b happens, then the angle $\alpha$ cannot be determined unequivocally. This is because the same distance $s$ refers to different directions of signal arrival. When the wavefront detected by $R_0$ is not the same as the wavefront detected by $R_1$, a correction must be added to the measured time $t_{01} = t_1 - t_0$. This correction is a multiple of the signal period $T_a$ and is different for different cases. For example, for the case presented in Fig. 3b it is equal to $1 \cdot T_a$, while for the case presented in Fig. 3c, it is $-2 \cdot T_a$. Considering the 2-D case presented in Fig. 3, the range of possible multiples for the full range $\boldsymbol{\alpha} = [-\frac{\pi}{2}, \frac{\pi}{2}]$ of arrival angles is limited to

$$\mathcal{I}_{\boldsymbol{\alpha}}(b_{01}) = \left[ \left\lfloor -\frac{b_{01}}{\lambda} \right\rfloor, \left\lfloor \frac{b_{01}}{\lambda} \right\rfloor \right],$$

where $\lambda$ is the wavelength of the signal and $\lfloor . \rfloor$ is the floor function. However, due to receivers directivity the range of arrival angles of the signal which can be detected is much smaller than $[-\frac{\pi}{2}, \frac{\pi}{2}]$. Assuming that the range is $\boldsymbol{\alpha} = [\alpha_{min}, \alpha_{max}]$, then the range of possible multiples can be expressed as follows

$$\mathcal{I}_{\boldsymbol{\alpha}}(b_{01}) = \left[ \left\lfloor \frac{b_{01} \sin \alpha_{min}}{\lambda} \right\rfloor, \left\lfloor \frac{b_{01} \sin \alpha_{max}}{\lambda} \right\rfloor \right].$$

It enables to construct the set of all possible values of time measurements for the same wavefront detected by $R_0$ and $R_1$ that is

$$\boldsymbol{\tau}_{\boldsymbol{\alpha}}(\tau_{01}) = \{\tau_i : \tau_i = \tau_{01} + kT_a, k \in \mathcal{I}_{\boldsymbol{\alpha}}(b_{01}) \wedge |v_a \tau_i| < b_{01}\}.$$

It is worth noting that

$$\forall \tau_i \in \boldsymbol{\tau}_{\boldsymbol{\alpha}}(\tau_j), \quad \boldsymbol{\tau}_{\boldsymbol{\alpha}}(\tau_i) = \boldsymbol{\tau}_{\boldsymbol{\alpha}}(\tau_j).$$

For simplicity, it can be assumed that the directivity pattern of all receivers is the same and has axial symmetry along the receiver acoustic axis. In the same way, it can be created the set $\boldsymbol{\tau}_\alpha(\tau_{02})$ for the second receiver pair of the system $\mathbb{R} = (R_0, R_1, R_2)$ presented in Fig. 2. In this way for this system is obtained

$$\boldsymbol{\tau}_\alpha^2(\tau_{01}, \tau_{02}) = \{(\tau_i, \tau_j) : \tau_i \in \boldsymbol{\tau}_\alpha(\tau_{01}) \;\wedge\; \tau_j \in \boldsymbol{\tau}_\alpha(\tau_{02})\}.$$

In the context of Eq. (2), these sets make it possible to create the set of parameters $(s_i, s_j)$

$$\mathfrak{s}(\tau_{01}, \tau_{02}) = \{(s_i, s_j) : s_i = v_a\tau_i, \tau_i \in \boldsymbol{\tau}_\alpha(\tau_{01}) \;\wedge\; s_j = v_a\tau_j, \tau_j \in \boldsymbol{\tau}_\alpha(\tau_{02})\}$$

In consequence, it determines a set of solutions of Eq. (3),

$$\mathfrak{n}^*(\tau_{01}, \tau_{02}) = \{(n_{x,1}, n_{y,1}, n_{z,1}), \ldots, (n_{x,k}, n_{y,k}, n_{z,k})\}.$$

The measured intervals $\tau_{01}$ and $\tau_{02}$ are determined by geometry of the receiver system $\mathbb{R}$ and the received signal $S$. Therefore, for further analysis it will be more convenient to use $\mathbb{R}$ and $S$ as the mapping arguments than the measurement values $\tau_{01}$ and $\tau_{02}$. Moreover, due to Eq. (1), to determine angles $\phi$ and $\theta$, it is enough to know $n_y$ and $n_z$. Thus, a bit more restricted set $\mathfrak{n}^2$ will be considered further and defined as follows

$$\mathfrak{n}^2(\mathbb{R}, S) = \{(n_{y,1}, n_{z,1}), \ldots, (n_{y,k}, n_{z,k})\}.$$

Having values of $n_{y,i}$ and $n_{y,i}$, then, using Eq. (4) the set of hypothetical signal arrival directions can be created

$$\mathcal{D}(\mathbb{R}, S) = \{(\phi_1, \theta_1), \ldots, (\phi_k, \theta_k)\}.$$

## 4 Ambiguity Removal

In order to select the proper angles, the 2-D case approach presented in [3] has been adapted. To explain the main idea in the simpler way, the 2-D case will be used again. In this case, it is enough to consider two receiver pairs whose inter-distances are greater than half the wavelength of the signal and they are different. When they receive the same signal, due to ambiguity, two different sets of designated arrival angles are obtained. The only common element is the correct angle of signal arrival (see Fig. 4). Thus, this feature can be exploited for selecting the final solution. To minimize the number of receivers, these two pairs can be integrated in one system consisted of three receivers, using one of the receivers as the common one. It is simple to notice that this approach fails when the inter-distances are the same. Unfortunately, this is not the only case. Therefore, the proper choice of receivers' inter-distances is crucial. A detailed analysis of this approach for the 2-D case is provided in [3]. It should be emphasized that

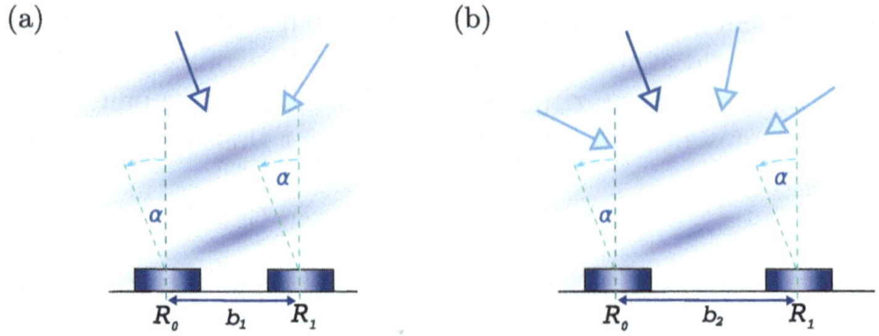


**Fig. 4.** Examples of possible incident angles for measurement data obtained while the real incident angle is equal to 20°. Directions were determined for the gap size b equal to a – 11 mm, b – 15 mm

this analysis and discussion presented above are true for the narrowband signal. This assumption is met, when popular piezoelectric ultrasonic transducers are used.

The main difference between the 2-D and 3-D cases is that the elementary receiver system has to consist of three receivers instead of two. Moreover, as said before, they cannot be collinear. Following the idea presented for the 2-D case, to apply it to the 3-D case the second receiver system is needed whose inter-distances of receivers are different than in the previous system (see Fig. 5a). Fortunately, these systems can be integrated in a similar way as it was done for the 2-D case. This time, two receivers are the common to both systems (see Fig. 5b). This means that only one additional receiver is needed to follow the approach outlined above. In this way, having the set of transducers $\boldsymbol{R} = \{R_0, R_1, R_2, R_3\}$, two receiver systems can be distinguished, i.e. $\mathbb{R}_1 = \{R_0, R_1, R_2\}$ and $\mathbb{R}_2 = \{R_1, R_2, R_3\}$. However, it is simple to notice, that also two additional systems can be found. In this case, these systems are $\mathbb{R}_3 = \{R_0, R_1, R_3\}$ and $\mathbb{R}_4 = \{R_0, R_2, R_3\}$.

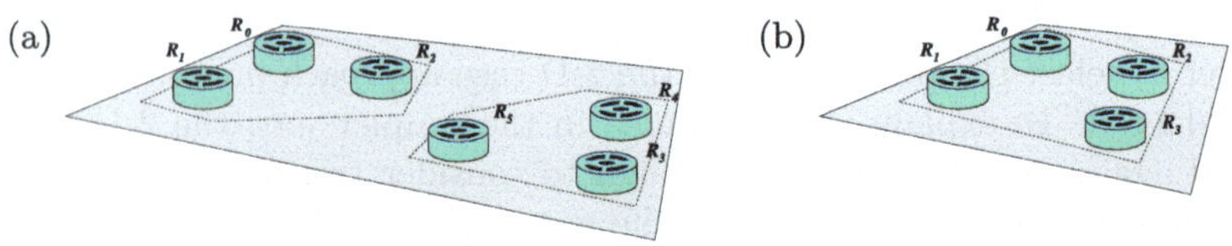


**Fig. 5.** Receiver system for the 3-D case that makes possible to solve the problem of ambiguity (a) two separate elementary receiver systems, (b) single integrated receiver system

## 5 Robustness

Because of time measurement errors, values of $n_y$ and $n_z$ are subjects to errors. Assuming that measurement errors of $\tau$ intervals can be approximated by the same value $\Delta\tau$ and taking into account (3), then

$$\Delta n_y = \left|\frac{z_1 - z_2}{y_1 z_2 - y_2 z_1}\right| v_a \Delta\tau, \qquad \Delta n_z = \left|\frac{y_2 - y_1}{y_1 z_2 - y_2 z_1}\right| v_a \Delta\tau, \tag{5}$$

It is obvious, but nevertheless it is worth noticing, that for each receiver system $\mathbb{R}_q$ errors of determining coordinates $n_y$ and $n_z$ are constant and strongly depend on the location of its receivers. Taking into account measurement errors, instead of a point set $\mathfrak{n}^2(\mathbb{R}_q, S)$, a set of rectangles is got. It can be defined as follows

$$\mathfrak{n}^2_\Delta(\mathbb{R}_q, S) = \{c_i = [n_{y,i} - \Delta_q n_y, n_{y,i} + \Delta_q n_y] \times [n_{z,i} - \Delta_q n_z, n_{z,i} + \Delta_q n_z] : \\ (n_{y,i}, n_{z,i}) \in \mathfrak{n}^2(\mathbb{R}_q, S)\}$$

Assuming that $(n_{x,S}, n_{y,S}, n_{z,S})$ are the vector coordinates of the direction of the $S$ signal propagation and taking into consideration Eq. (5), a *rectangle of uncertainty* can be obtained

$$c_S = [n_{y,S} - \Delta_q n_y, n_{y,S} + \Delta_q n_y] \times [n_{z,S} - \Delta_q n_z, n_{z,S} + \Delta_q n_z].$$

Following the idea presented in Fig. 4, for the direction $n_S$ the following condition must be met

$$\forall i \in [1, \ldots, n]\ \exists c_i \in \mathfrak{n}^2_\Delta(\mathbb{R}_i, S); \quad c_S \cap \left(\bigcap_{i=1}^{n} c_i\right) \neq \emptyset. \tag{6}$$

The ambiguity is removed when Condition (6) is met only for a single sequence of sets $c_i \in \mathfrak{n}^2_\Delta(\mathbb{R}_i, S)$. Estimating the direction $n_S$, it is assumed that when only a single sequence of sets $c_i \in \mathfrak{n}^2_\Delta(\mathbb{R}_i, S)$ exits, for which Condition (7) is met,

$$\forall i \in [1, \ldots, n]\ \exists c_i \in \mathfrak{n}^2_\Delta(\mathbb{R}_i, S); \quad \bigcap_{i=1}^{n} c_i \neq \emptyset. \tag{7}$$

then Condition (6) is also met. To estimate values of $n_{y,S}$ and $n_{z,S}$, uncertainty of their determination by subsequent receiver systems has to be taken into account. The uncertainty for $n_{y,i}$ can be defined as follows

$$p_{i,y} = \frac{\sum_{j=1}^{n} \Delta_j n_y - \Delta_i n_y}{(n-1)\sum_{j=1}^{n} \Delta_j n_y}.$$

It is simple noticing that

$$\sum_{i=1}^{n} p_{i,y} = 1.$$

To estimate $n_{y,S}$, a formula can be used

$$\hat{n}_{y,S} = \sum_{i=1}^{n} p_{i,y} n_{y,i}.$$

where $n_{y,i}$ determines $c_i$ which belongs to the sequence of sets meeting Condition (7). In the same way, the value of $n_{z,S}$ can be estimating. Having that, the estimation of $\phi$ and $\theta$ can be computed using Eq. (4).

## 6 Simulations

At the current stage of work, the preliminary implementation of the method was performed. In tests, the geometry of receivers arrangement presented in Fig. 6a was used. This corresponds to the geometry of the real sonar module which was constructed to implement the method of the 2-D case [3]. However, the geometry of receivers arrangement meets the condition which allows it to be used in the implementation of the method presented in this paper. It was estimated that when $(\phi, \theta) \in [-40°, 40°] \times [-40°, 40°]$, a time measurement error of at least 0.5 μs was correctly tolerated. It means that in this range of error values, ambiguities are successfully removed and the DoA is correctly determined.

(a)

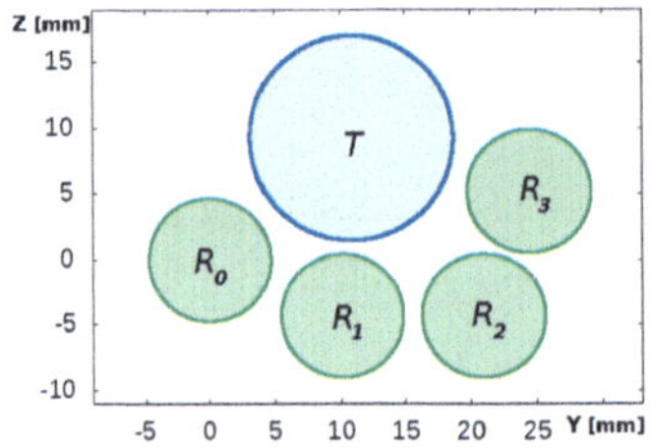


(b)

**Fig. 6.** Geometrical arrangement of sonar receivers and a transmitter (a) dimensioned positions of receivers and a transmitter (b) the real sonar module

## 7 Conclusion

The presented concept of the method of determining the direction of ultrasonic signal arrival is the extension of the previous version of the method for the 2-D case [3]. The main advantage of a such approach is that using larger inter-distances between receivers, the accuracy of estimation of the signal arrival direction is increased. It can be supposed that this feature will be preserved in extending the method to the 3-D case. At this moment, a preliminary implementation of the method was performed and it was tested in simulation for the geometry of receivers arrangement of the sonar module used in [3]. It was estimated that a time measurement error of at least 0.5 µs was correctly tolerated. This value is well below expectations. In the case of 2-D, this arrangement of receivers allows obtaining a tolerated measurement error value of up to 1.56 µs. Further examination should make it possible to explain the observed change. This should allow finding a way to increase the value of the tolerated error.

It is worth noting that the field of 3-D ultrasonic sensor construction for robotics is no longer just a field of scientific research. In 2019, Toposens introduced the TS3 sensor [11]. It can be expected that this will promote the use of such sensors in the precise navigation of mobile robots.

## References


1. Haardt, M., Nossek, J.A.: Unitary ESPRIT: how to obtain increased estimation accuracy with a reduced computational burden. IEEE Trans. Signal Process. **43**(5), 1232–1242 (1995)
2. Kreczmer, B.: Azimuth angle determination for the arrival direction for an ultrasonic echo signal. J. Autom. Mobile Robot. Intell. Syst. **11**(02), 31–41 (2017)
3. Kreczmer, B.: Estimation of the azimuth angle of the arrival direction for an ultrasonic signal by using indirect determination of the phase shift. Arch. Acoust. **44**(3), 585–601 (2019). http://acoustics.ippt.gov.pl/index.php/aa/article/view/2427
4. Roy, R., Kailath, T.: ESPRIT-estimation of signal parameters via rotational invariance techniques. IEEE Trans. Acoust. Speech Signal Process. **37**(7), 984–995 (1989)
5. Roy, R., Paulraj, A., Kailath, T.: Direction-of-arrival estimation by subspace rotation methods - ESPRIT. In: IEEE International Conference on Acoustics, Speech, and Signal Processing, ICASSP 1986, vol. 11, pp. 2495–2498, April 1986
6. Schmidt, R.: Multiple emitter location and signal parameter estimation. IEEE Trans. Antennas Propag. **34**(3), 276–280 (1986)
7. Steckel, J., Peremans, H.: Spatial sampling strategy for a 3D sonar sensor supporting BatSLAM. In: 2015 IEEE/RSJ International Conference on Intelligent Robots and Systems (IROS), pp. 723–728, September 2015
8. Steckel, J., Peremans, H.: Batslam: simultaneous localization and mapping using biomimetic sonar. PLOS ONE **8**(1), 1–11 (2013)
9. Sun, F., Lan, P., Gao, B.: Partial spectral search-based DOA estimation method for co-prime linear arrays. Electron. Lett. **51**(24), 2053–2055 (2015)
10. Tayem, N., Kwon, H.M.: Conjugate ESPRIT (C-SPRIT). In: IEEE Military Communications Conference, MILCOM 2003, vol. 2, pp. 1155–1160, October 2003
11. Toposens: TS3—3D ultrasonic sensor (2019). https://toposens.com/ts3/#specs
12. Zhang, D., Zhang, Y., Zheng, G., Feng, C., Tang, J.: Improved DOA estimation algorithm for co-prime linear arrays using root-MUSIC algorithm. Electron. Lett. **53**(18), 1277–1279 (2017)
13. Zhou, C., Shi, Z., Gu, Y., Shen, X.: DECOM: DOA estimation with combined MUSIC for coprime array. In: 2013 International Conference on Wireless Communications and Signal Processing, pp. 1–5, October 2013

# Investigation of GNSS Jamming Incidences on Main Transport Corridors in Central Europe

Artur Wieczyński(✉), Konrad Bożek, Maria Baczyńska-Wilkowska, and Arkadiusz Perski

ŁUKASIEWICZ Research Network – Industrial Research Institute for Automation and Measurements PIAP, Al. Jerozolimskie 202, 02-486 Warsaw, Poland
awieczynski@piap.pl

**Abstract.** In the article there are presented results of investigation on GNSS jamming incidences in different localizations on main transport corridors in Poland taking into account also traffic from neighbouring countries: Germany, Belarus and Lithuania. Localisations of the measurement sessions were carefully chosen to enable the general analysis of the GNSS jamming problem on the main transport corridors in Central Europe. Static and mobile measuring systems were designed and built in ŁUKASIEWICZ Research Network – Industrial Research Institute for Automation and Measurements PIAP based on software defined radio (SDR) platform. The system scans among all GNSS bands and seeks for the in-band interference signals. The detected and recorded jamming signals were further analysed. The analysis was done to assess the volume and types of GNSS jamming incidences. The collected data allow to estimate the scale of the problem and prepare a road map for the development of legislative documents and technical countermeasures. The research was done in the frame of "Evaluation of Galileo/EGNOS services jamming incidence in Central Europe" project founded by European Space Agency.

**Keywords:** GNSS · Jamming incidences · Radio interferences · SDR · Mobile measurement system · Statistical analysis

## 1 Introduction

Although the use of jamming devices is illegal in the EU, they are particularly often used to affect the vehicle tracking systems, both by rogue employees of transport companies (blocking vehicle location monitoring), as well as by organized crime groups with embezzlement and theft of cargo/vehicles [2, 3].

Considering the fact that the important roads and transport corridors are often located near objects especially sensitive for GNSS jamming turned on jammers emitting signals even with low EIRP power are able to destabilize or even totally block the critical safety infrastructure.

The goal of the work was to investigate how often Global Navigation Satellite System services could be exposed to interferences through the use of jamming devices


R. Szewczyk et al. (Eds.): AUTOMATION 2020, AISC 1140, pp. 368–378, 2020.
https://doi.org/10.1007/978-3-030-40971-5_34

available on the market. The system designed within the project was intended to be used as a detection and measuring equipment for GNSS interferences. Additionally the system had an ability to record detected interferences for further analysis.

Within the project a several specialized platforms were designed and built, and an analytical software prepared, allowing for conducting tests on selected locations and performing mobile research on trans-European transport corridors, national roads and local roads in the centers of large cities.

## 2 System Overview

In the frame of the project, ŁUKASIEWICZ – PIAP performed the following actions:

- development of unique mobile system architecture for detection of GNSS bands interference,
- preparation of a comprehensive solution, combining elements of digital signal processing, spectral analysis of signals [5], systems using GNSS positioning,
- implementation of software for statistical analysis,
- improvement of the knowledge about the scale of the threat in Central Europe in the area of potential terrorist attack disrupting critical for national security infrastructure based on GNSS.

## 3 System Development

The system was based on the SDR platform located in the measurement vehicle [1], enabling:

- analysis of GNSS in-band signals power, being a potential problem when GNSS receivers operate in such interfered environment,
- analysis of out-of-band signal power, which might cause saturation of receiver active elements (i.e. LNA).

The first of them was to scan among all GNSS bands and to seek the occurrence of in-band interference signal. If the power of received signal exceeds given threshold level, the signal samples were recorded into a file for further analysis, done in post-processing.

The second system functionality was to provide continuous recording of wideband signal power received from antenna. For that purpose dedicated wideband power detector and antenna were designed, manufactured and then used.

The measurements campaign envisaged within the project was preceded with the preliminary measurements campaign. It was done to gather practical experience on the type of interferences (their signals spectrum, powers, etc.) that could be detected in real life conditions. The preliminary measurements campaigns were performed in several stages. On the basis of performed measurements it was possible to tune up the system architecture to be more efficient for the final measurement campaign (Figs. 1 and 2).

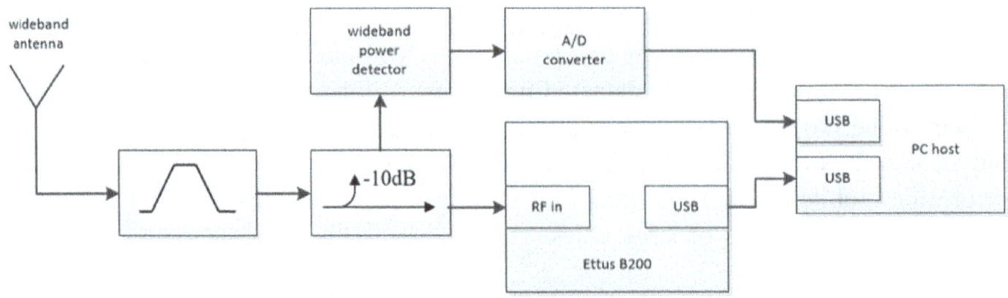


**Fig. 1.** Measurement system block diagram.

**Fig. 2.** Measurement ambulance equipped with proposed hardware solutions.

# 4 Measurement Campaigns

The measurement campaign lasted more than 49 days of measurement, performed in seven different locations. At each location at least six days of continuous measurement was done.

## 4.1 Location of Measurement Sessions

Locations of measurement sessions were distributed within Poland territory and located on different transport corridors to avoid measurement sessions on consecutive sections of the same road (Fig. 3).

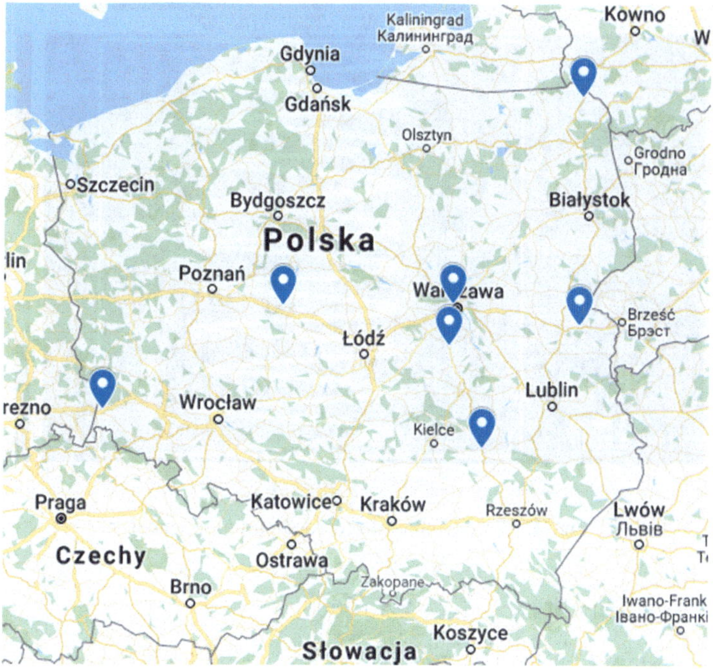


**Fig. 3.** Map of measurement sessions locations.

## 5 Types of Detected Jamming Incidents

During the measurement campaign different kinds of jamming incidents were detected and recorded. Within the post processing phase, all incidents were grouped into seven different categories, as follows:

- chirp – signal of a teeth saw shape, very popular for intentional jamming devices,
- continuous wave – a single continuous wave with some particular frequency,
- modulated informational – modulated signal that probably is a carrier of some information,
- modulated noninformational – modulated signal with probably no information,
- radar – signal that probably comes from a radar sources,
- other – wideband signals from unknown source,
- non jamming – detected and recorded signals which are outside but very close to GNSS bands (Fig. 4).

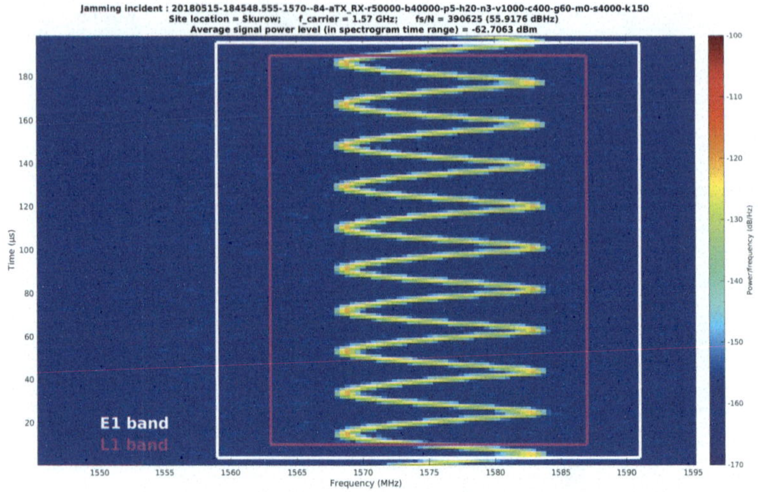


**Fig. 4.** An example of recorded jamming signal.

# 6 Gathered Results

Thanks to continuous method of measurements lasting seven days it was possible to gather information about jamming incidents from different days of week. The sections were chosen according the predominant type of traffic to make measurements sessions in different traffic conditions (different share of types of vehicles). In all measurement sessions the chosen section of a given road has high traffic in comparison to other sections on the same road. Additionally two sections were chosen in such a way to have a section with high share of trucks traffic (Szypliszki) and a section with low share of trucks traffic (Golina). In each measurement section except one (Warsaw), the measurement session was held for 7 days. The number of observed jamming incidents is presented in the Table 1 below.

**Table 1.** Observed jamming incidents in total.

| Jamming incidents | Warsaw | Skurów | Opatów | Biała Podlaska | Szypliszki | Golina | Żarska Wieś |
|---|---|---|---|---|---|---|---|
| Lower band | 1709 | 26 | 43 | 20 | 39 | 9 | 1137 |
| Upper band | 785 | 29 | 29 | 20 | 61 | 9 | 18 |
| Total | 2494 | 55 | 72 | 40 | 100 | 18 | 1155 |

# 7 Analysis of Detected Jamming Signals

## 7.1 Jamming Incidents in Lower and Upper Band

During the further analysis of detected jamming incidents it was assumed that chirp signals, continuous wave and modulated signals come from passing vehicles. The rest of detected incidents (radar signals and other signals from unknown sources) do not come from passing vehicles. The number of lower and upper band incidents from passing vehicles is shown in the Table 2 and a diagram below (Fig. 5).

**Table 2.** Observed jamming incidents from passing vehicles.

| Jamming signals | Warsaw | Skurów | Opatów | Biała Podlaska | Szypliszki | Golina | Żarska Wieś |
|---|---|---|---|---|---|---|---|
| Lower band | 25 | 11 | 14 | 2 | 20 | 5 | 10 |
| Upper band | 114 | 28 | 27 | 19 | 59 | 9 | 18 |
| Total | 139 | 39 | 41 | 21 | 79 | 14 | 28 |

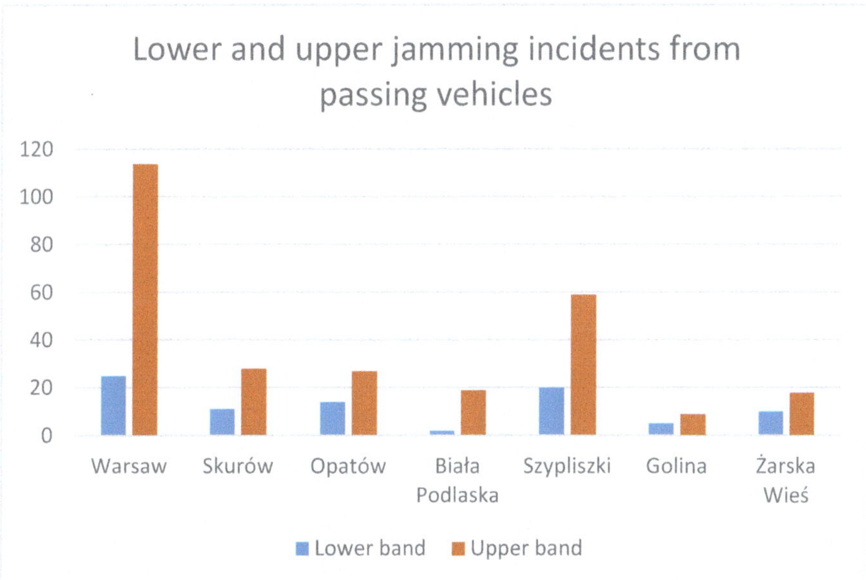


**Fig. 5.** Observed jamming incidents from passing vehicles.

## 7.2 Jamming Incidents in Different Parts of a Day

It is important to assess in which part of a day different jamming incidents occur mostly. For that purpose a day was divided into four different parts:

- night – from 00:00 am to 06:00 am,
- morning – from 06:00 am to 12:00 am,
- afternoon – from 12:00 am to 06:00 pm,
- evening – from 06:00 pm to 12:00 pm.

The Table 3 below presents all jamming incidents detected and recorded during each of the measurement session divided into four parts of a day according the time of their occurrence.

**Table 3.** Jamming incidents in different parts of a day.

| | Warsaw | Skurów | Opatów | Biała Podlaska | Szypliszki | Golina | Żarska Wieś |
|---|---|---|---|---|---|---|---|
| Night | 534 | 16 | 16 | 5 | 16 | 4 | 630 |
| Morning | 876 | 23 | 26 | 15 | 35 | 6 | 158 |
| Afternoon | 644 | 15 | 24 | 17 | 25 | 7 | 136 |
| Evening | 449 | 8 | 11 | 4 | 29 | 1 | 382 |
| Total per location | 2503 | 62 | 77 | 41 | 105 | 18 | 1306 |

The following diagram present the distribution of jamming incidents within different parts of a day for one of the measurements locations (Warsaw) (Fig. 6).

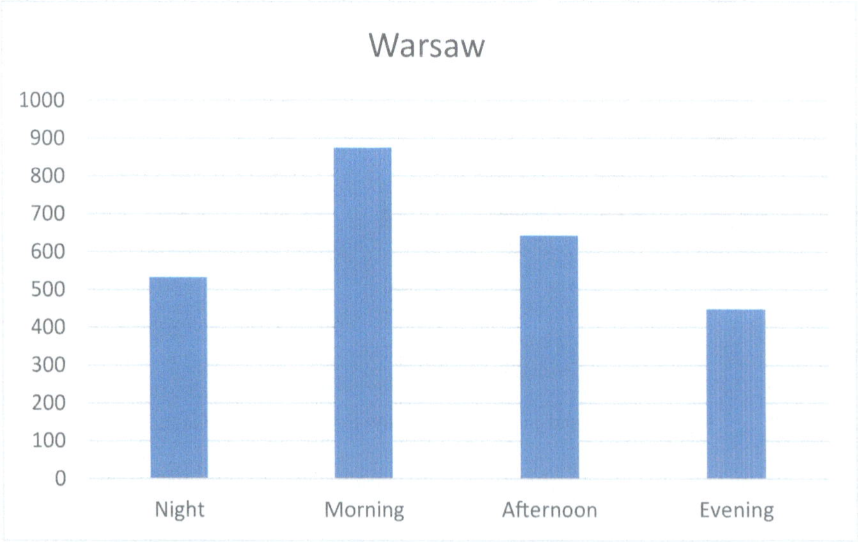


**Fig. 6.** Jamming incidents in different parts of a day in Warsaw.

### 7.3 Types of Detected Jamming Incidents by Measurements Locations

At each of the measurement session different jamming incidents were detected and recorded. They were categorized into seven different groups described above. The analysis was done for all locations but in this paper the results from only one location (Szypliszki) are presented below (Table 4).

**Table 4.** Jamming incidents in Szypliszki

| | Chirp | Continuous wave | Modulated informational | Modulated noninformational | Radar | Pulse | No jamming |
|---|---|---|---|---|---|---|---|
| Lower band | 1 | 1 | 15 | 3 | 4 | 15 | 0 |
| Upper band | 4 | 42 | 9 | 4 | 0 | 2 | 6 |
| Total | 5 | 43 | 24 | 7 | 4 | 17 | 6 |

The share of each type of jamming signals within all detected incidents is presented in a diagram below (Fig. 7).

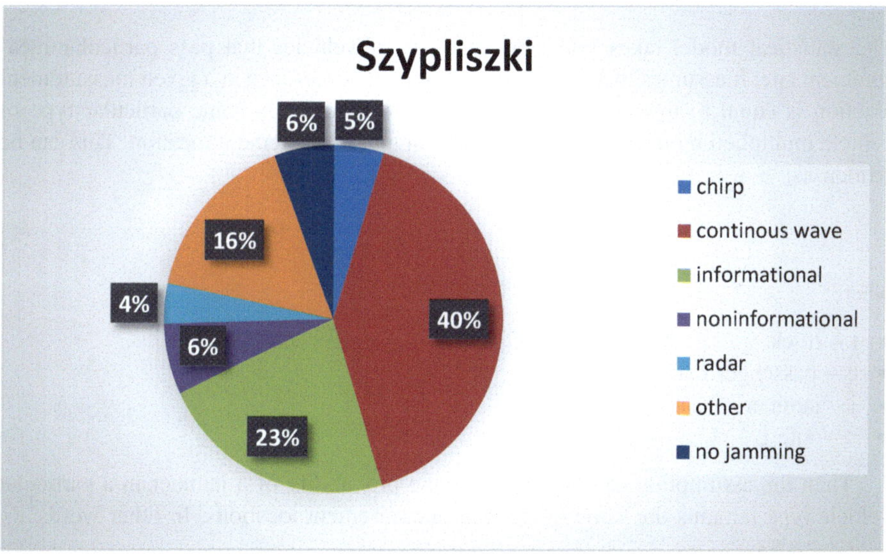


**Fig. 7.** Jamming incidents in Szypliszki.

Taking into account only jamming incidents that probably come from vehicles the diagram with different kinds of incidents is as follow (Fig. 8):

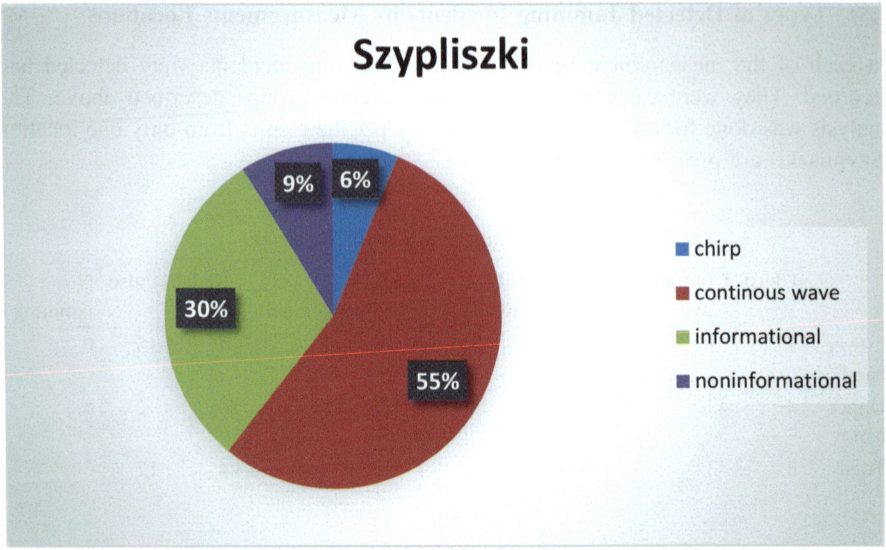


**Fig. 8.** Jamming incidents from vehicles in Szypliszki.

### 7.4 Statistical Model

The statistical model takes into account types of vehicles that pass particular measurement site. It assumes that probability of jammer occurrence in a given measurement location is equal to the sum of probabilities of a jammer in some particular type of vehicle multiplied by probability of a vehicle in that measurement location. This can be written as:

$$P(j|s_i) = P(j|t,s_i) \cdot P(t|s_i) + P(j|c,s_i) \cdot P(c|s_i) \quad (1)$$

where:

- t – truck,
- c – passenger car,
- j – jammer,
- $s_i$ – site i.

Then the assumption can be made that the probability of a jammer in a particular vehicle type remains the same in different measurement locations. In other words

$$P(j|t,s_i) = P(j|t), P(j|c,s_i) = P(j|c) \quad (2)$$

With the above assumption, the model can be rewritten as:

$$P(j|s_i) = P(j|t) \cdot P(t|s_i) + P(j|c) \cdot P(c|s_i) \text{ for } j = 1, \ldots, 7 \quad (3)$$

Having the set of equations from all measurement locations a simple linear regression can be done.

The mathematical model took into account trucks and passenger cars. The potential contribution of other road users like bikes or motors were omitted. The data about current traffic was taken from General Directorate of National Roads and Motorways. For this statistical modelling only chirp, continuous wave and modulated (both informational and noninformational) signals were taken, radar and other signals were omitted. It gives us the following number of jamming incidents taken into account (Table 5).

**Table 5.** Number of jamming incidents from vehicles

| | Warsaw | Skurów | Opatów | Biała Podlaska | Szypliszki | Golina | Żarska Wieś |
|---|---|---|---|---|---|---|---|
| Number of jamming signals from vehicles | 139 | 39 | 41 | 21 | 79 | 14 | 28 |

The simple linear regression give us the average number of jammers in a vehicle:

- average number of jammers in a passenger car: 0,0000 ± 0,0001,
- average number of jammers in a truck: 0,0017 ± 0,0003.

The first number give us the average number of jammers in a vehicle of a particular type while the second number is an error with which the average number is calculated.

## 8 Conclusions

Global Navigation Satellite System services are exposed to interferences through the use of jamming devices that are easily available [4]. For that purpose seven measurement sessions were carried out during which all signals jamming GNSS bands were detected and recorded. The post processing of gathered results allowed to investigate what were the jamming signals and how often they occurred.

The obtained results show that majority of jamming incidents that come from passing vehicles are caused by continuous wave signals. In five out of seven locations continuous wave jamming signals were more than half of all jamming signals from vehicles. In one of these location it exceeded 70%. At the same time detected chirp signals were present more rarely than expected. Additionally in some locations other jamming signals were detected very often. In Żarska Wieś many signals from a radar operating somewhere nearby were observed while in Warsaw many other signals in form of very short wideband pulses were noticed. It gives an important remark that not only intentional jamming signals can cause a disturbance but also many unintentional signals influence the appropriate reception of GNSS signals.

All jamming signals were grouped according they appearance in different parts of a day. It allowed to answer a question when jamming incidents always happen. It occurred that majority of jamming incidents happen during a day on mornings and afternoons. Less jamming incidents happen during evenings and night. It could be explained by higher traffic volume during daytime than during night-time. An exception is a location in Żarska Wieś but it was caused by a radar operating during nights.

Statistical model was also created to predict if potential jammers are located in trucks or in passenger cars. It was calculated for all jamming incidents from vehicles in total and also for each type of jamming incidents separately. All cases show that the probability of a jammer in truck is much higher than a probability of a jammer in passenger car.

The above analysis leads to the idea of further investigation of jamming incidents. The investigation should be carried out on different transit corridors to achieve more reliable data.

## References

1. Bożek, K., Perski, A., Wieczyński, A., Baczyńska-Wilkowska, M.: Detection of GNSS jamming incidence. In: Automation 2018. Advances in Automation, Robotics and Measurement Techniques, pp. 548–557. Springer (2018)
2. Dunn, J.E.: GPS jamming caused by moonlighting truck drivers, research suggests. Techworld (2013)
3. Kraus, T., Bauernfeind, R., Eissfeller, B.: Survey of in-car jammers – analysis and modelling of the RF signals and IF samples (suitable for active signal cancellation). In: 24th International Technical Meeting of the Satellite Division of the Institute of Navigation, Portland, OR, 19–23 September (2011)
4. Pullen, S., Gao, G.: GNSS Jamming in the Name of Privacy. Inside GNSS (2012)
5. Borio, D., Dovis, F., Kuusniemi, H., Presti, L.L.: Impact and detection of GNSS jammers on consumer grade satellite navigation receivers. Proc. IEEE **104**(6), 1233–1245 (2016)

# Microbial Fuel Cell-Based Toxicity Sensor

Juste Rozene[1(✉)], Antanas Zinovicius[2], Barbora Kacinskaite[1], Vytautas Bučinskas[1], Arunas Ramanavicius[2], and Inga Morkvenaite-Vilkonciene[1,3]

[1] Department of Mechatronics and Robotics, Vilnius Gediminas Technical University, J. Basanaviciaus 28, 03224 Vilnius, Lithuania
juste.rozene@vgtu.lt
[2] Vilnius University, Naugarduko 24, 03226 Vilnius, Lithuania
[3] Laboratory of Electrochemical Energy Conversion, Center for Physical Sciences and Technology, Sauletekio 3, 10257 Vilnius, Lithuania

**Abstract.** The increase in energy consumption also increases the damage to the environment and the toxicity of water still is one of the main problems in the world. Microbial fuel cells (MFCs) are one of the alternative energy sources. The study demonstrates the possibility to use yeast as a probe to assess redox and electrophile-based toxicities and to extend the life of a MFCs. But as it is known, the quinones cause the oxidative stress of the cells. That is why the main aim of our research was to create a real-time biomonitoring system using MFCs to detect the concentration of quinones in solution, using one and two redox mediator's systems. Atomic force microscopy (AFM) was used to analyse the topography of the modified and non-modified graphite electrode and Potentiostat/Galvanostat Autolab PGSTAT 30 was used for the electrochemical measurements, which results are described in this paper.

**Keywords:** Microbial fuel cell · Toxicity · Sensor · Quinones

## 1 Introduction

The increase in energy consumption also increases the damage to the environment. That is why alternative energy sources are needed. Also, clean and safe drinking water is necessary for every person in the world [1, 2].

That is the reason why during the last decade a lot of researches turns to biofuel cells. The microorganisms used for biofuel cells turn to biocatalysts for simultaneous electricity generation and although scientists predict that this technology will not replace other renewable sources, it can be useful in covering wastewater treatment costs [3, 4]. The biofuel cell becomes a device that converts chemical energy into electricity through electrochemistry and biochemistry [5]. Not so long ago, yeast cells such as Saccharomyces cerevisiae as the main material for the biofuel cells started to be used. The yeast is used for thousand years for different kinds of purposes: food or drinks industry, nutritional supplements, probiotics, industrial ethanol production, and most important scientific researches. These fields often generate a lot of wastewaters, which can be used with biofuel cells for purification [3].


R. Szewczyk et al. (Eds.): AUTOMATION 2020, AISC 1140, pp. 379–388, 2020.
https://doi.org/10.1007/978-3-030-40971-5_35

A lot of researches are done when for the measurements the baking yeast is used as the microorganisms. Because yeast is highly resistant to environmental impact (including anaerobic conditions and varying pH), inexpensive, easy to manipulate and the yeast genome is fully sequenced allowing facile manipulation of their genetics to predictably control their susceptibility/resistance to quinone toxicity [6, 7]. It is expected that using them will make it possible to extend the life of the MFCs. More importantly, the yeast serves as a tool for the discovery of other, as yet unestablished, mechanisms of quinone-based toxicity [6].

While the toxicity of water is one of the main problems in the world, the researchers studied it while using microorganisms as the indicators [8, 9]. Unicellular organisms, like yeast, is as an early toxicity warning device because of the yeast respiratory function [10]. It depends on oxygen concentration in the solution. As it is known, quinones cause oxidative stress which causes the death of the cell [11, 12]. Quinones can "redox" cycle while interacting with a biological reductant and dioxygen ($O_2$) to generate reactive oxygen species (ROS) such as superoxide ($O_2^-$) or hydrogen peroxide ($H_2O_2$) which can be pernicious [13, 14]. Production of ROS can cause severe oxidative stress within cells through the formation of oxidized cellular macromolecules, including lipids, proteins, and DNA [15]. In an attempt to validate established mechanisms of quinone toxicity in a whole-cell in vivo system, to determine the contribution of each mechanism on the overall toxicity, and to assess participation of new and/or alternative mechanisms of toxicity, we have begun to examine the effects of quinone exposure on the viability of the yeast Saccharomyces cerevisiae.

To ensure the proper electron transfer from the cells to the electrode, a two-mediator system is usually applied: one mediator is lipophilic, which interacts with the intracellular redox centers; another mediator is hydrophilic, which takes the electrons from the lipophilic mediator and passes them to the electrode [16, 17].

Lipophilic 9,10-phenanthrenequinone (PQ) can work intracellularly when it is reduced and diffuse back through the membrane into the solution [18]. To ensure effective electron transfer to the electrode, the second redox mediator, such as ferricyanide should be used [16]. The study demonstrates the possibility to use yeast as a probe to assess redox and electrophile-based toxicities. Two types of ETM are needed because one of them, intracellular redox mediators available to permeate across the cell membrane (can penetrate the cell membrane) and react with the enzymes to deliver electrons to another, externally acting ETM, which transmits electrons to the electrode [19, 20].

The main aim of our research was to create a real-time biomonitoring system using MFCs to detect the concentration of quinones in solution, using one and two redox mediator's systems.

## 2 Materials and Methods

### 2.1 Materials

All chemicals used in electrochemical experiments were purchased from Merck, Riedel and Sigma–Aldrich companies. Dried yeast was purchased from food supplier "Dr. Oetker Lietuva" (Vilnius, Lithuania).

Phosphate buffer solution was prepared with 0.05 M $CH_3COONa$; 0.05 M $NaH_2PO_4$; 0.05 M $Na_2HPO_4$, and 0.1 M KCl in distilled water. Glucose and potassium ferricyanide were prepared in phosphate buffer solution. Before investigations, glucose solutions were allowed to mutarotate overnight. 9,10-phenanthrenequinone (PQ) was dissolved in 97% ethanol.

### 2.2 Preparation of Yeast

1 g of YPD-broth was mixed with 20 ml distilled water to get medium with YPD-broth 50 g/l concentration. 500 mg of dried yeast was introduced to prepared suspension. A further culture was grown in shaking incubator at 200 rpm till yeast reaches Logarithmic Phase (20–24 h).

### 2.3 Graphite Electrode Preparation

Graphite electrode (ø3 mm) was sanded with paper with three different grinding bead sizes and washed with distilled water and 97% ethanol. The yeast 0.125 g/mL (2.5 µL drop) are immobilized on the dry electrode (see Fig. 1A). In case if PQ was used, it was immobilized on to the prepared graphite electrode before yeast immobilization with concentration 3.75 mM, 2.5 µL drop (see Fig. 1B).

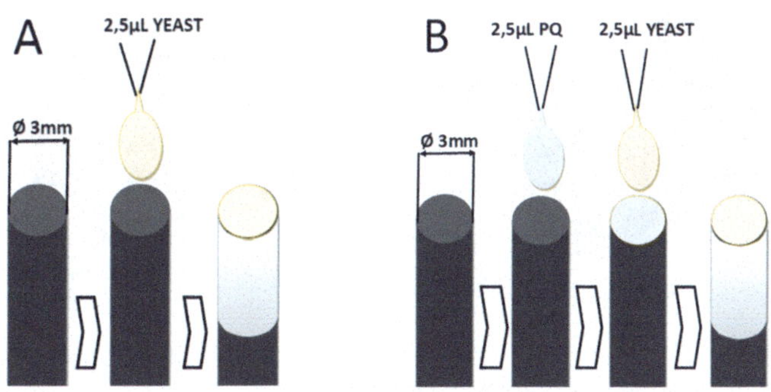


**Fig. 1.** Scheme of graphite electrode preparation: A. Immobilization of yeast; B. Immobilization of PQ and yeast.

### 2.4 Atomic Force Microscopy Measurements

The BioScope II AFM combined with optical microscope developed by Veeco Instruments Ltd (Santa Barbara, USA) and NP-D cantilever (Bruker, CA, USA) was used at contact mode in fluid at 0.15 Hz speed for the measurements of yeast to obtain high-resolution topographical images. Experiments were performed with the modified and non-modified graphite electrode to check the quality of yeast immobilization on the electrode. Contact mode in fluid was used in order to keep yeast cells active for a longer period. As a fluid phosphate buffer solution (pH = 6.77) was used. Raw data was analysed using Nanoscope Analysis 1.5 software.

### 2.5 Electrochemical Measurements

Electrochemical measurements were performed using an Autolab PGSTAT 30 Potentiostat/Galvanostat (Utrecht, Netherlands), and NOVA software. All experiments were carried out at ambient temperature (at 20 °C), while stirring in PBS, pH = 6.77, under aerobic conditions. Measurements were performed in three electrodes electrochemical cell, where the graphite electrode was connected as a working electrode, the platinum electrode as a counter electrode and Ag/AgCl electrode as a reference electrode. Cyclic voltammograms were recorded at the scan rate of 0.1 V/s in the range from –0.6 V to 0.8 V.

## 3 Results and Discussion

### 3.1 Electrodes Surface Analysis by AFM

Electrodes' surface was analysed before and after modification with yeast cells. AFM results show that the graphite electrode was successfully modified with yeast and it can be used for further investigation (see Fig. 2).

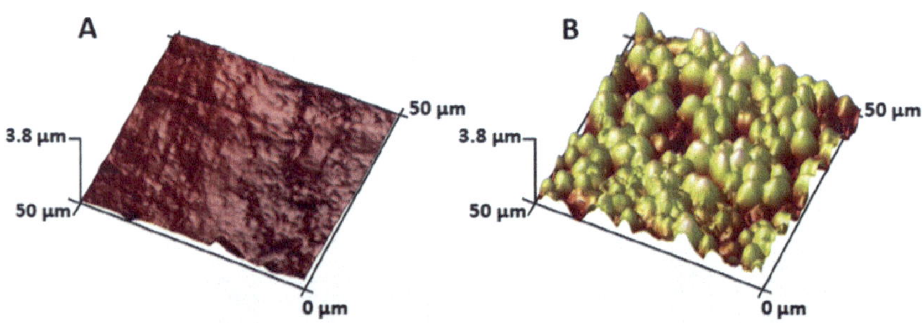


**Fig. 2.** Topographical images: A. Topography of the non-modified graphite electrode; B. Topography of the yeast-modified graphite electrode.

### 3.2 Investigation of Toxicity Sensor

**Experiments with Non-modified Electrode**

Cyclic voltamperograms were recorded at different concentrations of PQ in buffer solution (see Fig. 3). When PQ concentration increased, the current density values at both the cathodic (PQ reduction, at interval from –0.25 V to –0.15 V) and anodic (PQ oxidation, at interval from –0.12 V to –0.08 V) peaks increased accordingly. While the concentration of PQ increased from 0 to 0.75 mM the current density changed from –1 to –6.5 mA/cm$^2$ at –0.25 V potential and while the concentration of PQ increased from 0 to 0.75 mM the current density increased from 0.5 to 6 mA/cm$^2$ at –0.1 V potential.

The calibration curve – current density dependencies on PQ concentration at cathodic and anodic potentials are linear in the range of 0.11 mM at both cathodic and anodic peaks (see Fig. 5). Therefore, the microbial fuel cell can work as PQ toxicity sensor in current range, and both cathodic and anodic peaks of cyclic voltammograms can be used for this purpose.

When the concentration of PQ increased the current density grew up at –0.25 V and –0.1 V potentials. The increasing at both peaks stop when the concentration of 0.625 mM is reached (see Fig. 4). However, the effective sensor work can be ensured at lower (0.11 mM) range.

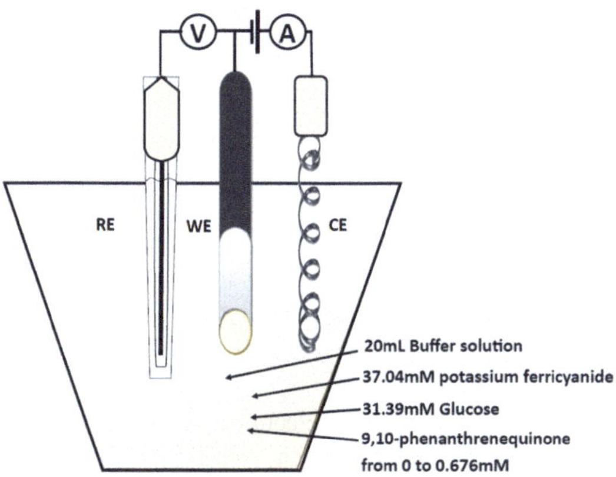


**Fig. 3.** The electrochemical cell with three electrodes in the buffer solution with potassium ferricyanide, the glucose and the 9,10-phenanthrenequinone. WE – working electrode (graphite electrode), RE – reference (Ag/AgCl) electrode, and CE – counter (platinum) electrode. Voltmeter V measures the potential difference between WE and RE; amperemeter A measures the current flowing through WE and CE.

**Experiments with Yeast-Modified Electrode**

The three electrodes electrochemical cell was used for measurements, as it is shown in Fig. 5. The cyclic voltammograms of the yeast-modified graphite electrode (see Fig. 1A) were recorded with the potassium ferricyanide and the glucose in the buffer solution with different PQ concentrations (see Fig. 5).

The current density increases more significantly after the concentration of the PQ reaches 0.098 mM at –0.15 V and –0.35 V potentials (see Fig. 6). But the peaks at 0.1 V and 0.4 V potentials starts to be unstable (see Fig. 7). The linear dependence of current density vs. concentration was found in the range from 0.176 mM to 0.676 mM. Comparing to non-modified electrode, the range of detection became 6 times higher

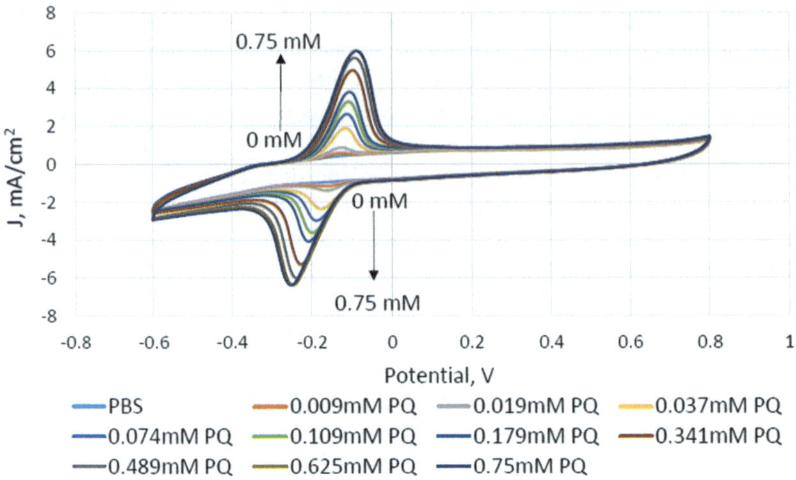


**Fig. 4.** Cyclic voltammograms recorded with non-modified graphite electrode at different PQ concentrations in the buffer solution. Scan rate was 0.1 V/s.

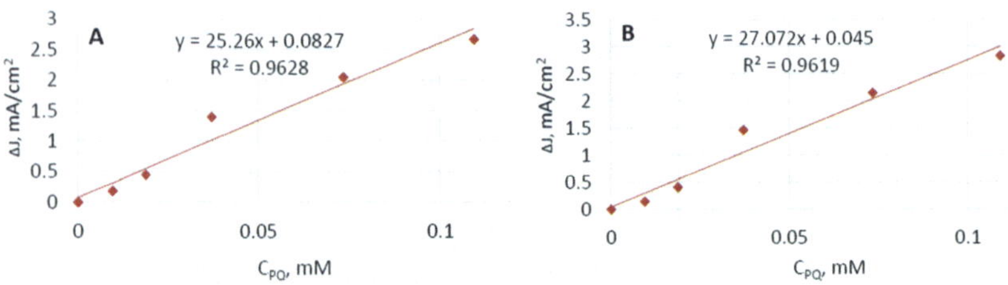


**Fig. 5.** The calibration curve of non-modified electrode: A. At cathodic potential; B. At anodic potential. The data for calibration curve was taken from Fig. 2.

(from 0.11 mM measured with non-modified electrode to 0.676 mM obtained at modified electrode).

## Experiments with PQ and Yeast-Modified Electrode

The graphite electrode modified with PQ and yeasts was used as a working electrode in electrochemical cell (see Fig. 1B) and cyclic voltamperograms were recorded without and with glucose. The steady-state voltammograms were taken into account (see Fig. 8). Cyclic voltammograms, observed in buffer solution and after adding of glucose, are similar. The PQ oxidation-reduction peaks was obtained, since PQ was immobilized on the electrode. In order to achieve better sensitivity of the sensor and increase the range of current density, we used second redox mediator potassium ferricyanide. As expected, it took 20 min to get steady-state conditions for yeast [16].

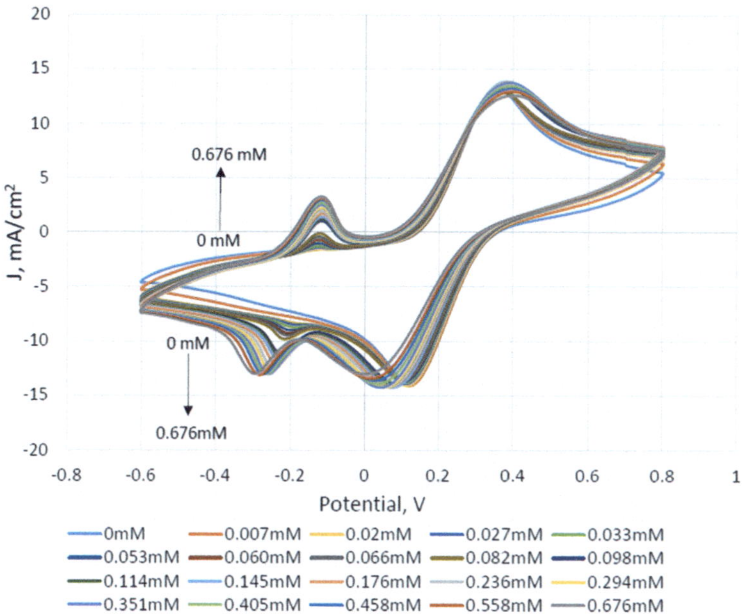


**Fig. 6.** Cyclic voltammograms of yeast-modified graphite electrode at different PQ concentrations. Measurements were performed in buffer solution with the potassium ferricyanide 37.04 mM and glucose 31.39 mM.

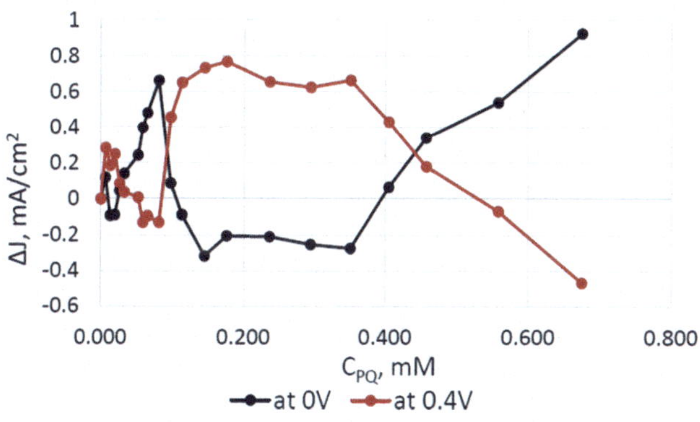


**Fig. 7.** The yeast-modified electrode used as a signal converter at cathodic potential (0 V) and at anodic potential (0.4 V) from Fig. 6.

The current density decreases at –0.1 V potential, but at the other three peaks, it increases. The maximum increment shows at 0.15 V and 0.35 V potentials, as it should be for potassium ferricyanide. When the potassium ferricyanide is mixed in, the current density increased in time: from 0.5 mM/cm$^2$ (at 0 min) to 8 mM/cm$^2$ (at 20 min) at 0.15 V potential; and from –0.5 mM/cm$^2$ (at 0 min) to 7 mM/cm$^2$ (at 20 min) at 0.25 V potential.

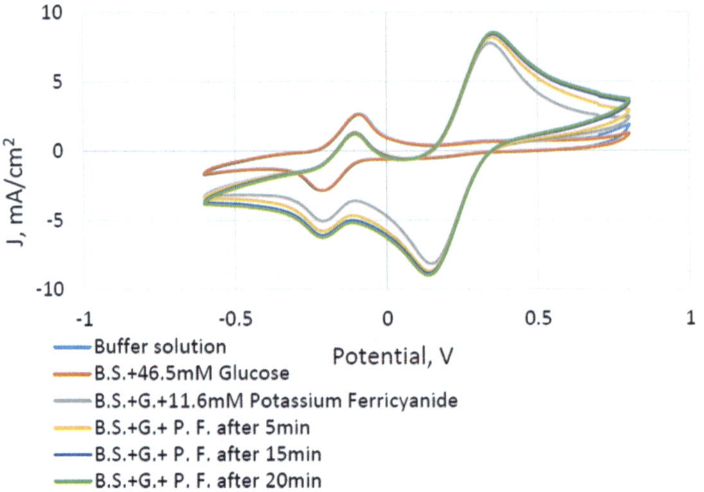


**Fig. 8.** Cyclic voltamperograms of graphite electrode with immobilized PQ and yeast cells before and after the glucose and later the potassium ferricyanide was mixed in.

## 4 Conclusions

The PQ toxicity sensor, based on microbial fuel cell was developed and evaluated. Non-modified graphite electrode and yeast-modified electrodes were used during the investigation. Experiments with non-modified electrode showed that while the concentration of PQ increased the current density changed and the dependence on PQ concentration is linear at both cathodic and anodic peaks. The sensor's calibration curves show that the device can detect the PQ in the range from 0 mM to 0.11 mM using the data at both cathodic and anodic peaks. The current density reaches steady-state value at the concentration of 0.625 mM. Also, it was important to investigate the modified graphite electrode surface with AFM to be sure that the modification was successful and it can be used for further investigation. Experiments with yeast-modified electrode showed that the current density increases more significantly after the concentration of the PQ reaches the average concentration measurement value – 0.098 mM. But the peaks at positive potentials start to be unstable. Experiments with PQ and yeast-modified electrode showed that the current density increases rapidly after the second ETM (potassium ferricyanide) is mixed in and after 20 min sensor reaches steady-state, where the PQ is used for electron transfer via the cell membrane and the second ETM is used for electron transfer in extracellular space. The increment noticeable not at all peaks, and the maximum values are reached only at positive potentials. Using yeast-modified electrode as a toxicity sensor, we found that PQ detection range is from 0.176 mM to 0.676 mM. In further our research, several quinones will be used as ETMs in the microbial fuel cells for toxicity investigations.

**Acknowledgments.** This Research was funded by European Social Fund (Project No 09.3.3.-LMT-K-712-16-0211).

## References


1. Du, Z., Li, H., Gu, T.: A state of the art review on microbial fuel cells: a promising technology for wastewater treatment and bioenergy. Biotechnol. Adv. **25**, 464–482 (2007). https://doi.org/10.1016/j.biotechadv.2007.05.004
2. Liu, H., Ramnarayanan, R., Logan, B.E.: Production of electricity during wastewater treatment using a single chamber microbial fuel cell. Environ. Sci. Technol. **38**, 2281–2285 (2004). https://doi.org/10.1021/es034923g
3. Hubenova, Y., Mitov, M.: Extracellular electron transfer in yeast-based biofuel cells: a review. Bioelectrochemistry **106**, 177–185 (2015). https://doi.org/10.1016/j.bioelechem.2015.04.001
4. Lei, Y., Chen, W., Mulchandani, A.: Microbial biosensors. Anal. Chim. Acta **568**, 200–210 (2006)
5. Bullen, R.A., Arnot, T.C., Lakeman, J.B., Walsh, F.C.: Biofuel cells and their development. Biosens. Bioelectron. **21**, 2015–2045 (2006)
6. Rodriguez, C.E., Shinyashiki, M., Froines, J., Yu, R.C., Fukuto, J.M., Cho, A.K.: An examination of quinone toxicity using the yeast Saccharomyces cerevisiae model system. Toxicology (2004). https://doi.org/10.1016/j.tox.2004.04.016
7. Baronian, K.H.R.: The use of yeast and moulds as sensing elements in biosensors. Biosens. Bioelectron. **19**, 953–962 (2004)
8. Campanella, L., Favero, G., Mastrofini, D., Tomassetti, M.: Further developments in toxicity cell biosensors. Sens. Actuators B Chem. **44**, 279–285 (1997)
9. Hollis, R.P., Killham, K., Glover, L.A.: Design and application of a biosensor for monitoring toxicity of compounds to eukaryotes. Appl. Environ. Microbiol. **66**, 1676–1679 (2000)
10. Stein, N.E., Hamelers, H.M.V., Van Straten, G., Keesman, K.J.: On-line detection of toxic components using a microbial fuel cell-based biosensor. J. Process Control **22**, 1755–1761 (2012). https://doi.org/10.1016/j.jprocont.2012.07.009
11. Jamieson, D.J.: Oxidative stress responses of the yeast Saccharomyces cerevisiae. Yeast **14**, 1511–1527 (1998)
12. Breitenbach, M., Weber, M., Rinnerthaler, M., Karl, T., Breitenbach-Koller, L.: Oxidative stress in fungi: its function in signal transduction, interaction with plant hosts, and lignocellulose degradation. Biomolecules **5**, 318–342 (2015)
13. O'Brien, P.J.: Molecular mechanisms of quinone cytotoxicity. Chem. Biol. Interact. **80**, 1–41 (1991)
14. Kumagai, Y., Koide, S., Taguchi, K., Endo, A., Nakai, Y., Yoshikawa, T., Shimojo, N.: Oxidation of proximal protein sulfhydryls by phenanthraquinone, a component of diesel exhaust particles. Chem. Res. Toxicol. **15**, 483–489 (2002). https://doi.org/10.1021/tx0100993
15. Bolton, J.L., Trush, M.A., Penning, T.M., Dryhurst, G., Monks, T.J.: Role of quinones in toxicology. Chem. Res. Toxicol. **13**, 135–160 (2000)
16. Morkvenaite-Vilkonciene, I., Ramanaviciene, A., Ramanavicius, A.: 9,10-Phenanthrenequinone as a redox mediator for the imaging of yeast cells by scanning electrochemical microscopy. Sens. Actuators B Chem. **228,** 200–206 (2016). https://doi.org/10.1016/j.snb.2015.12.102
17. Kisieliute, A., Popov, A., Apetrei, R.-M., Cârâc, G., Morkvenaite-Vilkonciene, I., Ramanaviciene, A., Ramanavicius, A.: Towards microbial biofuel cells: Improvement of charge transfer by self-modification of microorganisms with conducting polymer – Polypyrrole. Chem. Eng. J. **356**, 1014–1021 (2019). https://doi.org/10.1016/J.CEJ.2018.09.026

18. Rodriguez, C.E., Sobol, Z., Schiestl, R.H.: 9,10-Phenanthrenequinone induces DNA deletions and forward mutations via oxidative mechanisms in the yeast Saccharomyces cerevisiae. Toxicol. Vitro **22**, 296–300 (2008)
19. Allen, R.M., Bennetto, H.P.: Microbial fuel-cells: electricity production from carbohydrates. Appl. Biochem. Biotechnol. **39**, 27–40 (1993)
20. Hubenova, Y.V., Rashkov, R.S., Buchvarov, V.D., Arnaudova, M.H., Babanova, S.M., Mitov, M.Y.: Improvement of yeast – biofuel cell output by electrode modifications. Ind. Eng. Chem. Res. **50**, 557–564 (2010)

# Author Index




R. Szewczyk et al. (Eds.): AUTOMATION 2020, AISC 1140, pp. 389–390, 2020.
https://doi.org/10.1007/978-3-030-40971-5